STUDIES IN MICROBIOLOGY V

Principles of
Gene Manipulation

AN INTRODUCTION TO
GENETIC ENGINEERING

R. W. OLD MA, PhD
Department of Biological Sciences,
University of Warwick,
Coventry

S. B. PRIMROSE BSc, PhD
Amersham International plc,
Amersham,
Buckinghamshire

FOURTH EDITION

OXFORD
Blackwell Scientific Publications
LONDON EDINBURGH BOSTON
MELBOURNE PARIS BERLIN VIENNA

© 1980, 1981, 1985, 1989 by
Blackwell Scientific Publications
Editorial offices:
Osney Mead, Oxford OX2 OEL
25 John Street, London WC1N 2BL
23 Ainslie Place, Edinburgh EH3 6AJ
3 Cambridge Center, Cambridge,
  Massachusetts 02142, USA
54 University Street, Carlton
  Victoria 3053, Australia

Other Editorial Offices:
Arnette SA
2, rue Casimir-Delavigne
75006 Paris
France

Blackwell Wissenschafts-Verlag
Meinekestrasse 4
D-1000 Berlin 15
Germany

Blackwell MZV
Feldgasse 13
A-1238 Wien
Austria

First published 1980
Second edition 1981
Reprinted 1982, 1983 (twice)
Third edition 1985
Reprinted 1986 (twice), 1987, 1988
Fourth edition 1989
Reprinted 1991, 1992

Set by Setrite Typesetters, Hong Kong
Printed and bound in Spain by
Printek, S.A., Bilbao

DISTRIBUTORS

Marston Book Services Ltd
PO Box 87
Oxford OX2 0DT
(*Orders*: Tel: 0865 791155
         Fax: 0865 791927
         Telex: 837515)

USA
Blackwell Scientific Publications, Inc.
3 Cambridge Center
Cambridge, MA 02142
(*Orders*: Tel: (800) 759—6102
         (617) 225—0401)

Canada
Times Mirror Professional Publishing, Ltd
5240 Finch Avenue East
Scarborough, Ontario M1S 5A2
(*Orders*: Tel: (800) 268—4178
         (416) 298—1588)

Australia
Blackwell Scientific Publications
(Australia) Pty Ltd
54 University Street
Carlton, Victoria 3053
(*Orders:* Tel: (03) 347—0300)

British Library
Cataloguing in Publication Data
Old, R. W.
  Principles of gene manipulation. — 4th ed.
  1. Genetic engineering
  I. Title II. Series
  575.1
  ISBN 0—632—02608—1

Library of Congress
Cataloging in Publication Data
Old, R. W.
  Principles of gene manipulation: an
  introduction to genetic engineering/
  R. W. Old, S. B. Primrose. — 4th ed.
    p. cm. — (Studies in microbiology; v. 2)
  Includes bibliographical references.
  ISBN 0—632—02608—1:
    1. Genetic engineering. 1. Primrose, S.
  B. II. Title. III. Series: Studies in
  microbiology (Oxford, England); v. 2.
  QH442.042 1989
  660′.65—dc20

Principles of Gene Manipulation
AN INTRODUCTION TO
GENETIC ENGINEERING

0632026081

STUDIES IN MICROBIOLOGY

EDITORS

N. G. CARR

Department of Biochemistry
University of Liverpool

J. L. INGRAHAM

Department of Bacteriology
University of California at Davis

S. C. RITTENBERG

Department of Bacteriology
University of California at Los Angeles

Contents

Preface

It is about 10 years since the first edition of *Principles of Gene Manipulation* was published. In writing the first edition, our aim was to explain a new and rapidly growing technology. Our basic philosophy was to present the principles of gene manipulation, and its associated techniques, in sufficient detail to enable the non-specialist reader to understand them. We also intended that the scope of this technology, and its potential impact on virtually all areas of biology, would be evident. It was assumed that the reader would have a reasonable working knowledge of molecular biology.

The second and third editions were enlarged, to cope with the advances in technology which were quickly broadening the field. It had become apparent that, around a core of fundamental techniques concerning the manipulation of DNA *in vitro*, there was developing an ever-expanding repertoire of transformation techniques, library construction and screening methods, expression systems, and host–vector systems.

In this, the fourth, edition we have stuck to the basic philosophy of the previous ones, which we feel has been justified with the passage of time: we think it is valuable to identify and explain the basic principles. As before, the book is intended to be an introduction to the subject. But with the field having grown as much as it now has, we hope that the book will provide a useful overview of the state-of-the-art for researchers who already have experience of recombinant DNA work. We recognize that previous editions have been used widely as textbooks for undergraduates, and we want to continue to serve this readership. Therefore we have not changed the level at which the book is written, nor the general style. The book is, inevitably, larger than before. Despite this, we hope that it communicates the essential simplicity of gene manipulation.

Once again we should like to acknowledge the assistance of our colleagues, at the University of Warwick, at Amersham International plc, and elsewhere, in the impossible task of keeping up-to-date in this exciting field.

R. W. Old
S. B. Primrose

Abbreviations and Conversion Scale

amber (mutation)	*am*
covalently closed circles	CCC
dihydrofolate reductase	DHFR
gene for DNA ligase	*lig*
kilobases	kb
megadaltons	MDal.
molecular weight	mol. wt
plaque-forming unit	p.f.u.
purine	Pu
pyrimidine	Py
resistance (to an antibiotic)	R
sensitivity (to an antibiotic)	s
temperature-sensitive (mutation)	*ts*
thymidine kinase	TK

ABBREVIATIONS OF ANTIBIOTIC NAMES

ampicillin	Ap
chloramphenicol	Cm
kanamycin	Km
neomycin	Nm
streptomycin	Sm
sulphonamide	Su
tetracycline	Tc
trimethoprim	Tp

Scale for conversion between kilobase pairs of duplex DNA and molecular weight.

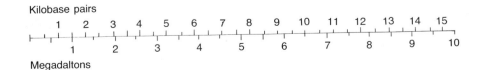

viii

Section 1
Introduction to Gene Manipulation

1 Basic Techniques

Introduction

Occasionally technical developments in science occur that enable leaps forward in our knowledge and increase the potential for innovation. Molecular biology and biomedical research have recently experienced just such a revolutionary change with the development of gene manipulation. The term gene manipulation can be applied to a variety of sophisticated *in-vivo* genetics as well as to *in-vitro* techniques. In fact, in most Western countries there is a precise *legal* definition of gene manipulation as a result of Government legislation to control it. In the United Kingdom gene manipulation is defined as 'the formation of new combinations of heritable material by the insertion of nucleic acid molecules, produced by whatever means outside the cell, into any virus, bacterial plasmid or other vector system so as to allow their incorporation into a host organism in which they do not naturally occur but in which they are capable of continued propagation.' The definitions adopted by other countries are similar and all adequately describe the subject matter of this book.

This legal definition emphasizes the propagation of foreign nucleic acid molecules (the nucleic acid is nearly always DNA) in a different, host organism. The ability to cross natural species barriers and place genes from any organism in an unrelated host organism is one important feature of gene manipulation. A second important feature is the fact that a defined and relatively small piece of DNA is propagated in the host organism. As we shall see, this has far-reaching consequences, for it is then possible to obtain a pure DNA fragment in bulk. This opens the door to a range of molecular biological opportunities including nucleotide sequence determination, site-directed mutagenesis, and manipulation of gene sequences to ensure very high-level expression of an encoded polypeptide in a host organism. In addition, the DNA fragment provides a molecular hybridization probe of absolute sequence purity, totally uncontaminated by other sequences from the donor organism.

The initial impetus for gene manipulation *in vitro* came about in the early 1970s with the simultaneous development of techniques for:
1 transformation* of *Escherichia coli*,
2 cutting and joining DNA molecules, and
3 monitoring the cutting and joining reactions.

---

* The sudden change of an animal cell possessing normal growth properties into one with many of the growth properties of the cancer cell is called *growth transformation*. Growth transformation is mentioned in Chapter 13 and should not be confused with bacterial transformation which is described here.

In order to explain the significance of these developments we must first consider the essential requirements of a successful gene manipulation procedure.

The basic problems

Before the advent of modern gene manipulation methods there had been many early attempts at transforming pro- and eukaryotic cells with foreign DNA. But, in general, little progress could be made. The reasons for this are as follows. Let us assume that the exogenous DNA is taken up by the recipient cells. There are then two basic difficulties. First, where detection of uptake is dependent on gene expression, failure could be due to lack of accurate transcription or translation. Second, and more importantly, the exogenous DNA may not be maintained in the transformed cells. If the exogenous DNA is integrated into the host genome, there is no problem. The exact mechanism whereby this integration occurs is not clear and it is usually a rare event. However this occurs, the result is that the foreign DNA sequence becomes incorporated into the host cell's genetic material and subsequently will be propagated as part of that genome. If, however, the exogenous DNA fails to be integrated, it will probably be lost during subsequent multiplication of the host cells. The reason for this is simple. In order to be replicated, DNA molecules must contain an *origin of replication*, and in bacteria and viruses there is usually only one per genome. Such molecules are called *replicons*. Fragments of DNA are not replicons and in the absence of replication will be diluted out of their host cells. It should be noted that even if a DNA molecule contains an origin of replication this may not function in a foreign host cell.

There is an additional, subsequent problem. If the early experiments were to proceed, a method was required for assessing the fate of the donor DNA. In particular, in circumstances where the foreign DNA was maintained because it had become integrated in the host DNA, a method was required for mapping the foreign DNA and the surrounding host sequences.

THE SOLUTIONS: BASIC TECHNIQUES

If fragments of DNA are not replicated, the obvious solution is to attach them to a suitable replicon. Such replicons are known as *vectors* or *cloning vehicles*. Small plasmids and bacteriophages are the most suitable vectors for they are replicons in their own right, their maintenance does not necessarily require integration into the host genome and their DNA can be isolated readily in an intact form. The different plasmids and phages which are used as vectors are described in detail in Chapters 3 and 4. Suffice it to say at this point that initially plasmids and phages suitable as vectors were only found in *Escherichia coli*. An important consequence follows from the use of a vector to carry the foreign DNA: this is because relatively simple methods are available for purifying the vector molecule, complete with its foreign DNA insert, from transformed host cells. Thus not only does the vector provide the replicon function, but it also permits the easy bulk preparation of the foreign DNA sequence, free from host cell DNA.

Composite molecules in which foreign DNA has been inserted into a vector molecule are sometimes called DNA *chimaeras* because of their analogy with the Chimaera of mythology—a creature with the head of a lion, body of a goat and the tail of a serpent. The construction of such composite or *artificial recombinant* molecules has also been termed *genetic engineering* or *gene manipulation* because of the potential for creating novel genetic combinations by biochemical means. The process has also been termed *molecular cloning* or *gene cloning* because a line of genetically identical organisms, all of which contain the composite molecule, can be propagated and grown in bulk, hence *amplifying* the composite molecule and *any gene product whose synthesis it directs*.

Although conceptually very simple, cloning of a fragment of foreign, or *passenger*, or *target* DNA in a vector demands that the following can be accomplished.

1 The vector DNA must be purified and cut open.

2 The passenger DNA must be inserted into the vector molecule to create the artificial recombinant. DNA joining reactions must therefore be performed. Methods for cutting and joining DNA molecules are now so sophisticated that they warrant a chapter of their own (Chapter 2).

3 The cutting and joining reactions must be readily monitored. This is achieved by the use of gel electrophoresis.

4 Finally, the artificial recombinant must be transformed into E. coli, or other host cell. Further details on the use of gel electrophoresis and transformation of E. coli are given in the next section. As we have noted, the necessary techniques became available at about the same time and quickly led to many cloning experiments, the first of which were reported in 1972 (Jackson et al. 1972, Lobban & Kaiser 1973).

Agarose gel electrophoresis

The progress of the first experiments on cutting and joining of DNA molecules was monitored by velocity sedimentation in sucrose gradients. However, this has been entirely superseded by gel electrophoresis. Gel electrophoresis is not only used as an analytical method, it is routinely used preparatively for the purification of specific DNA fragments. The gel is composed of polyacrylamide or agarose. Agarose is convenient for separating DNA fragments ranging in size from a few hundred to about 20 kb. Polyacrylamide is preferred for smaller DNA fragments. For very large DNA fragments, up to 1000–2000 kb, eletrophoresis in agarose with pulsed electrical fields, or field inversion, has been developed, see Chapter 7.

A gel is a complex network of polymeric molecules. DNA molecules are negatively charged, and under an electric field DNA molecules migrate through the gel at rates dependent upon their sizes: a small DNA molecule can thread its way through the gel easily and hence migrates faster than a larger molecule (Fig. 1.1). Aaij & Borst (1972) showed that the migration

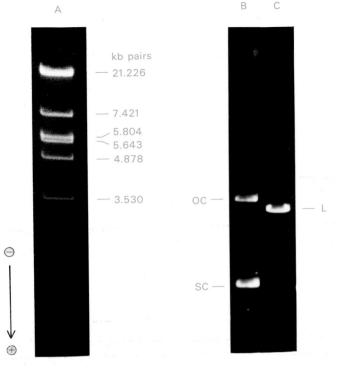

Fig. 1.1 Electrophoresis of DNA in agarose gels. The direction of migration is indicated by the arrow. DNA bands have been visualized by soaking the gel in a solution of ethidium bromide (which complexes with DNA by intercalating between stacked base pairs) and photographing the orange fluorescence which results upon ultraviolet irradiation. (A) Phage λ DNA restricted with *Eco* RI and then electrophoresed in a 1% agarose gel. The λ restriction map is given in Fig. 4.4. (B) Open circular (OC) and super-coiled (SC) forms of a plasmid of 6.4 kb pairs. Note that the compact super-coils migrate considerably faster than open circles. (C) Linear plasmid (L) DNA produced by treatment of the preparation shown in lane B with *Eco* RI for which there is a single target site. Under the conditions of electrophoresis employed here, the linear form migrates just ahead of the open circular form.

rates of the DNA molecules were inversely proportional to the logarithms of the molecular weights. More recently, Southern (1979a,b) has shown that plotting fragment length or molecular weight against the reciprocal of mobility gives a straight line over a wider range than the semi-logarithmic plot. In any event, gel electrophoresis is frequently performed with marker DNA fragments of known size which allow accurate size determination of an unknown DNA molecule by interpolation. A particular advantage of gel electrophoresis is that the DNA bands can be readily detected at high sensitivity. The bands of DNA in the gel are stained with the intercalating dye ethidium bromide (Fig. 1.2), and as little as 0.05 μg of DNA in one band can be detected as visible fluorescence when the gel is illuminated with ultraviolet light.

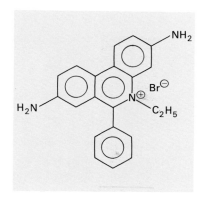

Fig. 1.2 Ethidium bromide.

In addition to resolving DNA fragments of different lengths, gel electro-phoresis separates the different molecular configurations of a DNA mol-ecule. The covalently closed circular, the nicked (relaxed) circular and linear forms of a DNA molecule have different mobilities (Fig. 1.1).

Southern blotting; Northern and Western blotting

Frequently it is necessary to know what sequences in a DNA restriction fragment are transcribed into RNA, or to be able to map sequences by hybridization to restriction fragments. Clearly it would be helpful to have a method of detecting fragments in an agarose gel that are complementary to a given RNA or DNA sequence. This can be done by a neat method described by Southern (1975, 1979b).

This method, often referred to as *Southern* blotting, has since been extended to the analysis of RNA and proteins, and these methods have acquired the jargon terms Northern and Western blotting (see below). The essence of these blotting techniques is to transfer the macromolecules from the gel, through which they have been electrophoretically separated, on to the surface of a membrane. Once transformed, the macromolecules are, or can be, immobilized or 'fixed' more or less permanently on the membrane. The membranes are relatively easy to handle and can be subjected to a variety of analytical techniques. As a result, membranes have become widely used in the detection and analysis of nucleic acids and proteins.

The prototype of these methods, Southern blotting, is shown in Fig. 1.3. DNA restriction fragments in an agarose gel are denatured into single-standard form by alkali treatment, and the gel is then laid on top of buffer-saturated filter paper. The top surface of the gel is covered with a nitrocellulose filter membrane and this membrane is itself overlain with dry filter paper. Many further layers of dry filter paper or absorbent tissue are stacked on top. Buffer passes through the gel, drawn by the capillary action accompanying the progressive wetting of the dry filter stack, and in so doing elutes the denatured DNA from the gel. When the single-stranded DNA comes in contact with the nitrocellulose, it binds there. This blotting process takes several hours to complete and results in the transfer of the

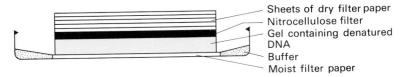

Fig. 1.3 The Southern blot technique. See text for details.

DNA from the gel to the membrane so as to give a spatial pattern of bands on the surface of the membrane which resembles the original gel fractionation pattern, with minimal loss of resolution. The stack can then be disassembled. The DNA is then permanently immobilized on the membrane by baking at 80°C *in vacuo*.

The filter is then placed in a solution of radioactive RNA, single-stranded DNA, or oligodeoxynucleotide which is complementary in sequence to the blot-transferred DNA band or bands which is to be detected. Conditions are chosen so that the radioactive nucleic acid hybridizes with the DNA on the membrane. Since the radioactive nucleic acid is used to detect and locate the complementary sequence, it is called the *probe*. Non-specific binding of the probe to the membrane must not occur, and this is particularly troublesome with DNA probes because single-stranded DNA, unlike RNA, has a high affinity for the membrane. Denhardt (1966) showed that the non-specific binding can be eliminated by treating the membrane carrying the fixed DNA with a solution containing 0.2% each of Ficoll (an artificial polymer of sucrose), polyvinylpyrrolidone and bovine serum albumin. Denhardt's mixture of macromolecules is usually included in the hybridization reaction itself as well as during a pretreatment step. The mixture is often supplemented with an 'irrelevant' nucleic acid such as tRNA, and probably acts by occupying all the available non-specific binding sites for macromolecules on the membrane. Usually, conditions are chosen which maximize the rate of hybridization, compatible with a low background of non-specific binding on the membrane. After the hybridization reaction has been carried out, the membrane is washed to remove unbound radioactivity and regions of hybridization are detected autoradiographically by placing the membrane in contact with X-ray film. A common approach is to carry out the hybridization under conditions of relatively low stringency which permit a high rate of hybridization, followed by a series of post-hybridization washes of increasing stringency (i.e. higher temperature or, more commonly, lower ionic strength). Autoradiography following each washing stage will reveal any DNA bands that are related to, but not perfectly complementary with, the probe and will also permit an estimate of the degree of mismatching to be made.

The Southern blotting methodology can be extremely sensitive. It can be applied to mapping restriction sites around a single-copy gene sequence in a complex genome such as that of man (Fig. 1.4), and when a 'mini-satellite' probe is used it can be applied forensically to minute amounts of DNA (see Chapter 15).

Southern's technique has been of enormous value, but it was thought that it could not be applied directly to the blot-transfer of RNAs separated

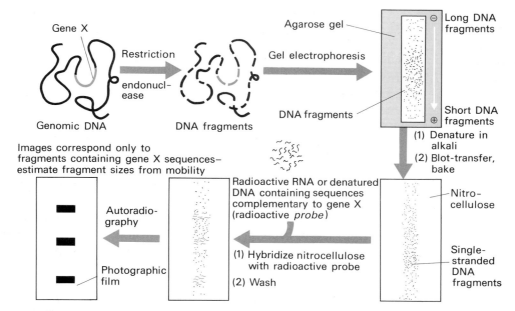

Fig. 1.4 Mapping restriction sites around a hypothetical gene sequence in total genomic DNA by the Southern blot method.

Genomic DNA is cleaved with a restriction endonuclease into hundreds of thousands of fragments of various sizes. The fragments are separated according to size by gel electrophoresis and blot-transferred on to nitrocellulose paper. Highly radioactive RNA or denatured DNA complementary in sequence to gene X is applied to the nitrocellulose paper bearing the blotted DNA. The radiolabelled RNA or DNA will hybridize with gene X sequences and can be detected subsequently by autoradiography, so enabling the sizes of restriction fragments containing gene X sequences to be estimated from their electrophoretic mobility. By using several restriction endonucleases singly and in combination, a map of restriction sites in and around gene X can be built up.

by gel electrophoresis, since RNA was found not to bind to nitrocellulose. Alwine *et al.* (1979), therefore, devised a procedure in which RNA bands are blot-transferred from the gel onto chemically reactive paper, where they are bound covalently. The reactive paper is prepared by diazotization of aminobenzyloxymethyl paper (creating diazobenzyloxymethyl, DBM, paper), which itself can be prepared from Whatman 540 paper by a series of uncomplicated reactions. Once covalently bound, the RNA is available for hybridization with radiolabelled DNA probes. As before, hybridizing bands are located by autoradiography. Alwine's method thus extends that of Southern and for this reason it has acquired the jargon term *Northern* blotting.

Because of the firm covalent binding of the RNA to the paper, such blot-transfers are reusable; the probe from previous hybridization reactions having been eluted by washing at a temperature at which hybrids are not stable. Although originally devised for the transfer of RNA bands, the chemically reactive paper is equally effective in binding denatured DNA. In fact, small DNA fragments are more efficiently transferred to the diazotized paper derivative than to nitrocellulose.

More recently it has been found that RNA bands can indeed be blotted onto nitrocellulose membranes under appropriate conditions (Thomas 1980). In addition, nylon membranes have been developed for both Southern and Northern blotting.

The new membrane materials have advantages of physical robustness and reusability (after the probe has been removed by high-temperature washing or some other denaturing procedure). Nylon membranes have the additional advantage that nucleic acids can be cross-linked, and therefore permanently fixed, to the membrane by brief exposure to ultraviolet light, instead of by the more time-consuming baking step required for nitrocellulose. Because of the convenience of these more recent methods, which do not require freshly activated paper, the use of DBM paper has been superseded.

Boxing the compass continued with the arrival of the term 'Western' blotting (Burnette 1981). This refers to a procedure which does not directly involve nucleic acids, but which is of importance in gene manipulation. It involves the transfer of electrophoresed protein bands from a polyacrylamide gel onto a membrane of nitrocellulose or nylon, to which they bind strongly (Gershoni & Palade 1982, Renart & Sandoval 1984). The bound proteins are then available for analysis by a variety of specific protein—ligand interactions. Most commonly, antibodies are used to detect specific antigens. Lectins have been used to identify glycoproteins. In these cases the probe may itself be labelled with radioactivity, or some other 'tag' may be employed. Often, however, the probe is, unlabelled and is itself detected in a 'sandwich' reaction using a second molecule which is labelled, for instance a species-specific second antibody, or protein A of *Staphylococcus aureus* (which binds to certain subclasses of IgG antibodies), or streptavidin (which binds to antibody probes that have been biotinylated). These second molecules may be labelled in a variety of ways with radioactive, enzyme or fluorescent tags. An advantage of the sandwich approach is that a single preparation of labelled second-molecule can be employed as a general detector for different probes. For example, an antiserum may be raised in rabbits which reacts with a range of mouse immunoglobins. Such a rabbit anti-mouse (RAM) antiserum may be radiolabelled and used in a number of different applications to identify polypeptide bands probed with different, specific, monoclonal antibodies, each monoclonal antibody being of mouse origin. The sandwich method may also give a substantial increase in sensitivity, owing to the multivalent binding of antibody molecules.

The original protein blotting technique employed capillary blotting, but nowadays the blotting is usually accomplished by electrophoretic transfer of polypeptides from an SDS-polyacrylamide gel onto the membrane (Towbin *et al.* 1979). This electroblot technique is faster and more efficient than capillary blotting, and has in its turn been applied in some laboratories to Southern and Northern blotting. Because nucleic acids bind to nitrocellulose only under conditions of high ionic strength, which would lead to high electrical currents and hence overheating during electrophoresis, nylon membranes are preferred for such electroblotting.

Early attempts to achieve transformation of *E. coli* were unsuccessful and it was generally believed that *E. coli* was refractory to transformation. However, Mandel and Higa (1970) found that treatment with $CaCl_2$ allowed *E. coli* cells to take up DNA from bacteriophage λ. A few years later Cohen *et al*. (1972) showed that $CaCl_2$-treated *E. coli* cells are also effective recipients for plasmid DNA. Almost any strain of *E. coli* can be transformed with plasmid DNA, albeit with varying efficiency, whereas it was thought that only *recBC⁻* mutants could be transformed with linear bacterial DNA (Cosloy & Oishi 1973). Later, Hoekstra *et al*. (1980) showed that *recBC⁺* cells can be transformed with linear DNA, but the efficiency is only 10% of that in otherwise isogenic *recBC⁻* cells. Transformation of *recBC⁻* cells with linear DNA is only possible if the cells are rendered recombination proficient by the addition of a *sbc*A or *sbc*B mutation. The fact that the *recBC* gene product is an exonuclease explains the difference in transformation efficiency of circular and linear DNA in *recBC⁺* cells.

As will be seen from the next chapter, many bacteria contain restriction systems which can influence the efficiency of transformation. Although the complete function of these restriction systems is not known yet, one role they do play is the recognition and degradation of foreign DNA. For this reason it is usual to use a restriction-deficient strain of *E. coli* as a tranformable host.

Since transformation of *E. coli* is an essential step in many cloning experiments it is desirable that it be as efficient as possible. Several groups of workers have examined the factors affecting the efficiency of transformation. It has been found that *E. coli* cells and plasmid DNA interact productively in an environment of calcium ions and low temperature (0−5°C), and that a subsequent heat shock (37−45°C) is important, but not strictly required. Several other factors, especially the inclusion of metal ions in addition to calcium, have been shown to stimulate the process.

A very simple, moderately efficient transformation procedure for use with *E. coli* involves resuspending logphase cells in ice-cold 50 mM calcium chloride at about 10^{10} cells/ml and keeping them on ice for about 30 min. Plasmid DNA (0.1 mg) is then added to a small aliquot (0.2 ml) of these now *competent* (i.e. competent for transformation) cells, and the incubation on ice continued for a further 30 min, followed by a heat shock of 2 min at 42°C. The cells are then usually transferred to nutrient medium and incubated for some time (30 min to 1 hour) to allow phenotypic properties conferred by the plasmid to be expressed e.g. antibiotic resistance commonly used as a selectable marker for plasmid-containing cells. (This so-called *phenotypic lag* may not need to be taken into consideration with high-level ampicillin resistance. With this marker significant resistance builds up very rapidly, and ampicillin exerts its effect on cell-wall biosynthesis only in cells which have progressed into active growth.) Finally the cells are plated out on selective medium. Just why such a transformation procedure is effective is not fully understood. The calcium chloride affects the cell wall and may also be responsible for binding DNA to the cell surface. The actual uptake of DNA is stimulated by the brief heat shock.

Hanahan (1983) has re-examined factors that affect the efficiency of transformation, and has devised a set of conditions for optimal efficiency (expressed as transformants per µg plasmid DNA) applicable to most *E. coli* K12 strains. Typically, efficiencies of 10^7 or 10^8 transformants/µg can be achieved. Large DNAs transform less efficiently, on a molar basis, than small DNAs. Even with such improved transformation procedures certain potential gene cloning experiments requiring large numbers of clones are not reliable. One approach which can be used to circumvent the problem of low transformation efficiencies is to package recombinant DNA into virus particles *in vitro*. A particular form of this approach, the use of cosmids, is described in detail in Chapter 4.

Transformation of other organisms

Although *E. coli* often remains the host organism of choice for cloning experiments, many other hosts are now used, and with them transformation may still be a critical step. In the case of Gram-positive bacteria the two most important groups of organisms are *Bacillus* spp. and actinomycetes. That *B. subtilis* is naturally competent for transformation has been known for a long time and hence the genetics of this organism are fairly advanced. For this reason *B. subtilis* is a particularly attractive alternative prokaryotic cloning host. The significant features of transformation with this organism are detailed in Chapter 10. Of particular relevance here is that it is possible to transform protoplasts of *B. subtilis*, a technique which leads to improved transformation frequencies. A similar technique is used to transform actino-mycetes, and recently it has been shown that the frequency can be increased considerably by first entrapping the DNA in liposomes that then fuse with the host cell membrane.

In later chapters we discuss ways in which cloned DNA can be introduced into eukaryotic cells. With animal cells there is no great problem as only the membrane has to be crossed. In the case of yeast, protoplasts are required (Hinnen *et al.* 1978). With higher plants one strategy that has been adopted is either to package the DNA in a plant virus or to use a bacterial plant pathogen as the donor. It has also been shown that proto-plasts prepared from plant cells are competent for transformation, and that a very low but finite level of transformation can be achieved by injecting large amounts of DNA into the vascular system of rye plants at a critical time when the gametes are forming (see Chapter 12). A further remarkable approach that has been demonstrated with plants is the use of micropro-jectiles shot from a gun (Chapter 12).

A relatively new, rapid and simple method for introducing cloned genes into a wide variety of cells, including animal cell lines, yeast proto-plasts, plant protoplasts, and bacterial protoplasts, is *electroporation*. This technique depends upon the original observation by Zimmerman *et al.* (1983) that high-voltage electric pulses can induce cell plasma membranes to fuse. Subsequently it was found that when subjected to electric shock (typically a brief exposure to a voltage gradient of 4000 to 8000 V/cm), the cells take up exogenous DNA from the suspending solution, apparently through holes momentarily created in the plasma membrane. A proportion of these cells become stably transformed (Newman *et al.* 1982, Potter *et al.* 1984), and can be selected if a suitable marker gene is carried on the transforming DNA. Potter *et al.* (1984) also found that treating eukaryotic cells with colcemid before electroporation increases the transformation efficiency. This is probably because the cells, which are arrested in metaphase by the drug, lack a nuclear membrane or have an unusually permeable membrane. These workers also found that linear DNA transforms more efficiently than supercoiled DNA. This technique appears to be a general one. It is successful with animal cell types that have not been amenable to other approaches (Chu *et al.* 1987).

Animal cells, and protoplasts of yeast, plant and bacterial cells are susceptible to transformation by liposomes (Deshayes *et al.* 1985). A simple transformation system has been developed which makes use of liposomes prepared from a cationic lipid (Felgner *et al.* 1987). Small unilamellar (single bilayer) vesicles are produced. DNA in solution spontaneously and efficiently complexes with these liposomes (in contrast to previously employed liposome encapsidation procedures involving non-ionic lipids). The positively charged liposomes not only complex with DNA, but also bind to cultured animal cells and are efficient in transforming them, probably by fusion with the plasma membrane. The use of liposomes as a transformation or transfection system is called *lipofection*.

2 Cutting and Joining DNA Molecules

CUTTING DNA MOLECULES

It is worth recalling that prior to 1970 there was simply no method available for cutting a duplex DNA molecule into discrete fragments. DNA biochemistry was circumscribed by this impasse. It became apparent that the related phenomena of host-controlled restriction and modification might lead towards a solution to the problem when it was discovered that restriction involves specific endonucleases. The favourite organisms of molecular biologists, *E. coli* K12, was the first to be studied in this regard, but turned out to be an unfortunate choice. Its endonuclease is perverse in the complexity of its behaviour. The breakthrough in 1970 came with the discovery in *Haemophilus influenzae* of an enzyme that behaves more simply. Present-day DNA technology is totally dependent upon our ability to cut DNA molecules at specific sites with restriction endonucleases. An account of host-controlled restriction and modification therefore forms the first part of this chapter.

Host-controlled restriction and modification

Host-controlled restriction and modification are most readily observed when bacteriophages are transferred from one bacterial host strain to another. If a stock preparation of phage λ, for example, is made by growth upon *E. coli* strain C and this stock is then titred upon *E. coli* C and *E. coli* K, the titres observed on these two strains will differ by several orders of magnitude, the titre on *E. coli* K being the lower. The phage are said to be *restricted* by the second host strain (*E. coli* K). When those phage that do result from the infection of *E. coli* K are now replated on *E. coli* K they are no longer restricted; but if they are first cycled through *E. coli* C they are once again restricted when plated upon *E. coli* K (Fig. 2.1). Thus the efficiency with which phage λ plates upon a particular host strain depends upon the strain on which it was last propagated. This non-heritable change conferred upon the phage by the second host strain (*E. coli* K) that allows it to be replated on that strain without further restriction is called modification.

The restricted phages adsorb to restrictive hosts and inject their DNA normally. When the phage are labelled with ^{32}P it is apparent that their DNA is degraded soon after injection (Dussoix & Arber 1962) and the endonuclease that is primarily responsible for this degradation is called a *restriction endonuclease* or restriction enzyme (Lederberg & Meselson 1964). The restrictive host must, of course, protect its own DNA from the potentially lethal effects of the restriction endonuclease and so its DNA must be

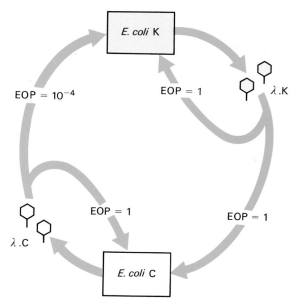

Fig. 2.1 Host-controlled restriction and modification of phage λ in *E. coli* strain K, analysed by efficiency of plating (EOP). Phage propagated by growth on strains K or C (i.e. λK or λC) have EOPs on the two strains as indicated by arrows.

appropriately modified. Modification involves methylation of certain bases at a very limited number of sequences within DNA which constitute the recognition sequences for the restriction endonuclease. This explains why phage that survive one cycle of growth upon the restrictive host can subsequently reinfect that host efficiently; their DNA has been replicated in the presence of the modifying methylase and so it, like the host DNA, becomes methylated and protected from the restriction system.

Although phage infection has been chosen as our example to illustrate restriction and modification, these processes can occur whenever DNA is transferred from one bacterial strain to another. Conjugation, transduction, transformation and transfection are all subject to the constraint of host-controlled restriction. The genes that specify host-controlled restriction and modification systems may reside upon the bacterial chromosome itself or may be located on a plasmid or prophage such as P1.

The restriction endonuclease of *E. coli* K was the first to be isolated and studied in detail. Meselson and Yuan (1968) achieved this by devising an ingenious assay in which a fractionated cell extract was incubated with a mixture of unmodified and modified phage λ DNAs which were differentially radiolabelled — one with ^{3}H, the other with ^{32}P — so that they could be distinguished. After incubation, the DNA mixture was analysed by sedimentation through a sucrose gradient, where the appearance of degraded unmodified DNA in the presence of undegraded modified DNA indicated the activity of restriction endonuclease.

The enzyme from *E. coli* K, and the similar one from *E. coli* B, were found to have unusual properties. In addition to magnesium ions, they require the cofactors ATP and S-adenosylmethionine, and DNA degradation *in vitro* is accompanied by hydrolysis of the ATP in amounts greatly exceeding the stoichiometry of DNA breakage (Bickle *et al.* 1978). In addition, the enzymes are now known to interact with an unmodified *recognition* sequence in duplex DNA and then, surprisingly, to track along the DNA molecule. After travelling for a distance corresponding to between 1000 and 5000 nucleotides the enzyme cleaves one strand only of the DNA at an apparently random site, and makes a gap about 75 nucleotides in length by releasing acid-soluble oligonucleotides. There is no evidence that the enzyme is truly catalytic, and having acted once in this way a second enzyme molecule is required to complete the double-strand break (Rosamond *et al.* 1979). Enzymes with these properties are now known as type I restriction endonucleases. Like all restriction endonucleases they recognize specific nucleotide sequences. However, they are not particularly useful for gene manipulation since their cleavage sites are non-specific. Their biochemistry still presents many puzzles; for instance, the precise role of S-adenosylmethionine remains unclear.

While these bizarre properties of type I restriction enzymes were being unravelled, a restriction endonuclease from *H. influenzae* Rd was discovered (Kelly & Smith 1970, Smith & Wilcox 1970) that was to become the prototype of a large number of restriction endonucleases—now known as type II enzymes—that have none of the unusual properties displayed by type I enzymes and which are fundamentally important in the manipulation of DNA. The type II enzymes recognize a particular target sequence in a duplex DNA molecule and break the polynucleotide chains within, or near to, that sequence to give rise to discrete DNA fragments of defined length and sequence. In fact, the activity of these enzymes is often assayed and studied by gel electrophoresis of the DNA fragments which they generate (see Fig. 1.1). As expected, digests of small plasmid or viral DNAs give characteristic simple DNA band patterns.

Very many type II restriction endonucleases have now been isolated from a wide variety of bacteria. Several hundred restriction enzymes have been identified and at least partially characterized (Kessler *et al.* 1985), and the number continues to grow as more bacterial genera are surveyed for their presence. It is worth noting that many so-called restriction endonucleases have not formally been shown to correspond with any genetically identified restriction and modification system of the bacteria from which they have been prepared: it is usually assumed that a site-specific endonuclease which is inactive upon host DNA and active upon exogenous DNA is, in fact, a restriction endonuclease.

Nomenclature

The discovery of a large number of restriction enzymes called for a uniform nomenclature. A system based upon the proposals of Smith & Nathans (1973) has been followed for the most part. The proposals were as follows.
1 The species name of the host organism is identified by the first letter of the genus name and the first two letters of the specific epithet to form a

three-letter abbreviation in italic: for example, *Escherichia coli* = *Eco* and *Haemophilus influenzae* = *Hin*.

2 Strain or type identification is written as a subscript, e.g. *Eco*$_K$. In cases where the restriction and modification system is genetically specified by a virus or plasmid, the abbreviated species name of the host is given and the extrachromosomal element is identified by a subscript, e.g. *Eco*$_{PI}$, *Eco*$_{RI}$.

3 When a particular host strain has several different restriction and modification systems, these are identified by roman numerals, thus the systems from *H. influenzae* strain Rd would be *Hin*$_d$ I, *Hin*$_d$ II, *Hin*$_d$ III, etc. These roman numerals should not be confused with those in the classification of restriction enzymes into type I, etc.

4 All restriction enzymes have the general name endonuclease R, but, in addition, carry the system name, e.g. endonuclease R.*Hin*$_d$ III. Similarly, modification enzymes are named methylase M followed by the system name. The modification enzyme from *H. influenzae* Rd corresponding to endonuclease R.*Hin*$_d$ III is designated methylase M.*Hin*$_d$ III.

In practice this system of nomenclature has been simplified further: (a) subscripts are typographically inconvenient: the whole abbreviation is now usually written on the line, e.g. *Hind* III, and (b) where the context makes it clear that restriction enzymes only are involved, the designation endonuclease R is omitted. This is the system used in Table 2.1, which lists some of the more commonly used restriction endonucleases. A more extensive list is given in Appendix 1.

Target sites

The vast majority of, but not all, type II restriction endonucleases recognize and break DNA within particular sequences of tetra-, penta-, hexa- or hepta-nucleotides, which have a twofold axis of *rotational symmetry*; for example, *Eco* RI cuts at the positions indicated by arrows in the sequence

<pre>
          axis of symmetry
                  * |
       5′ — G↓A Å | T T C —
       3′ — C T T | A A↑G —
                  | *
</pre>

giving rise to termini bearing 5′-phosphate and 3′-hydroxyl groups. Such sequences are sometimes said to be *palindromic* by analogy with words that read alike backwards and forwards. (However, this term has also been applied to sequences such as

<pre>
       5′ — A G C C G A —
       3′ — T C G G C T —
</pre>

which are palindromic *within one strand*, yet do not have an axis of rotational symmetry.)

The structure of the enzyme–DNA complex has been determined by X-ray crystallography (McClarin *et al.* 1986). It is evident that the endonuclease acts as a dimer of identical subunits, and that the palindromic

Table 2.1 Target sites for some restriction endonucleases.

Anabaena variabilis	*Ava* I	$C\downarrow\binom{C}{T}CG\binom{A}{G}G$	
Bacillus amyloliquefaciens H	*Bam* HI	$G\downarrow GATCC$	
Bacillus globigii	*Bgl* II	$A\downarrow GATCT$	
Escherichia coli RY13	*Eco* RI	$G\downarrow A\overset{*}{A}TTC$	1,4
Escherichia coli R245	*Eco* RII	$\downarrow CC\binom{A}{T}GG$	2
Haemophilus aegyptius	*Hae* III	$GG\downarrow \overset{*}{C}C$	
Haemophilus gallinarum	*Hga* I	$GACGC$	3
Haemophilus haemolyticus	*Hha* I	$G\overset{*}{C}G\downarrow C$	
Haemophilus influenzae Rd	*Hind* II	$GT\binom{C}{T}\downarrow\binom{A}{G}\overset{*}{A}C$	
	Hind III	$\overset{*}{A}\downarrow AGCTT$	
Haemophilus parainfluenzae	*Hpa* I	$GTT\downarrow AAC$	
	Hpa II	$C\downarrow \overset{*}{C}GG$	
Klebsiella pneumoniae	*Kpn* II	$GGTAC\downarrow C$	
Moraxella bovis	*Mbo* I	$\downarrow GATC$	
Providencia stuartii	*Pst* I	$CTGCA\downarrow G$	
Serratia marcescens	*Sma* I	$CCC\downarrow GGG$	
Streptomyces stanford	*Sst* I	$GAGCT\downarrow C$	
Xanthomonas malvacearum	*Xma* I	$C\downarrow CCGGG$	

Source: Kessler *et al.* (1985). Recognition sequences are written from 5′→3′, only one strand being given, and the point of cleavage is indicated by an arrow. Bases written in parentheses signify that either base may occupy that position. Where known, the base modified by the corresponding specific methylase is indicated by an asterisk. $\overset{*}{A}$ is N^6-methyladenine, $\overset{*}{C}$ is 5-methylcytosine.

Notes
1, 2 The names of these two enzymes are anomalous. The genes specifying the enzymes are borne on two resistance transfer factors which have been classified separately. Hence RI and RII.
3 *Hga* I is a type II restriction endonuclease, cleaving as indicated:

    5′ GACGCNNNNN ↓
    3′ CTGCGNNNNN NNNNN ↑

where N is any nucleotide.
4 Under certain conditions (low ionic strength, alkaline pH or 50% glycerol) the *Eco* RI specificity is reduced so that only the internal tetranucleotide sequence of the canonical hexanucleotide is necessary for recognition and cleavage. This is so-called *Eco* RI∗ (RI-star) activity. It is inhibited by parachloromercuribenzoate, whereas *Eco* RI activity is insensitive (Tikchonenko *et al.* 1978). Many other enzymes exhibit star activity, i.e. reduced specificity, under suboptimal conditions.

nature of the target sequence reflects the twofold rotational symmetry of the dimeric protein. If the target sequence is modified by methylation so that 6-methyladenine residues are found at *one* or *both* of the positions indicated by asterisks, then the sequence is resistant to endonuclease R. *Eco* RI. The resistance of the half-methylated site protects the bacterial host's own duplex DNA from attack immediately after semiconservative replication of the fully methylated site until the modification methylase can once again restore the daughter duplexes to the fully methylated state.

We can see that *Eco* RI makes single-strand breaks four nucleotide pairs apart in the opposite strands of its target sequence, and so generates fragments with protruding 5′-termini. These DNA fragments can associate by hydrogen bonding between overlapping 5′-termini, or the fragments can circularize by intramolecular reaction, and for this reason the fragments are said to have *sticky* or *cohesive* ends (Fig. 2.2). In principle, DNA fragments from diverse sources can be joined by means of the cohesive ends, and it is possible, as we shall see later, to seal the remaining nicks in the two strands to form an intact *artificially recombinant* duplex DNA molecule.

It is clear from Table 2.1 that not all type II enzymes cleave their target sites like *Eco* RI. Some enzymes (e.g. *Pst* I) produce fragments bearing 3′-cohesive ends. Others (e.g. *Hae* III) make even cuts giving rise to flush- or blunt-ended fragments with no cohesive end at all. Some enzymes recognize tetranucleotide sequences, others recognize longer sequences, and this of course determines the average fragment length produced. We would expect any particular tetranucleotide target to occur about once every 4^4 (i.e. 256) nucleotide pairs in a long random DNA sequence, assuming all bases are equally frequent. Any particular hexanucleotide target would be expected to occur once in every 4^6 (i.e. 4096) nucleotide pairs. Some enzymes (e.g. *Sau* 3AI) recognize a tetranucleotide sequence that is included within the hexanucleotide sequence recognized by a different enzyme (e.g. *Bam* HI). The cohesive termini produced by these enzymes are such that fragments produced by *Sau* 3AI will cohere with those produced by *Bam* HI. If the fragments are then covalently joined, the 'hybrid site' so produced will be

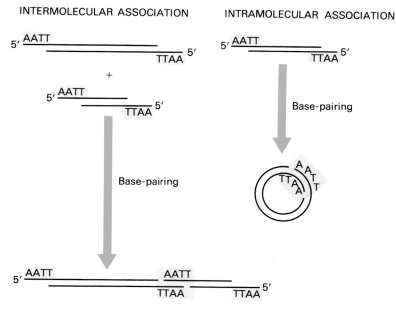

Fig. 2.2 Cohesive ends of DNA fragments produced by digestion with *Eco* RI.

once again sensitive to *Sau* 3AI, but may not constitute a target for *Bam* HI; this will depend upon the nucleotides adjacent to the original *Sau* 3AI site (Fig. 2.3). Several other combinations of enzymes have this property.

From Table 2.1 we can also see that *Hind* II, the first type II enzyme to be discovered, is an example of an enzyme recognizing a sequence with some ambiguity; in this case all three sequences corresponding to the structure given in Table 2.1 are substrates. There are also several known examples of enzymes from different sources which recognize the same target. They are *isoschizomers*. Some pairs of isoschizomers cut their target at different places (e.g. *Sma* I, *Xma* I).

In our discussion of the phenomena of restriction and modification of phage λ by *E. coli* K, we saw that methylation was the basis of modification in that system. In the wide variety of type II restriction enzymes now known there are some curious and useful examples of the influence of methyl groups at restriction sites. The enzymes *Hpa* II and *Msp* I are isoschizomers with the target sequence CCGG. *Hpa* II will not cut the target when it contains 5-methylcytosine as indicated by the asterisk CĊGG (indeed this is the product of M.*Hpa* II). However, *Msp* I is known to be indifferent to methylation at this nucleotide: it cleaves whether or not this C residue is methylated. Now it has been found that over 90% of the methyl groups in genomic DNA of many animals, including vertebrates and echinoderms, occur as 5-methylcytosine in the sequence CG. Many of these methyl groups occur at *Msp* I sites, and their presence can be detected by comparing digests of the DNA generated by *Hpa* II and *Msp* I. Indeed methylation at *Msp* I sites around a single gene can be investigated in detail by combining the use of these two enzymes with the Southern-blot technique (Bird & Southern 1978, Razin & Riggs 1980).

A potentially troublesome effect of DNA methylation concerns the *dam* and *dcm* methylation activities of many *E. coli* strains. The *dam* (DNA adenine methylase) activity methylates adenine, creating 6-methyladenine, in the sequence GATC. The *dcm* (DNA cytosine methylase) activity methylates the internal cytosine, creating 5-methylcytosine, in the sequence CC(A/T)GG. The *dam* methylation appears to play a role in strand discrimination for mismatch repair (Lu *et al.* 1983, Radman & Wagner 1984). Further postulated functions for the adenine and cytosine methylation in

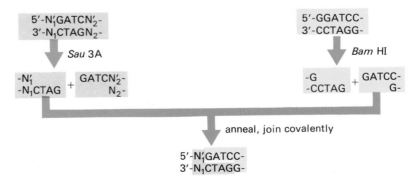

Fig. 2.3 Production of a hybrid site by cohesion of complementary sticky ends generated by *Sau* 3A and *Bam* HI.

the initiation of DNA replication or recombination are less well understood. From the point of view of the gene manipulator, the main importance of these methylation functions is their effect of limiting the ability of certain restriction endonucleases to act on DNAs from *dam*[+] and *dcm*[+] *E. coli* strains: two examples follow. *Dam* methylation of the sequence GATC renders it resistant to restriction by *Mbo* I. This problem is overcome by the availability of the *Sau* 3AI isoschizomer, which cuts the sequence whether or not the adenine residue is methylated. It is mainly for this reason that *Sau* 3AI has replaced *Mbo* I in the catalogues of commercially available restriction endonucleases. *Dcm* methylation of the CC(A/T)GG sequence prevents restriction by *Eco* RII. In this example also there is isoschizomer available, *BstN* I, which is indifferent to the *dcm* methylation. An alternative solution to the problem in this case, and in several other examples where no methylation−indifferent isoschizomer is available, is to produce the DNA which is to be restricted in a *dam*[−] or *dcm*[−] *E. coli* host. In fact, *dcm*[−] *dam*[−] double mutant strains are available for this application (Marinus *et al.* 1983). (Refer to Chapter 7 for the involvement of *dam* methylation in cleavage by *Dpn* I.)

A further kind of type II restriction endonuclease has been identified. Enzymes of this type make breaks in the two strands at *measured distances* to one side of their asymmetric target sequence (e.g. *Hga* I; see Table 2.1). Finally, type III restriction enzymes have been classified as those which cleave DNA at well-defined sites, require ATP and Mg^{2+}, but which have only a partial requirement for (i.e. are stimulated by) S-adenosylmethionine. In these respects they have properties intermediate between type I and type II enzymes (Table 2.2).

The wide variety of properties exhibited by restriction endonucleases described in the preceding paragraphs has provided great scope for ingenious and resourceful gene manipulators. This will be apparent from examples in following chapters.

What is the function of restriction endonucleases *in vivo*? Clearly host-controlled restriction acts as a mechanism by which bacteria distinguish self from non-self. It is analogous to an immunity system. Restriction is moderately effective in preventing infection by some bacteriophages. It may be for this reason that the T-even phages (T2, T4 and T6) have evolved with glucosylated hydroxymethylcytosine residues replacing cytosine in their DNA, so rendering it resistant to many restriction endonucleases. The restriction and glucosylation modification of T-even phage DNA is beyond the scope of this book. For a detailed discussion the reader is referred to Kornberg (1980). However, it is worth noting that a mutant strain of T4 is available which does have cytosine residues in its DNA and is therefore amenable to conventional restriction methodology (Velten *et al.* 1976, Murray *et al.* 1979, Krisch & Selzer 1981). As an alternative to the unusual DNA structure of the T-even phages, other mechanisms appear to have evolved in T3 and T7 for overcoming restriction *in vivo* (Spoerel *et al.* 1979, Bandyopadhyay *et al.* 1985, Kruger *et al.* 1985). In spite of this evidence we may be mistaken in concluding that immunity to phage infection is the sole or main function of all restriction endonucleases in nature.

Table 2.2 Characteristics of restriction endonucleases (Yuan 1981, Hadi *et al.* 1982, Iida *et al.* 1982, Kessler *et al.* 1985).

	Type I	Type II	Type III
Restriction and modification activities	single multifunctional enzyme	separate endonuclease and methylase	separate enzymes with a subunit in common
Protein structure of restriction endonuclease	3 different subunits	simple	2 different subunits
Requirements for restriction	ATP, Mg^{2+} S-adenosylmethionine	Mg^{2+}	ATP, Mg^{2+} (S-adenosylmethionine)
Sequence of host specificity sites	*Eco* B: TGAN$_8$TGCT *Eco* K: AACN$_6$GTGC	rotational symmetry (usually, but not in all examples)	*Eco* P1: AGACC *Eco* P15: CAGCAG
Cleavage sites	possibly random, at least 1000 bp from host specificity site	at or near host specificity site	24–26 bp to 3′ of host specificity site
Enzymatic turnover	no	yes	yes
DNA translocation	yes	no	no
Site of methylation	host specificity site	host specificity site	host specificity site

N = any nucleotide.

The importance of eliminating restriction systems in *E. coli* strains used as hosts for recombinant molecules

If foreign DNA is introduced into an *E. coli* host it may be attacked by restriction systems active in the host cell. An important feature of these systems is that the fate of the incoming DNA in the restrictive host depends not only on the sequence of the DNA but also upon its history: the DNA sequence may or may not be restricted, depending upon its source immediately prior to transforming the *E. coli* host strain. As we have seen, post-replication modifications of the DNA, usually in the form of methylation of particular adenine or cytosine residues in the target sequence, protects against cognate restriction systems but not, in general, against different restriction systems. Such is the case with the *E. coli* K restriction and modification system, where methylation of the sequence 5'-AAC-(N_6)-GTGC3' or 5'-GCAC-(N_6)-GTT3' in the second or first adenines respectively protects against K restriction.

Because restriction provides a natural defence against invasion by foreign DNA, it is usual to employ a K restriction-deficient *E. coli* K12 strain as a host in transformation with newly created recombinant molecules. Thus where, for example, mammalian DNA has been ligated into a plasmid vector, transformation of the *Eco* K restriction-deficient host eliminates the possibility that the incoming sequence will be restricted, even if the mammalian sequence contains an unmodified *Eco* K target site. If the host happens to be *Eco* K restriction-deficient but *Eco* K modification-*proficient*, propagation on the host will confer modification methylation and hence allow subsequent propagation of the recombinant in *Eco* K restriction-proficient strains, if desired.

It has recently been discovered that *Eco* K is not the only restriction-like activity in *E. coli* K12 strains. Two new activities have been found in many laboratory strains of *E. coli* K12. By contrast with the examples above, the targets for restriction by these activities are *methylated* DNAs. In particular, DNA containing 5-methylcytosine is restricted by these strains. The restriction is due to two genetically distinct systems, that differ in their sequence specificities and are named McrA and McrB (for *modified cytosine restriction*) (Raleigh & Wilson 1986, Raleigh *et al.* 1988). It had been known for many years that *E. coli* restricts DNA with an unusual cytosine modification (5-hydroxymethylation of cytosine residues, as in restriction of *glucoseless* T-even phages, conferring sensitivity to the *E. coli* RglA and RglB systems). The newly discovered restriction of DNA containing 5-methylcytosine, a much more common modification, by the McrA and McrB systems is controlled by the genetic loci, *mcrA* and *mcrB*, which appear to be the same as those that govern the previously known Rgl restriction (Raleigh *et al.* 1988).

The best-studied of these systems is McrB. If DNA is methylated at the cytosine of 5'-GC 3' dinucleotide sequences it will be subject to restriction by the McrB system, and the degree of restriction is a function of the number of methylated sites. Many bacterial species contain methylases that produce DNA that will be a substrate for McrB if that DNA is transformed into an McrB$^+$ *E. coli* strain. Modification methylases M. *Pvu*

II (C-A-G-^{m}C-T-G) and M.*Hae* III (G-G-^{m}C-C) are examples of such methylases. Of course bacteria are not the only organisms that contain 5-methylcytosine in their DNA. In plants, as much as 40−50% of all the C may be methylated, including the majority of C-G, C-A-G and C-T-G sequences. In vertebrates, the overall amount of methylation is less (about 10% of all C residues in mammals), the great majority is in the C-G dinucleotide sequence, and the cell-type specific methylation appears to be involved in gene regulation. The C methylation in DNA from such sources may give rise to a methylated G-C dinucleotide (e.g. in mammals where the sequence G-^{m}C-G occurs) and therefore render the DNA susceptible to restriction by the McrB activity. The McrA activity is less well understood at present, but is known to restrict DNA with the following methylated sequence, C-^{m}C-G-G, such as that produced by M.*Hpa* II. The main importance of these newly discovered systems, for the gene manipulator, is their possible influence in leading to the under-representation of certain genomic sequences during the creation of genomic libraries. It is therefore wise to use for such cloning an E. *coli* host strain that not only lacks the three familiar restriction systems (*Eco* K, *Eco* B and the prophage P1-specific *Eco* P1 system) found in many laboratory strains of E. *coli*, but which also lacks the Mcr systems. Table 2.3 shows the restriction properties of some commonly used host strains. Note that the foreign methylation pattern will be lost upon replication of the cloned DNA in E. *coli*, so that after the critical transformation the cloned DNAs can be freely moved into Mcr$^+$ strains.

Mechanical shearing of DNA

In addition to digesting DNA with restriction endonucleases to produce discrete fragments, there are a variety of treatments which result in non-specific breakage. Non-specific endonucleases and chemical degradation can be used but the only method that has been much applied to gene manipulation involves mechanical shearing.

The long, thin threads which constitute duplex DNA molecules are sufficiently rigid to be very easily broken by shear forces in solution. Intense sonication with ultrasound can reduce the length to about 300 nucleotide pairs. More controlled shearing can be achieved by high-speed stirring in a blender. Typically, high mol. wt DNA is sheared to a population of molecules with a mean size of about 8 kb pairs by stirring at 1500 rev/min for 30 min (Wensink *et al.* 1974). Breakage occurs essentially at random with respect to DNA sequence. The termini consist of short, single-stranded regions which may have to be taken into account in subsequent joining procedures.

JOINING DNA MOLECULES

Having described the methods available for cutting DNA molecules we must consider the ways in which DNA fragments can be joined to create artificially recombinant molecules. There are currently three methods for

Table 2.3 Restriction properties of some commonly used *E. coli* host strains.

E. coli strain	Notes	Restriction properties			
		Eco K or *Eco* B	*Eco* P1	*mcr*A	*mcr*B
HB101	K strain carrying hsd−mcrB region from *E. coli* B. Is *hsd* 20 and thus phenotypically R⁻M⁻ for *Eco* K and *Eco* B		−	+	−
DH1	*rec A gyr A*	$R_k^- M_k^+$	−	+	+
DH5	derived from DH1	$R_k^- M_k^+$	−	+	+
JM101	M13 host	$R_k^+ M_k^+$	−	+	?
JM107	M13 host	$R_k^+ M_k^+$	−	+	+
Y1090	λgt11 host	$R_k^- M_k^+$	−	−	+
LE392	*E sup F*	$R_k^- M_k^+$	−	−	+
MC1061		$R_k^- M_k^+$	nt	−	
K802		$R_k^- M_k^+$	−	−	−
K803		$R_k^- M_k^-$	−	−	−
GM2163	*dam⁻ dcm⁻*	$R_k^- M_k^+$	−	−	−

Sources: Raleigh & Wilson (1980), Raleigh *et al.* (1988).

joining DNA fragments *in vitro*. The first of these capitalizes on the ability of DNA ligase to join covalently the annealed cohesive ends produced by certain restriction enzymes. The second depends upon the ability of DNA ligase from phage T4-infected *E. coli* to catalyse the formation of phospho-diester bonds between blunt-ended fragments. The third utilizes the enzyme terminal deoxynucleotidyltransferase to synthesize homopolymeric 3'-single-stranded tails at the ends of fragments. We can now look at these three methods a little more deeply.

DNA ligase

E. coli and phage T4 encode an enzyme, DNA ligase, which seals single-stranded nicks between adjacent nucleotides in a duplex DNA chain (Olivera *et al.* 1968, Gumport & Lehman 1971). Although the reactions catalysed by the enzymes of *E. coli* and T4-infected *E. coli* are very similar, they differ in their cofactor requirements. The T4 enzyme requires ATP, whilst the *E. coli* enzyme requires NAD^+. In each case the cofactor is split and forms an enzyme-AMP complex. The complex binds to the nick, which must expose a 5'-phosphate and 3'-OH group, and makes a covalent bond in the phosphodiester chain as shown in Fig. 2.4.

When termini created by a restriction endonuclease that creates cohesive

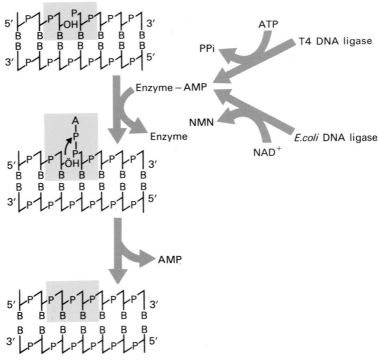

Fig. 2.4 Action of DNA ligase. An enzyme-AMP complex binds to a nick bearing 3'-OH 5'-P groups. The AMP reacts with the phosphate group. Attack by the 3'-OH group on this moiety generates a new phosphodiester bond which seals the nick.

ends associate, the joint has nicks a few base pairs apart in opposite strands. DNA ligase can then repair these nicks to form an intact duplex. This reaction, performed *in vitro* with purified DNA ligase, is fundamental to many gene manipulation procedures such as that shown in Fig. 2.5.

The optimum temperature for ligation of nicked DNA is 37°C, but at this temperature the hydrogen-bonded joint between the sticky ends is unstable. *Eco* RI-generated termini associate through only four AT base pairs and these are not sufficient to resist thermal disruption at such a high temperature. The optimum temperature for ligating the cohesive termini is therefore a compromise between the rate of enzyme action and association of the termini, and has been found by experiments to be in the range 4–15°C (Dugaicyzk *et al.* 1975, Ferretti & Sgaramella 1981).

The ligation reaction can be performed so as to favour the formation of recombinants. First, the population of recombinants can be increased by performing the reaction at a high DNA concentration; in dilute solutions *circularization* of linear fragments is relatively favoured because of the reduced frequency of intermolecular reactions. Second, by treating linearized plasmid vector DNA with alkaline phosphatase to removed 5'-terminal phosphate groups, both recircularization and plasmid dimer

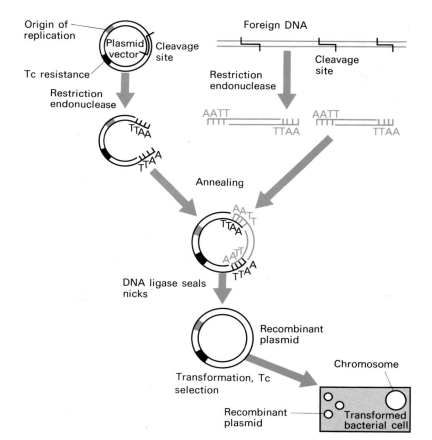

Fig. 2.5 Use of DNA ligase to create a covalent DNA recombinant joined through association of termini generated by *Eco* RI.

formation are prevented (Fig. 2.6). In this case, circularization of the vector can occur only by insertion of non-phosphatase-treated foreign DNA which provides one 5'-terminal phosphate at each join. One nick at each join remains unligated, but after transformation of host bacteria, cellular repair mechanisms reconstitute the intact duplex.

Joining DNA fragments with cohesive ends by DNA ligase is a relatively efficient process which has been used extensively to create artificial re-combinants. A modification of this procedure depends upon the ability of T4 DNA ligase to join blunt-ended DNA molecules (Sgaramella 1972). The *E. coli* DNA ligase will not catalyse blunt ligation except under special reaction conditions of macromolecular crowding (Zimmerman & Pheiffer 1983). Blunt ligation is most usefully applied to joining blunt-ended fragments via *linker* molecules; in an early example of this, Scheller *et al.* (1977) synthesized self-complementary decameric oligonucleotides, which contain sites for one or more restriction endonucleases. One such molecule is shown in Fig. 2.7. The molecule can be ligated to both ends of the foreign DNA to be cloned, and then treated with restriction endonuclease to

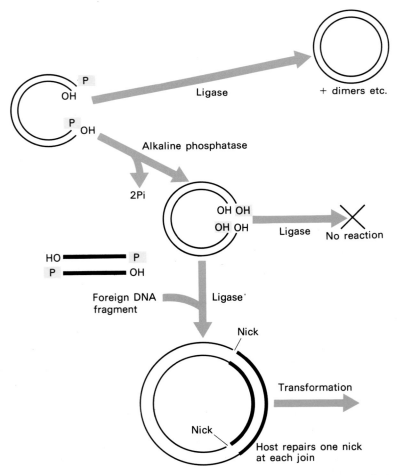

Fig. 2.6 Application of alkaline phosphatase treatment to prevent recircularization of vector plasmid without insertion of foreign DNA.

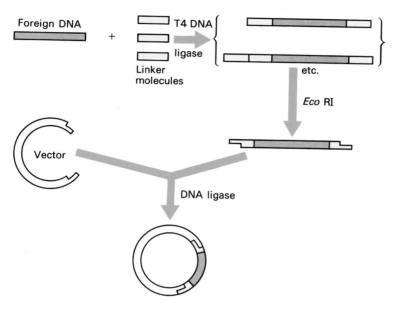

Fig. 2.7 A decameric linker molecule containing an *Eco* RI target site is joined by T4 DNA ligase to both ends of flush-ended foreign DNA. Cohesive ends are then generated by *Eco* RI. This DNA can then be incorporated into a vector that has been treated with the same restriction endonuclease.

produce a sticky-ended fragment which can be incorporated into a vector molecule that has been cut with the same restriction endonuclease. Insertion by means of the linker creates restriction enzyme target sites at each end of the foreign DNA and so enables the foreign DNA to be excised and recovered after cloning and amplification in the host bacterium.

Double-linkers

Plasmid vectors have been derived which contain a set of closely clustered cloning sites. An example of such a vector is pUC8, which is described in more detail on page 58. This vector has been used to clone duplex cDNA molecules by the double-linker approach (Kurtz & Nicodemus 1981, Helfman *et al*. 1983), in which *different* linker molecules are added to the opposite ends of the cDNA (Fig. 2.8). This has the following advantages.
1 The problem of vector reclosure without insertion of foreign DNA is overcome. Partly for this reason the method is efficient, i.e. cDNAs have been cloned which were derived from rare mRNA molecules in the starting population.

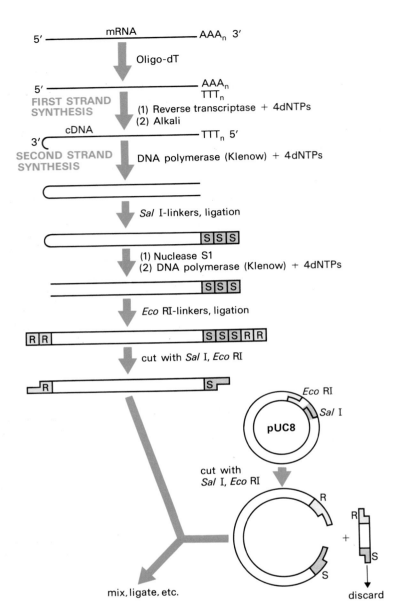

Fig. 2.8 Double-linkers. mRNA is copied into double-stranded cDNA and *Sal* I-linkers (S) are added. The hairpin loop formed by self-priming of second strand synthesis is removed by nuclease S1. In this procedure, any raggedness left by nuclease S1 (i.e. short, single-strand projections at the terminus) is removed by polishing with Klenow polymerase. *Eco* RI-linkers are then ligated to the duplex molecule, and cohesive termini revealed by restriction with *Sal* I and *Eco* RI. The cDNA plus linkers is then ligated into a vector cut with the same two enzymes.

2 The use of linkers, rather than homopolymers, is desirable when expression from a vector-borne promoter is sought (see Chapter 7).

3 The orientation of the inserted DNA is fixed.

Adaptors

It may be the case that the restriction enzyme used to generate the cohesive ends in the linker will also cut the foreign DNA at internal sites. In this situation the foreign DNA will be cloned as two or more subfragments. One solution to this problem is to choose another restriction enzyme, but there may not be a suitable choice if the foreign DNA is large and has sites for several restriction enzymes. Another solution is to methylate internal restriction sites with the appropriate modification methylase. An example of this is described in Chapter 7. Alternatively, a general solution to the problem is provided by chemically synthesized adaptor molecules which have a *preformed* cohesive end (Wu *et al.* 1978). Consider a blunt-ended foreign DNA containing an internal *Bam* HI site (Fig. 2.9), which is to be cloned in a *Bam* HI-cut vector. The *Bam* adaptor molecule has one blunt end, bearing a 5'-phosphate group, and a Bam cohesive end which is not phosphorylated. The adaptor can be ligated to the foreign DNA ends, without any risk of adaptor self-polymerization. The foreign DNA plus added adaptors is then phosphorylated at the 5'-termini and ligated into the *Bam* HI site of the vector. If the foreign DNA

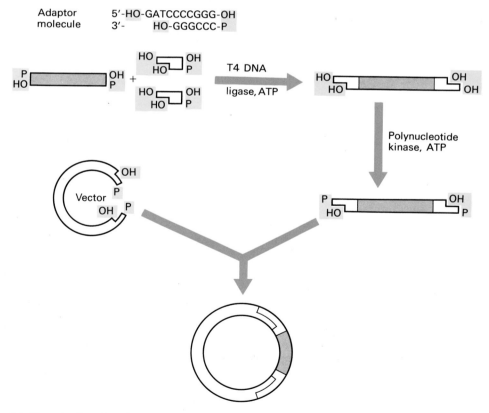

Fig. 2.9 Use of a *Bam* HI adaptor molecule. A synthetic adaptor molecule is ligated to the foreign DNA. The adaptor is used in the 5'-hydroxyl form to prevent self-polymerization. The foreign DNA plus ligated adaptors is phosphorylated at the 5'-termini and ligated into the vector previously cut with *Bam* HI.

were to be recovered from the recombinant with *Bam* HI, it would be obtained in two fragments. However, the adaptor is designed to contain two other restriction sites (*Sma* I, *Hpa* II) which may enable the foreign DNA to be recovered intact.

Homopolymer tailing

A general method for joining DNA molecules makes use of the annealing of complementary homopolymer sequences. Thus, by adding oligo (dA) sequences to the 3' ends of one population of DNA molecules, and oligo (dT) blocks to the 3' ends of another population, the two types of molecule can anneal to form mixed dimeric circles (Fig. 2.10).

An enzyme purified from calf-thymus, terminal deoxynucleotidyl-transferase, provides the means by which the homopolymeric extensions can be synthesized, for if presented with a single deoxynucleotide tri-phosphate it will repeatedly add nucleotides to the 3'-OH termini of a population of DNA molecules (Chang & Bollum 1971). DNA with exposed 3'-OH groups, such as arise from pretreatment with phage λ exonuclease or restriction with an enzyme such as *Pst* I, is a very good substrate for the transferase. However, conditions have been found in which the enzyme will extend even the shielded 3'-OH of 5'-cohesive termini generated by *Eco* RI (Roychoudhury *et al.* 1976, Humphries *et al.* 1978).

The terminal transferase reactions have been recently characterized in detail with regard to their use in gene manipulation (Deng & Wu 1981, Michelson & Orkin 1982). Typically, 10−40 homopolymeric residues are added to each end.

In 1972, Jackson *et al.* were among the first to apply the homopolymer method when they constructed a recombinant in which a fragment of phage λ DNA was inserted into SV40 DNA. In their experiments, the single-stranded gaps which remained in the two strands at each join were repaired *in vitro* with DNA polymerase and DNA ligase so as to produce covalently closed circular molecules, which were then used to transfect susceptible mammalian cells (see Chapter 13).

Subsequently, the homopolymer method, employing either dA.dT or dG.dC homopolymers, has been applied extensively in constructing re-combinant plasmids for cloning in *E. coli*. Commonly, the annealed circles are used directly for transformation with repair of the gaps occurring *in vivo*. In the example which follows we shall see how homopolymer tailing can be applied to cloning DNA copies of eukaryotic messenger RNA and how a careful choice of which homopolymers are used can be important.

Cloning cDNA by homopolymer tailing

If we wish to construct a clone containing sequences derived from eukaryotic mRNA, we must first obtain the sequence in DNA form. We can do this by making a complementary (cDNA) copy of the mRNA, using the enzyme reverse transcriptase, which is a type of DNA polymerase found in retroviruses, and whose function is to synthesize DNA upon an RNA template.

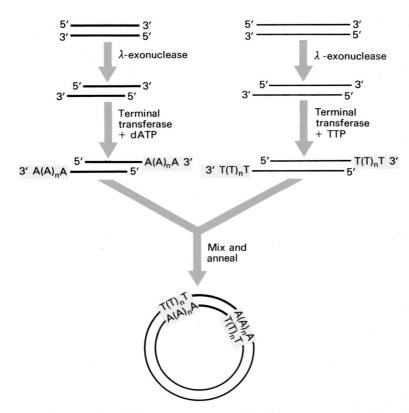

Fig. 2.10 Use of calf-thymus terminal deoxynucleotidyl-transferase to add complementary homopolymer tails to two DNA molecules.

Like other true DNA polymerases, reverse transcriptase can only synthesize a new DNA strand if provided with a growing point in the form of a pre-existing primer which is base-paired with the template and bears a free 3'-OH group. Fortunately, most eukaryotic mRNAs occur naturally in a polyadenylated form with up to 200 adenylate residues at their 3'-termini and hence we can provide a primer simply by hybridizing a short oligo (dT) molecule with this poly (A) sequence. The primer is then suitably located for synthesis of a complete cDNA by reverse transcriptase in the presence of all four deoxynucleoside triphosphates (Fig. 2.11).

The immediate product of the reaction is an RNA–DNA hybrid. The RNA strand can then be destroyed by alkaline hydrolysis, to which DNA is resistant, leaving a single-stranded cDNA which can be converted into the double-stranded form in a second DNA polymerase reaction. This reaction depends upon the observation that cDNAs can form a transient self-priming structure in which a hairpin loop at the 3'-terminus is stabilized by enough base-pairing to allow initiation of second-strand synthesis. Once initiated, subsequent synthesis of the second strand stabilizes the hairpin (Efstratiadis *et al.* 1976, Higuchi *et al.* 1976). The hairpin and any single-stranded DNA at the other end of the cDNA molecule are then trimmed away by treatment with the single-strand-specific nuclease S1, giving rise to a fully duplex molecule.

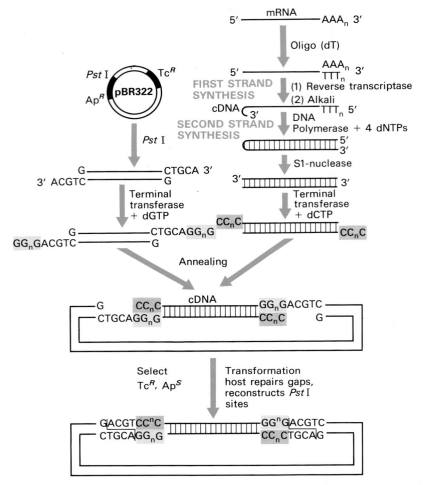

Fig. 2.11 Synthesis of a cDNA copy of a polyadenylated mRNA and insertion into a vector molecule by homopolymer tailing. See text for explanation.

In our example (Fig. 2.11) the duplex cDNA is tailed with oligo (dC) and annealed with the pBR322 vector which has been cut open with *Pst* I and tailed with oligo (dG). It will be seen that these homopolymers have been chosen so that *Pst* I target sites are reconstructed in the recombinant molecule, thus providing a simple means for excising the inserted sequences after amplification (Smith *et al.* 1979). This can be accomplished in another way, by constructing dA.dT joins. In that case the homopolymeric regions will have a lower melting temperature than the rest of the recombinant molecule, and so under partially denaturing conditions can be cleaved by nuclease S1 to release the inserted sequence.

Full-length cDNA cloning

In the two cDNA schemes illustrated in this chapter, second-strand synthesis is self-primed, resulting in the formation of a duplex cDNA with a hairpin loop that is subsequently removed by nuclease S1. This step necessarily leads to the loss of a certain amount of sequence corresponding to the 5′ end of the mRNA, and unless the nuclease S1 is very pure, there can be adventitious damage to the duplex cDNA. For this reason self-priming of second-strand synthesis is now rarely carried out; improved methods have been developed and are discussed in Chapter 7.

Section 2
Cloning in *E. coli*

3 Plasmids as Cloning Vehicles for Use in *E. coli*

Basic properties of plasmids

Plasmids are widely used as cloning vehicles, but before discussing their use in this context it is appropriate to review some of their basic properties. Plasmids are replicons which are stably inherited in an extrachromosomal state. It should be emphasized that extrachromosomal nucleic acid molecules are not necessarily plasmids, for the definition given above implies genetic homogeneity, constant monomeric unit size, and the ability to replicate independently of the chromosome. Thus the heterogeneous circular DNA molecules which are found in *Bacillus megaterium* (Carlton & Helinski 1969) are not necessarily plasmids. The definition given above, however, does include the prophages of those temperate phages, e.g. P1, which are maintained in an extrachromosomal state, as opposed to those such as λ (see Chapter 4) which are maintained by integration into the host chromosome. Also included are the replicative forms of the filamentous coliphages which specify the continued production and release of phage particles without concomitant cell lysis.

Most plasmids exist as double-stranded circular DNA molecules. If both strands of DNA are intact circles the molecules are described as covalently closed circles or CCC DNA (Fig. 3.1). If only one strand is intact, then the molecules are described as open circles or OC DNA. When isolated from cells, covalently closed circles often have a deficiency of turns in the double helix such that they have a supercoiled configuration. The enzymatic interconversion of supercoiled, relaxed CCC DNA* or OC DNA is shown in Fig. 3.1. Because of their different structural configurations, supercoiled and OC DNA separate upon electrophoresis in agarose gels. Addition of an intercalating agent, such as ethidium bromide, to supercoiled DNA causes the plasmid to unwind. If excess ethidium bromide is added the plasmid will rewind in the opposite direction (Fig. 3.2). Use of this fact is made in the isolation of plasmid DNA (see p. 43).

In recent years a number of linear plasmids have been described in bacteria as diverse as the actinomycete *Streptomyces* (Kinashi *et al.* 1987) and the spirochete *Borrelia* (Plasterk *et al.* 1985). At least in the case of the *Borrelia* plasmid the use of the term 'linear' may be a misnomer, for the plasmid is in reality a circular molecule of single-stranded DNA (Fig. 3.3) in which there is extensive internal base-pairing (Barbour & Garon 1987). Although linear plasmids have not been isolated from *E. coli* this does not

---

* The reader should not be confused by the terms *relaxed circle* and *relaxed plasmid*. Relaxed circles are CCC DNA that does not have a supercoiled configuration. Relaxed plasmids are plasmids with multiple copies per cell.

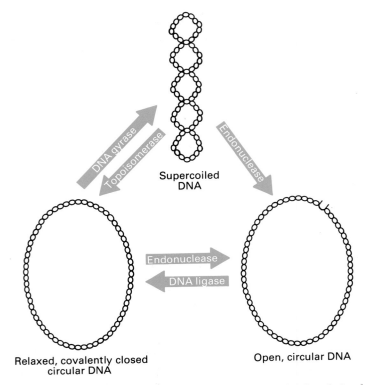

Fig. 3.1 The interconversion of supercoiled, relaxed covalently closed circular DNA, and open circular DNA.

mean that they do not exist. In the case of *Streptomyces* (Kinashi *et al.* 1987), linear plasmids only could be detected by means of the recently developed technique of orthogonal-field-alternation gel electrophoresis (Carle & Olson 1984).

Plasmids are widely distributed throughout the prokaryotes, vary in size from less than 1×10^6 daltons to greater than 200×10^6, and are generally dispensable. Some of the phenotypes which these plasmids confer on their host cells are listed in Table 3.1. Plasmids to which phenotypic traits have not yet been ascribed are called *cryptic* plasmids.

Plasmids can be categorized into one of two major types—conjugative or non-conjugative—depending upon whether or not they carry a set of transfer genes, called the *tra* genes, that promotes bacterial conjugation. Plasmids can also be categorized on the basis of their being maintained as multiple copies per cell (*relaxed* plasmids) or as a limited number of copies per cell (*stringent* plasmids). Generally, conjugative plasmids are of relatively high mol. wt and are present as one to three copies per chromosome whereas non-conjugative plasmids are of low mol. wt and present as multiple copies per chromosome (Table 3.2). An exception is the conjugative plasmid R6K which has a mol. wt of 25×10^6 daltons and is maintained as a relaxed plasmid.

Often it is important to measure plasmid copy number, and although a number of techniques are available they are not without their limitations.

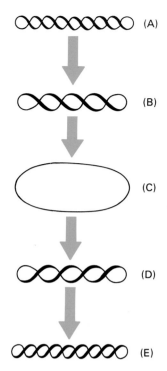

Fig. 3.2 Effect of intercalation of ethidium on supercoiling of DNA. As the amount of intercalated ethidium increases, the double helix untwists with the result that the supercoiling decreases until the open form of the circular molecule is produced. Further intercalation introduces excess turns in the double helix resulting in supercoiling in the opposite sense (note the direction of coiling at B and D). For clarity, only a single line represents the double helix.

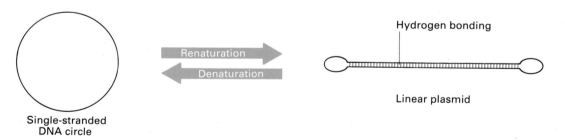

Fig. 3.3 The interconversion of a circular, single-stranded DNA molecule and a linear double-stranded plasmid.

A commonly used method (Twigg & Sherratt 1980) is to extract total DNA from a cell paste and to separate the chromosomal DNA from the plasmid DNA by agarose gel electrophoresis. After photographing the gel, the photographic negative is scanned with a densitometer. The density of the plasmid DNA relative to the chromosomal DNA is a measure of plasmid copy number. An alternative method is to quantify the plasmid DNA by high-performance liquid chromatography (Coppella *et al.* 1987). Both these

Table 3.1 Phenotypic traits exhibited by plasmid-carried genes.

antibiotic resistance	heavy-metal resistance
antibiotic production	bacteriocin production
degradation of aromatic compounds	induction of plant tumours
	hydrogen sulphide production
haemolysin production	Host-controlled restriction and modification
sugar fermentation	
enterotoxin production	

Table 3.2 Properties of some conjugative and non-conjugative plasmids of Gram-negative organisms.

Plasmid	Size (Mdal.)	Conjugative	No. of plasmid copies/ chromosome equivalent	Phenotype
Col E1	4.2	no	10–15	Col* E1 production
RSF 1030	5.6	no	20–40	ampicillin resistance
clo DF13	6	no	10	cloacin production
R6K	25	yes	13–38	ampicillin and streptomycin resistance
F	62	yes	1–2	—
RI	62.5	yes	3–6	multiple drug resistance
Ent P 307	65	yes	1–3	enterotoxin production

methods suffer from the disadvantage that errors can arise if there is differential or incomplete release of plasmid DNA from cell extracts.

An alternative procedure for estimating plasmid copy number is to measure the amount of a plasmid-encoded gene product. For example, β-lactamase activity can be measured if the plasmid specifies ampicillin resistance. Before this method can be used it is essential to show that the selected gene is expressed constitutively under all physiological conditions. Klotsky & Schwartz (1987) have shown that this is indeed the case for the ampicillin-resistance gene present on many cloning vectors used with *E. coli*.

Plasmid *incompatibility* is the inability of two different plasmids to coexist in the same host cell in the absence of selection pressure. The term incompatibility can only be used when it is certain that entry of the second plasmid has taken place and that DNA restriction is not involved. Groups of plasmids which are mutually incompatible are considered to belong to the same incompatibility group. Currently, over 30 incompatibility groups have been defined among plasmids of *E. coli* and 13 for plasmids of *Staphylococcus aureus*. Plasmids belonging to incompatibility classes P, Q, and W are termed *promiscuous* for they are capable of promoting their own transfer to a wide range of Gram-negative bacteria (group P and W) and of being stably maintained in these diverse hosts. Such promiscuous plasmids thus offer the potential of readily transferring cloned DNA molecules into a wide range of genetic environments (see p. 171).

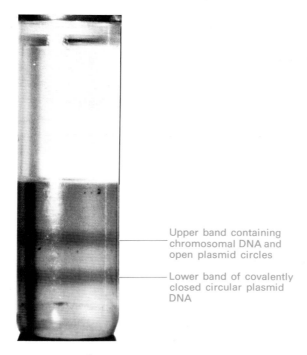

Upper band containing chromosomal DNA and open plasmid circles

Lower band of covalently closed circular plasmid DNA

Fig. 3.4 Purification of ColEI kan^R plasmid DNA by isopycnic centrifugation in a CsCl-EtBr gradient. (Photograph by courtesy of Dr G. Birnie.)

An extremely useful article explaining the terminology used in plasmid genetics is that of Novick *et al.* (1976). A much fuller discussion of the topics outlined above is provided by Falkow (1975) and Broda (1979).

The purification of plasmid DNA

An obvious prerequisite for cloning in plasmids is the purification of the plasmid DNA. Although a wide range of plasmid DNAs are now routinely purified, the methods used are not without their problems. Undoubtedly the trickiest stage is the lysis of the host cells; both incomplete lysis and total dissolution of the cells result in greatly reduced recoveries of plasmid DNA. The ideal situation occurs when each cell is just sufficiently broken to permit the plasmid DNA to escape without too much contaminating chromosomal DNA. Provided the lysis is done gently most of the chromosomal DNA released will be of high mol. wt and can be removed, along with cell debris, by high-speed centrifugation to yield a *cleared lysate*. The production of satisfactory cleared lysates from bacteria other than *E. coli*, particularly if large plasmids are to be isolated, is frequently a combination of skill, luck and patience.

Many methods are available for isolating pure plasmid DNA from cleared lysates but only two will be described here. The first of these is the 'classical' method and is due to Vinograd (Radloff *et al.* 1967). This method involves isopycnic centrifugation of cleared lysates in a solution of CsCl containing ethidium bromide (EtBr). EtBr binds by intercalating between the DNA base pairs, and in so doing causes the DNA to unwind. A CCC DNA molecule such as a plasmid has no free ends and can only unwind to a limited extent, thus limiting the amount of EtBr bound. A linear DNA molecule, such as fragmented chromosomal DNA, has no such topological constraints and can therefore bind more of the EtBr molecules. Because the density of the DNA/EtBr complex decreases as more EtBr is bound, and because more EtBr can be bound to a linear molecule than a covalent circle, the covalent circle has a higher density at saturating concentrations of EtBr. Thus covalent circles (i.e. plasmids) can be separated from linear chromosomal DNA (Fig. 3.4).

Currently the most popular method of extracting and purifying plasmid DNA is that of Birnboim and Doly (1979). This method makes use of the observation that there is a narrow range of pH (12.0–12.5) within which denaturation of linear DNA, but not covalently closed circular DNA, occurs. Plasmid-containing cells are treated with lysozyme to weaken the cell wall and then lysed with sodium hydroxide and sodium dodecyl sulphate. Chromosomal DNA remains in a high mol. wt form but is denatured. Upon neutralization with acidic sodium acetate, the chromosomal DNA renatures and aggregates to form an insoluble network. Simultaneously, the high concentration of sodium acetate causes precipitation of protein-SDS complexes and of high mol. wt RNA. Provided the pH of the alkaline denaturation step has been carefully controlled the CCC plasmid DNA molecules will remain in a native state and in solution while the contaminating macromolecules co-precipitate. The precipitate can be removed by centrifugation and the plasmid concentrated by ethanol precipitation. If necessary, the plasmid DNA can be purified further by gel filtration.

Although most cloning vehicles are of low mol. wt (see next section) it sometimes is necessary to use the much larger conjugative plasmids. Although these high mol. wt plasmids can be isolated by the methods just described, the yields often are very low. Either there is inefficient release of the plasmids from the cells as a consequence of their size or there is physical destruction caused by shear forces during the various manipulative steps. A number of alternative procedures have been described (Gowland & Hardmann 1986) many of which are a variation on that of Eckhardt (1978). Bacteria are suspended in a mixture of Ficoll and lysozyme and this results in a weakening of the cell walls. The samples then are placed in the slots of an agarose gel where the cells are lysed by the addition of detergent. The plasmids subsequently are extracted from the gel following electrophoresis. The use of agarose which melts at low temperature facilitates extraction of the plasmid from the gel.

Desirable properties of plasmid cloning vehicles

An ideal cloning vehicle would have the following three properties:
1 low mol. wt,
2 ability to confer readily selectable phenotypic traits on host cells,
3 single sites for a large number of restriction endonucleases, preferably in genes with a readily scorable phenotype.

The advantages of a low mol. wt are several. First, the plasmid is much easier to handle, i.e. it is more resistant to damage by shearing, and is readily isolated from host cells. Second, low mol. wt plasmids are usually present as multiple copies (see Table 3.2), and this not only facilitates their isolation but leads to gene dosage effects for all cloned genes. Finally, with a low mol. wt there is less chance that the vector will have multiple substrate sites for any restriction endonuclease (see below).

After a piece of foreign DNA is inserted into a vector the resulting chimaeric molecules have to be transformed into a suitable recipient. Since the efficiency of transformation is so low it is essential that the chimaeras have some readily scorable phenotype. Usually this results from some gene, e.g. antibiotic resistance, carried on the vector, but could also be produced by a gene carried on the inserted DNA.

One of the first steps in cloning is to cut the vector DNA and the DNA to be inserted with either the same endonuclease or ones producing the same ends. If the vector has more than one site for the endonuclease, more than one fragment will be produced. When the two samples of cleaved DNA are subsequently mixed and ligated, the resulting chimaeras will, in all probability, lack one of the vector fragments. It is advantageous if insertion of foreign DNA at endonuclease-sensitive sites inactivates a gene whose phenotype is readily scorable, for in this way it is possible to distinguish chimaeras from cleaved plasmid molecules which have self-annealed. Of course, readily detectable insertional inactivation is not essential if the vector and insert are to be joined by the homopolymer tailing method (see p. 32) or if the insert confers a new phenotype on host cells.

Some examples will be presented which illustrate the points raised above, but first we shall consider how some of the common plasmids rate as cloning vehicles.

Usefulness of 'natural' plasmids as cloning vehicles

The term 'natural' is used loosely in this context to describe plasmids which were not constructed *in vitro* for the sole purpose of cloning. Col E1 is a naturally occurring plasmid which specifies the production of a bacteriocin, colicin E1. By necessity this plasmid also carries a gene which confers on host cells immunity to colicin E1. RSF 2124 is a derivative of Col E1 which carries a transposon specifying ampicillin resistance. The exact origin of pSC101 is not clear, but for the purposes of this discussion we shall consider it to be a 'natural' plasmid. Details of these plasmids are shown in Table 3.3.

To clone DNA in pSC101, the plasmid DNA and the DNA to be inserted are digested with *Eco* RI, for example, and treated with DNA ligase. The ligated molecules are then used to transform a suitable recipient to tetracycline resistance. Unfortunately there is no easy genetical method of distinguishing chimaeras from reconstituted vector DNA unless the insert confers a new phenotype on the transformants. Two examples of the use of pSC101 for cloning DNA are presented in the next section. When using *Eco* RI, cloning with Col E1 as the vector is a little simpler. Transformants are selected on the basis of immunity to colicin E1 and chimaeras recognized by their inability to produce colicin E1. Unfortunately, screening

Table 3.3 Properties of some 'natural' plasmids used for cloning DNA.

Plasmid	Size (Mdal.)	Single sites for endonucleases	Marker for selecting transformants	Insertional inactivation of
pSC101	5.8	Xho I, Eco RI Pvu II, Hinc II Hpa I	tetracycline resistance	—
		Hind III, Bam HI Sal I	—	tetracycline resistance
Col E1	4.2	Eco RI	immunity to colicin E1	colicin E1 production
RSF 2124	7.4	Eco RI, Bam HI	ampicillin resistance	colicin E1 production

for immunity to colicin E1 is not technically simple, and plasmid RSF 2124 is more useful in this respect since transformants are selected by virtue of their ampicillin resistance.

Col E1 and plasmids derived from it (see later) have two distinct advantages over pSC101. They have a higher copy number and they can be enriched with chloramphenicol. When chloramphenicol is added to a late log-phase culture of a Col E1-containing strain of E. coli, chromosome replication ceases because of the need for continued protein synthesis. However, the cessation of protein synthesis has no effect on Col E1 replication, such that after 10–12 hours over 50% of the DNA in the cells is plasmid DNA (Hershfield et al. 1974). Since there may be 1000–3000 copies of the plasmid in each cell it is easy to see why chloramphenicol enrichment is a useful step in plasmid isolation.

Examples of the use of pSC101 for cloning
1. Expression of *Staphylococcus* plasmid genes in *E. coli*

For this experiment Chang and Cohen (1974) considered S. aureus plasmid p1258 (mol. wt 20×10^6) as being particularly appropriate for experiment involving interspecies genome construction, since it carries several different genetic determinants that were potentially detectable in E. coli. Moreover, agarose gel electrophoresis indicated that this plasmid is cleaved by the Eco RI restriction endonuclease into four easily identifiable fragments. Molecular chimaeras containing DNA derived from both Staphylococcus and E. coli were constructed by ligation of a mixture of Eco RI-cleaved pSC101 and p1258 DNA and then were used to transform a restrictionless strain of E. coli (Fig. 3.5). E. coli transformants that expressed the ampicillin resistance determinant carried by the Staphylococcus plasmid were selected and checked for tetracycline resistance.

Caesium chloride gradient analysis of one ampicillin-resistant, tetracycline-resistant chimaera showed that its buoyant density was intermediate to the buoyant densities of the parental plasmids. In addition, treatment of this chimaera with Eco RI produced two fragments, one the

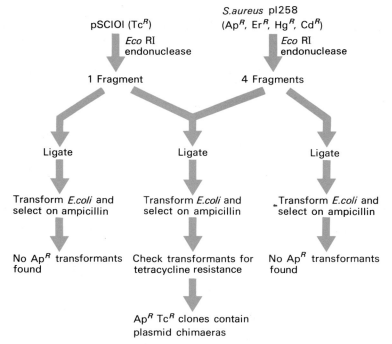

Fig. 3.5 Insertion of an *S. aureus* gene specifying ampicillin resistance into plasmid pSC101.

size of *Eco* RI cleaved pSC101 and the other the size of one of the *Eco* RI fragments of p1258.

Example of the use of pSC101 for cloning
2. Cloning of *Xenopus* DNA in *E. coli*

This experiment by Morrow *et al.* (1974) involved the construction *in vitro* of plasmid chimaeras composed of both prokaryotic and eukaryotic DNA, and the recovery of recombinant DNA molecules from transformed *E. coli* in the absence of selection for genetic determinants carried by the eukaryotic DNA. The amplified ribosomal DNA from *Xenopus laevis* oöcytes was used as the source of eukaryotic DNA for these experiments, since this DNA can be purified readily and had been well characterized. In addition, the repeat unit of *X. laevis* rDNA is susceptible to cleavage by *Eco* RI, resulting in the production of discrete fragments that can be linked to the pSC101 vector.

In this experiment (Fig. 3.6) a mixture of *Eco* RI-cleaved pSC101 DNA and *X. laevis* rDNA was ligated and was used to transform *E. coli* to tetracycline resistance. Fifty-five separate transformants were selected and their plasmid DNA extracted and analysed. All 55 plasmids gave a fragment of mol. wt 5.8×10^6 on *Eco* RI digestion and 13 of them yielded additional fragments corresponding in size to *Eco* RI-produced fragments of *X. laevis* rDNA. It was indeed fortuitous that such a high percentage (23.6%) of clones contained chimaeric molecules.

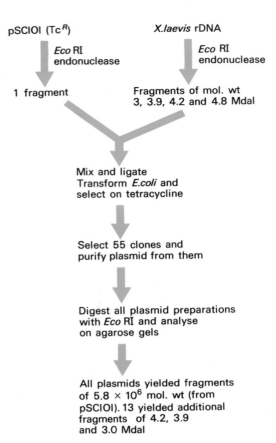

pSCIOI (Tc^R)   X.laevis rDNA

Eco RI endonuclease   *Eco* RI endonuclease

1 fragment

Fragments of mol. wt
3, 3.9, 4.2 and 4.8 Mdal

Mix and ligate
Transform *E.coli* and
select on tetracycline

Select 55 clones and
purify plasmid from them

Digest all plasmid preparations
with *Eco* RI and analyse
on agarose gels

All plasmids yielded fragments
of 5.8×10^6 mol. wt (from
pSCIOI). 13 yielded additional
fragments of 4.2, 3.9
and 3.0 Mdal

Fig. 3.6 Cloning of genes from *Xenopus laevis* in *Escherichia coli* with the aid of pSC101.

Construction and characterization of a new cloning vehicle: pBR322

Although pSC101, Col E1 and RSF 2124 can be used to clone DNA they suffer from a number of disadvantages as outlined above. For this reason considerable effort has been expended on constructing, *in vitro*, superior cloning vehicles. Undoubtedly the most versatile and widely used of these artificial plasmid vectors is pBR322. Plasmid pBR322 contains the Ap^R and Tc^R genes of RSF 2124 and pSC101 respectively, combined with replication elements of pMB1, a Col E1-like plasmid (Fig. 3.7a). The origins of pBR322, and its progenitor pBR313, are shown in Fig. 3.7b, and details of its construction can be found in the papers of Bolivar *et al.* (1977a, b).

Plasmid pBR322 has been completely sequenced. The original published sequence (Sutcliffe 1979) was 4362 base pairs long. Position O of the sequence was arbitrarily set between the A and T residues of the *Eco* RI recognition sequence (GAATTC). More recently (Backman & Boyer 1983, Peden 1983) the sequence has been revised by the inclusion of an additional CG base pair at position 526, thus increasing the size of the plasmid to 4363 base pairs. The most useful aspect of the DNA sequence is that it totally characterizes pBR322 in terms of its restriction sites such that the exact length of every fragment can be calculated. These fragments can serve as DNA markers for sizing any other DNA fragment in the range of several base pairs up to the entire length of the plasmid.

There are over 20 enzymes with unique cleavage sites on the pBR322 genome (Fig. 3.8). The target sites of seven of these enzymes (*Eco* RV, *Nhe* I, *Bam* HI, *Sph* I, *Sal* I, *Xma* III and *Nru* I) lie within the Tc^R gene, and there are sites for a further two (*Cla* I and *Hind* III) within the promoter of that gene. There are unique sites for three enzymes (*Pst* I, *Pvu* I and *Sca* I) within the Ap^R gene. Thus, cloning in pBR322 with the aid of any one of those 12 enzymes will result in insertional inactivation of either the Ap^R or the Tc^R markers. However, cloning in the other unique sites does not permit the easy selection of recombinants because neither of the antibiotic resistance determinants is inactivated.

Following manipulation *in vitro*, *E. coli* cells transformed with plasmids with inserts in the Tc^R gene can be distinguished from those cells transformed with recircularized vector. The former are Ap^R and Tc^S whereas the latter are both Ap^R and Tc^R. In practice, transformants are selected on the basis of their Ap resistance and then replica-plated onto Tc-containing media to identify those that are Tc^S. Cells transformed with pBR322 derivatives carrying inserts in the Ap^R gene can be identified more readily (Boyko & Ganschow 1982). Detection is based upon the ability of β-lactamase produced by Ap^R cells to convert penicillin to penicilloic acid, which in turn binds iodine. Transformants are selected on rich medium containing soluble starch and Tc. When colonized plates are flooded with an indicator solution of iodine and penicillin β-lactamase-producing (Ap^R), colonies clear the indicator solution whereas Ap^S colonies do not.

The *Pst* I site in the Ap^R gene is particularly useful because the 3'-tetranucleotide extensions formed on digestion are ideal substrates for terminal transferase. Thus this site is excellent for cloning by the homopolymer tailing method described on the previous chapter (see p. 32). If oligo (dG.dC) tailing is used, the *Pst* I site is regenerated (see Fig. 2.11) and the insert may be cut out with that enzyme. In addition, the *Pst* I site is particularly useful for obtaining expression of cloned genes and this aspect is covered in detail in Chapter 9 (p. 148).

Plasmid pBR322 is the most widely used cloning vehicle. In addition, it has been widely used as a model system for the study of prokaryotic transcription and translation as well as investigation of the effects of topological changes on DNA conformation. The popularity of pBR322 is a direct result of the availability of an extensive body of information on its structure and function. This in turn is increased with each new study. The reader wishing more detail on the structural features, transcriptional signals, replication, amplification, stability and conjugal mobility of pBR322 should consult the review of Balbas *et al.* (1986).

Example of the use of plasmid pBR322 as a vector: isolation of DNA fragments which carry promoters

Cloning into the *Hind* III site of pBR322 generally results in loss of tetracycline resistance. However, in some recombinants Tc^R is retained or even

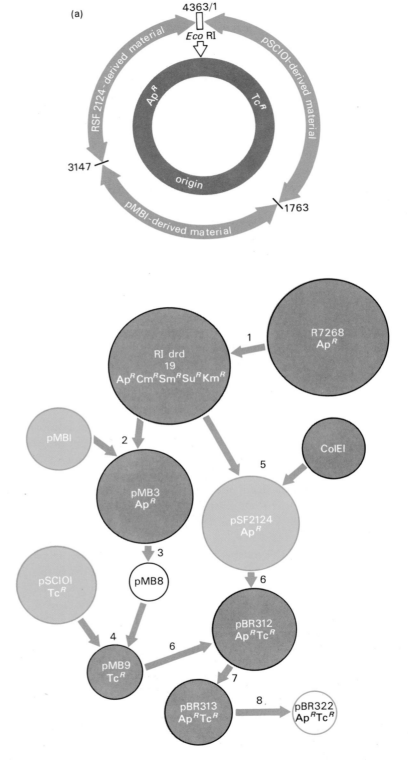

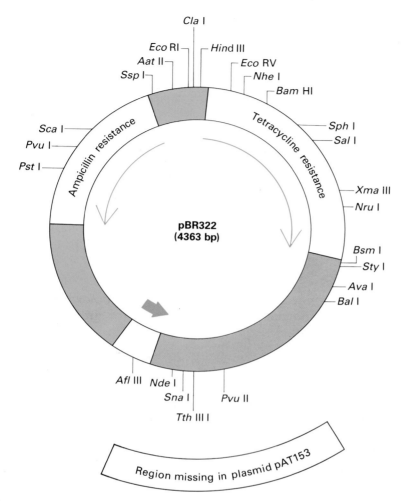

Fig. 3.8 The structure of pBR322 showing the unique cleavage sites. The thin arrows inside the circle show the direction of transcription of the Ap^R and Tc^R genes. The thick arrow shows the direction of DNA replication. Also shown is the region of pBR322 missing in plasmid pAT153 (see p. 56).

Fig. 3.7 *(Facing Page)* The origins of plasmid pBR322. (a) The boundaries between the pSC101, pMB1 and RSF 2124-derived material. The numbers indicate the positions of the junctions in base pairs from the unique *Eco* RI site. (b) The molecular origins of plasmid pBR322. R7268 was isolated in London in 1963 and later renamed R1. **1** A variant, R1*drd*19, which was derepressed for mating transfer, was isolated. **2** The Ap^R transposon, Tn3, from this plasmid was transposed onto pMB1 to form pMB3. **3** This plasmid was reduced in size by *Eco* RI* rearrangement to form a tiny plasmid, pMB8, which carries only colicin immunity. **4** *Eco* RI* fragments from pSC101 were combined with pMB8 opened at its unique Eco RI site and the resulting chimaeric molecule rearranged by *Eco* RI* activity to generate pMB9. **5** In a separate event, the Tn3 of R1*drd*19 was transposed to Col E1 to form pSF2124. **6** The Tn3 element was then transposed to pMB9 to form pBR312. **7** *Eco* RI* rearrangement of pBR312 led to the formation of pBR313, from which, **8**, two separate fragments were isolated and ligated together to form pBR322. During this series of constructions, R1 and Col E1 served only as carries for Tn3. (Reproduced by courtesy of Dr G. Sutcliffe and Cold Spring Harbor Laboratory.)

increased. This is because the *Hin*d III site lies within the promoter rather than the coding sequence. Thus whether or not insertional inactivation occurs depends on whether the cloned DNA carries a promoter-like sequence able to initiate transcription of the Tc^R gene. Widera *et al.* (1978) have used this technique to search for promoter-containing fragments.

Four structural domains can be recognized within *E. coli* promoters (see Chapter 9). These are:

1 position 1, the purine initiation nucleotide from which RNA synthesis begins,
2 position −6 to −12, the Pribnow box,
3 the region around base pair −35,
4 the sequence between base pairs −12 and −35.

Although the *Hin*d III site lies within the Pribnow box (Rodriguez *et al.* 1979) the box is re-created on insertion of a foreign DNA fragment (Fig. 3.9). Thus, when insertional inactivation occurs it must be the region from −13 to −40 which is modified.

Example of the use of plasmid pBR322 as a vector: expression in *E. coli* of a chemically synthesized gene for the hormone somatostatin

There are two reasons for selecting this particular early experiment to illustrate the use of pBR322 as a vector. First, the fact that gene manipulation had been successfully used to obtain a functional gene product from a chemically synthesized gene indicated that the potential of genetic engineering was not just a dream but had already been realized. Second, success was dependent on the use of a number of elegant 'tricks' and some of these warrant detailed examination. Although it does not make easy reading for the novice, the original paper is worth studying in detail for it is one of the classics of gene manipulation.

The rationale of the experiment was in fact to show that recombinant DNA technology can be used to fuse chemically synthesized genes to plasmid elements for expression in *E. coli* or other bacteria. As a model, Itakura *et al.* (1977) designed and synthesized a gene for the small hormone

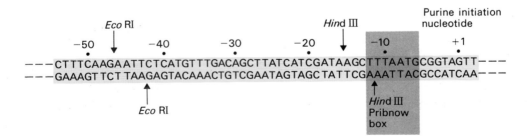

Fig. 3.9 DNA base sequence of the promoter of the tetracycline-resistance gene. The bases are numbered on the basis of the purine initiation nucleotide being position + 1. (In the conventional pBR322 map the bases are numbered from the *Eco* RI site.) The arrows indicate the positions of cleavage by restriction endonucleases *Eco* RI and *Hin*d III.

somatostatin. The somatostatin 'gene' was chosen because somatostatin is a small polypeptide of known amino acid composition, there were sensitive radioimmune and biological assays, and, being a hormone it was of intrinsic biological interest.

The somatostatin gene was synthesized chemically (see p. 9) such that there was an *Eco* RI site at one end and a *Bam* HI site at the other (Fig. 3.10). Also, a methionine codon preceded the normal NH$_2$-terminal amino acid of somatostatin and the COOH-terminal amino acid was followed by two stop codons.

In the first part of the experiment three new plasmids were created from pBR322. The control region of the *lac* operon, comprising the *lac* promoter, catabolite gene activator-protein binding site, the operator, the ribosome binding site, and the first seven triplets of the β-galactosidase structural gene, were inserted into the *Eco* RI site of pBR322 to create pBH10. Plasmid pBH10 has two *Eco* RI sites and one of these was removed to generate pBH20. Finally, the synthetic somatostatin gene was inserted next to the *lac* control gene to yield pSOM I (Fig. 3.11).

The DNA sequence of pSom I indicated that the clone carrying this plasmid should produce a peptide containing somatostatin, but no somatostatin was found. However, in reconstruction experiments it was observed that exogenous somatostatin was degraded rapidly in E. coli extracts. Thus the failure to find somatostatin activity could be accounted for by intracellular degradation by endogenous proteolytic enzymes. Such proteolytic degradation might be prevented by attachment of the somatostatin to a large protein, e.g. β-galactosidase. The β-galactosidase structural gene has an *Eco* RI site near the COOH-terminus and the available data on the amino acid sequence of this protein suggested that it would be possible to insert the synthetic gene into this site and still maintain the proper reading frame. In order to do this, two new plasmids pSom II and pSom II-3 were created.

The formation of pSom II (Fig. 3.11). Plasmid pSom I was digested with *Eco* RI and *Pst* I and the larger fragment, which contains the synthetic somatostatin gene, purified by gel electrophoresis. This fragment has lost the *lac* control region and part of the Ap^R gene. The Ap^R gene was restored by ligating this large fragment to the small fragment produced by digesting pBR322 with *Eco* RI and *Pst* I. The ligated mixture was used to transform E. coli to ampicillin resistance and the plasmid from one Ap^R clone selected and called pSom II.

The formation of pSom II-3 (Fig. 3.12). Bacteriophage λ*plac* 5 which carries the *lac* control region and the entire β-galactosidase gene was digested

Eco RI																*Bam* HI
	Met	Ala	Gly	Cys	Lys	Asn	Phe	Phe	Trp	Lys	Thr	Phe	Thr	Ser	Cys	Stop	Stop
5' AATTC	ATG	GCT	GGT	TGT	AAG	AAC	TTC	TTT	TGG	AAG	ACT	TTC	ACT	TCG	TGT	TGA	TAG
G	TAC	CGA	CCA	ACA	TTC	TTG	AAG	AAA	ACC	TTC	TGA	AAG	TGA	AGC	ACA	ACT	ATCCTAG 5'

Fig. 3.10 Sequence of the chemically synthesized somatostatin gene.

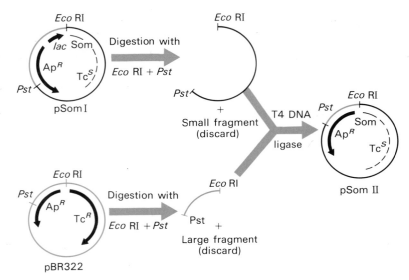

Fig. 3.11 Formation of pSom II. For clarity, only one strand of the plasmid DNA is shown.

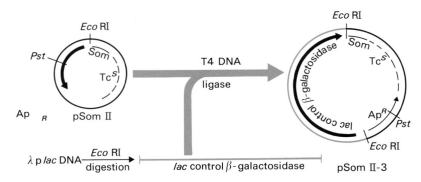

Fig. 3.12 Formation of pSom II-3 by insertion of *lac* region from λ*plac* into pSom II. For clarity, only one strand of DNA is shown.

with *Eco* RI and the mixture ligated with *Eco* RI-treated pSom II. The ligated mixture was used to transform *E. coli* and selection was made for blue colonies on medium containing ampicillin and the chromogenic substrate 5-bromo-4-chloro-3-indolyl-β-D-galactoside (Xgal). The rationale for this was as follows. Xgal is not an inducer of β-galactosidase but is cleaved by β-galactosidase, releasing a blue indolyl derivative. Since Xgal is not an inducer, only mutants constitutive for β-galactosidase produce blue colonies on medium containing Xgal. Plasmid pSom II-3 is maintained as a relaxed plasmid, i.e. multiple copies per cell. Thus, cells carrying pSom II-3 have multiple copies of the *lac* control region and can titrate out all the repressor produced by the single chromosomal *lac*I gene leading to a constitutive phenotype.

Approximately 2% of the transformants were blue and analysis of the plasmid from them showed the presence of a 4.4 Mdal. fragment identical to that found by digesting λ*plac*5 with *Eco* RI. Two orientations of the *Eco* RI *lac* fragment of λ*plac*5 are possible but only one of these would maintain the proper reading frame into the somatostatin gene. When a number of independent clones were examined, approximately 50% produced detectable somatostatin radioimmune activity and all of these had the desired orientation of the *lac* operon. The non-producing clones were found to have the opposite orientation.

It should be noted that no somatostatin radioimmune activity was detected prior to cyanogen-bromide cleavage of the total cellular protein. Since the antiserum used in the radioimmune assay requires a free NH_2-terminal alanine, no activity was expected prior to cleavage. Methionine residues are the site of cyanogen-bromide cleavage and it is for this reason that a methionine codon was included in the synthetic somatostatin gene (Figs 3.10 and 3.13).

Cyanogen bromide can only be used to cleave fusion proteins when the wanted portion has no internal methionine residues. Alternative methods for releasing the desired protein include the use of trypsin (Smith *et al.* 1984) and citraconic anhydride (Shine *et al.* 1980).

Improved vectors derived from pBR322

Over the years many different derivatives of pBR322 have been constructed, many to fulfil special-purpose cloning needs (Table 3.4). A compilation of the properties of these plasmids has been provided by Balbas *et al.* (1986). The design principles behind the improved general-purpose vectors are outlined below. Details of the more specialist vectors will be presented in later chapters.

As noted earlier, insertional inactivation frequently is used as a means of detecting the formation of recombinant molecules. The fact that cloning in pBR322 using *Eco* RI does not result in insertional inactivation is

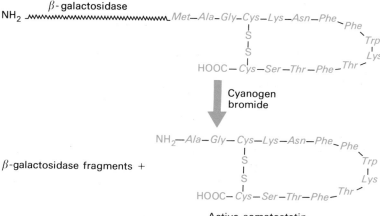

Fig. 3.13 Cleavage of the chimaeric protein by cyanogen bromide to yield active somatostatin. The somatostatin can readily be purified from cyanogen bromide and the fragments of β-galactosidase.

Table 3.4 Summary of modified vectors derived from pBR322.

Function of modified vector	Reference
vectors to facilitate expression of proteins	Chapter 9
vectors for the identification of regulatory signals	This Chapter
vectors for the direct selection of recombinants	This Chapter
vectors with additional restriction sites	This Chapter
vectors with different, additional or improved selective markers	This Chapter
vectors with increased stability	Chapter 9
vectors with altered copy numbers	This Chapter, Chapter 9
plasmids for DNA sequencing	Chapter 6
plasmids to permit secretion of proteins	Chapter 9
gene-fusion vectors to facilitate protein purification	Chapter 9
vectors for use in *E. coli* and unrelated organisms (shuttle vectors)	Chapters 10, 11, 12 and 13

unfortunate for *Eco* RI is one of the most widely used enzymes. Bolivar (1978) has constructed two different pBR322-derived plasmids containing unique *Eco* RI sites in selectable markers. One of these plasmids, pBR324, carries the Col E1 structural and immunity genes derived from pMB9. The position of a unique *Eco* RI site and a unique *Sma* I site in the Col E1 structural gene provides a relatively easy selection for *Eco* RI and *Sma* I endonuclease-generated •DNA fragments. The second plasmid, pBR325, carries the chloramphenicol-resistance gene derived from phage P1Cm. This plasmid has a unique *Eco* RI site in the Cm^R gene. Recovery of cells harbouring *Eco* RI-derived recombinant DNA molecules is facilitated by virtue of their $Ap^R Tc^R Cm^S$ phenotype.

Another derivative of pBR322, which now is widely used as a cloning vector, is pAT153. Twigg & Sherratt (1980) observed that when a particular *Hae* II fragment was removed from Col E1 the plasmid copy number in *E. coli* increased five- to sevenfold. The corresponding *Hae* II fragment was removed from pBR322 to produce pAT153 (Fig. 3.14). Cells containing pAT153 have only an increase in plasmid content of 1.5 to 3 times compared with those containing pBR322. Although this increase is smaller than for the Col E1 deletion derivatives, pAT153 is still a useful cloning vector because of greater levels of plasmid and plasmid-specified gene products. In terms of biological containment pAT153 has a great advantage over pBR322. Although pBR322 is not self-transmissible, it can be mobilized at a frequency of 10^{-1} from cells containing a conjugative plasmid plus plasmid Col K. However, the *Hae* II fragment removed from pBR322 during the formation of pAT153 contains a DNA sequence (called *nic* or *bom*) essential for conjugal transfer. As a consequence pAT153 cannot be mobilized and this provides a means of biological containment.

A whole series of vectors, the pUC series, has been developed from the mp series of M13-based sequencing vectors (Vieira & Messing, 1982, 1987, Yanisch-Perron *et al.* 1985). Since these vectors are described in the next chapter, only the barest details will be given here. The essence of the pUC vectors is that they carry a *polylinker* or *multiple cloning site*. A polylinker is a short stretch of DNA carrying sites for many different

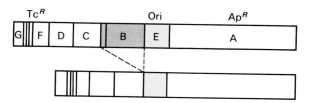

Fig. 3.14 The origin of pAT153. Plasmid pBR322 is cut into 11 fragments on complete digestion with *Hae* II. These fragments are assigned letters in alphabetical order on the basis of size. To construct pAT153, pBR322 DNA was partially digested with *Hae* II and religated. According to Twigg & Sherratt (1980) only fragment B was removed, but it is now known that the small fragment immediately to the left of B was also lost. See also Fig. 3.8.

restriction endonucleases (Fig. 3.15). The polylinker is placed within a gene such that gene function is maintained unless exogenous DNA is inserted at any of the restriction cleavage sites.

Direct selection vectors

Most plasmid vectors contain at least two selectable markers. One marker is kept intact and is used to select transformants. The transformants are then screened to detect inactivation of the second marker, indicating the insertion of foreign DNA. It would be more convenient if recombinant plasmids could be selected directly after transformation, and consequently a number of workers have constructed direct selection vectors.

Schumann (1979) has described a plasmid (pKN80) that carries a fragment of phage Mu DNA encoding a killing function that is expressed efficiently upon transformation of pKN80 into Mu-sensitive bacteria. Cloning of DNA fragments at the unique *Hpa* I site of pKN80 results in insertional inactivation of the killing function. Since then, many different direct selection vectors have been described (Roberts *et al.* 1980, Dean 1981, Hennecke *et al.* 1982, Hashimoto-Gotoh *et al.* 1986). Many of these vectors suffer from some or all of the following limitations: (a) requirement for specialized strains or media for selection, (b) limited availability of cloning sites, (c) inability to achieve regulated expression of the inserted fragments, and (d) high level of false positives. These problems have been overcome by Stevis & Ho (1987) who made use of the fact that high levels of xylose isomerase activity in wild-type *E. coli* strains results in a Xyl⁻ phenotype. They constructed plasmid pLX100, in which a gene for xylose isomerase, containing contiguous unique sites for *Hin*d III, *Pst* I, *Bam* HI and *Xho* I, is placed under the control of the *lac* promoter. *E. coli* transformants containing pLX100 cannot grow in minimal medium with xylose

Low copy number plasmid vectors

High copy number plasmid vectors such as pBR322 are widely used because they are easily purified and, via a gene dosage effect, they can direct the synthesis of high levels of cloned gene products. However there

pUC19

| 1 THR ACC (ATG) | 2 MET ATG | 3 ILE ATT | 4 THR ACG | 1 pro CCA | 2 ser AGC | 3 leu TTG *(Hind III)* | 4 his CAT *(Sph I)* | 5 ala GCC | 6 cys TGC *(Pst I)* | 7 arg AGG | 8 ser TCG *(Sal I / Acc I / Hinc II)* | 9 thr ACT | 10 leu CTA *(Xba I)* | 11 glu GAG | 12 asp GAT *(Bam HI)* | 13 pro CCC | 14 arg CGG *(Xma I / Sma I)* | 15 val GTA *(Kpn I)* | 16 pro CCG | 17 ser AGC *(Sst I)* | 18 ser TCG | 5 ASN AAT *(Eco RI)* | 6 SER TCA | 7 LEU CTG | 8 ALA GCC *(Hae III)* |

pUC18

| 1 THR ACC (ATG) | 2 MET ATG | 3 ILE ATT | 4 THR ACG | 5 ASN AAT *(Eco RI)* | 6 SER TCC | 1 ser AGC *(Sst I)* | 2 ser AGC | 3 leu TTG *(Hind III)* | 13 arg AGG | 14 cys TGC *(Pst I)* | 15 ala GCA | 16 ser AGC | 17 leu TTG | 6 ala GCA | 7 leu TTG | 8 ala GCA *(Hae III)* |

pUC13

| 1 THR ACC (ATG) | 2 MET ATG | 3 ILE ATT | 4 THR ACG | 5 ASN AAT *(Eco RI)* | 6 SER TCC | 1 pro CCA | 2 ser AGC | 3 leu TTG *(Hind III)* | 4 gly GGC | 5 cys TGC *(Pst I)* | 6 arg AGG | 7 ser TCG *(Sal I / Acc I / Hinc II)* | 8 thr ACT | 9 leu CTA *(Xba I)* | 10 glu GAG | 11 asp GAT *(Bam HI)* | 12 pro CCC | 13 arg CGG *(Xma I / Sma I)* | 14 ala GCG | 15 ser AGC *(Sst I)* | 16 his CAT | 17 leu TTG | 18 ser TCG *(Hind III)* |

pUC12

| 1 THR ACC (ATG) | 2 MET ATG | 3 ILE ATT | 4 THR ACG | 5 ASN AAT *(Eco RI)* | 6 SER TCC | 1 ser AGC *(Sst I)* | 2 ala GCA | 3 pro CCC | 4 gly GAT *(Bam HI)* | 5 pro GAT | 6 gly GGG | 7 ser TCG *(Sal I / Acc I / Hinc II)* | 8 pro CCT | 9 ser TCG | 10 thr ACC | 11 asp GAT *(Sal I / Acc I / Hinc II)* | 12 ser AGC | 13 pro CCA | 14 ser AGC | 15 ser AGC *(Sst I)* | 16 ser TCG *(Hind III)* | 17 leu TTG | 8 ALA GCC *(Hae III)* |

pUC9

| 1 THR ACC (ATG) | 2 MET ATG | 3 ILE ATT | 4 THR ACG | 1 pro CCA | 2 ser AGC *(Sst I)* | 3 leu TTG | 4 ala GCT | 5 ala GCA *(Pst I)* | 6 arg CGA | 7 arg CGG | 8 arg CGG *(Bam HI)* | 9 ile ATC | 10 pro CCC *(Xma I / Sma I)* | 11 gly GGG | 6 SER TCA | 7 LEU CTG | 8 ALA GCC *(Hae III)* |

pUC8

| 1 THR ACC (ATG) | 2 MET ATG | 3 ILE ATT | 4 THR ACG | 5 ASN AAT *(Eco RI)* | 6 SER TCC | 1 arg CGG *(Xma I / Sma I)* | 2 gly GGA | 3 asp GAT *(Bam HI)* | 4 val GTC | 5 asp GAC | 6 leu CTG *(Pst I)* | 7 cys TGC | 8 arg AGG | 9 pro CCA | 10 pro CCA *(Sal I / Acc I / Hinc II)* | 11 ala GCA | 7 LEU CTG | 8 ALA GCC *(Hae III)* |

pUC7

| 1 THR ACC (ATG) | 2 MET ATG | 3 ILE ATT | 4 THR ACG | 5 ASN AAT *(Eco RI)* | 1 ser AGC | 2 pro CCC | 3 asp GAT *(Bam HI)* | 4 pro CCG | 5 ser TCG *(Sal I / Acc I / Hinc II)* | 6 thr ACC | 7 cys TGC *(Pst I)* | 8 arg AGG *(Pst I)* | 9 ser TCG *(Sal I / Acc I / Hinc II)* | 10 ser ACG | 11 asp GAT *(Bam HI)* | 12 pro CCG | 13 gly GGG | 14 asn AAT | 6 SER TCA *(Eco RI)* |

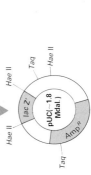

Circular plasmid map:
Center: pUC(-1.8 Mdal.)
Segments: lac Z', Amp^R
Sites: Hae II, Taq, Hae II, Taq, Hae II, Hae II

are some genes which cannot be cloned on high copy number vectors
because their presence seriously disturbs the normal physiology of the
cell. Examples include genes which encode surface structural proteins, e.g.
*omp*A (Beck & Bremer 1980), or proteins that regulate basic cellular metabolism, e.g. *pol*A (Murray & Kelley 1979). One strategy for cloning such
genes is to use low copy number vectors. Two suitable low copy number
vectors are phage λ in its lysogenic state (see Chapter 4) and plasmid
pSC101. Although pSC101 was the first plasmid to be used for cloning *in
vitro* (p. 46), it is seldom used because it carries only a single marker, and
insertional inactivation cannot be used to screen for recombinant clones.

Two sets of vectors have been constructed which retain the low copy
number of pSC101 but have two more antibiotic-resistance markers and
unique cleavage sites for many restriction endonucleases (Hashimoto-
Gotoh *et al.* 1981, Stoker *et al.* 1982).

Runaway plasmid vectors

One reason for cloning a gene on a multicopy plasmid is to increase
greatly its expression and hence facilitate purification of the protein it
encodes. However, as indicated above, some genes cannot be cloned on
high copy number vectors because excess gene product is lethal to the cell.
The use of low copy number vectors avoids cell killing, but this may be
self-defeating since expression of the cloned gene will be reduced. A
solution to this problem is to use runaway plasmid vectors.

The first runaway plasmid vectors were described by Uhlin *et al.*
(1979). At 30°C the plasmid vector is present in a moderate number of
copies per cell. Above 35°C all control of plasmid replication is lost and the
number of plasmid copies per cell increases continuously. Cell growth and
protein synthesis continue at the normal rates for two—three hours at the
higher temperature. During this period products from genes on the plasmid
are overproduced. Eventually inhibition of cell growth occurs and the cells
lose viability, but at this stage plasmid DNA may account for 50% of the
DNA in the cell. The two runaway vectors described by Uhlin *et al.* (1979),
now called pBEU1 and pBEU2, each carry a unique *Bam* HI site and a
single antibiotic-resistance marker.

More recently, improved runaway vectors have been constructed. Yasuda
and Takagi (1983) have described a runaway vector carrying a Km^R marker
and having unique recognition sites for 5 restriction endonucleases, and
Uhlin *et al.* (1983) have described one encoding resistance to Ap and Tc,
the latter marker with unique sites to permit insertional inactivation.
Larsen *et al.* (1984) have constructed cloning vectors that are present in
one copy per chromosome at temperatures below 37°C and that display
uncontrolled replication at 42°C. In addition they carry a partitioning
function (*par*, see p. 164) which stabilizes the plasmid at low temperatures
when grown in the absence of selection pressure.

Fig 3.15 (*Facing page*) Genetic maps of some pUC plasmids. The multiple
cloning site (MCS) is inserted into the *lacZ* gene but does not interfere with gene
function. The additional codons present in the *lacZ* gene as a result of the
polylinker are labelled with lower-case letters. These polylinker regions (MCS) are
identical to those of the M13 mp series of vectors (see p. 80).

4   Bacteriophage and Cosmid Vectors for *E. coli*

Essential features of bacteriophage λ

Bacteriophage λ is a genetically complex but very extensively studied virus of *E. coli*. Because it has been the object of so much molecular genetical research it was natural that, right from the beginnings of gene manipulation, it should have been investigated and developed as a vector. The DNA of phage λ, in the form in which it is isolated from the phage particle, is a linear duplex molecule of about 48.5 kb pairs. The entire DNA sequence has been determined (Sanger *et al.* 1982). At each end are short single-stranded 5'-projections of 12 nucleotides, which are complementary in sequence and by which the DNA adopts a circular structure when it is injected into its host cell, i.e. λ DNA naturally has cohesive termini which associate to form the *cos site*.

Functionally related genes of phage λ are clustered together on the map, except for the two positive regulatory genes *N* and *Q*. Genes on the left of the conventional linear map (Fig. 4.1) code for head and tail proteins of the phage particle. Genes of the central region are concerned with recombination (e.g. *red*) and the process of lysogenization in which the circularized chromosome is inserted into its host chromosome and stably replicated along with it as a prophage. Much of this central region, including these genes, is not essential for phage growth and can be deleted or replaced without seriously impairing the infectious growth cycle. Its dispensability is crucially important, as will become apparent later, in the construction of vector derivatives of the phage. To the right of the central region are genes

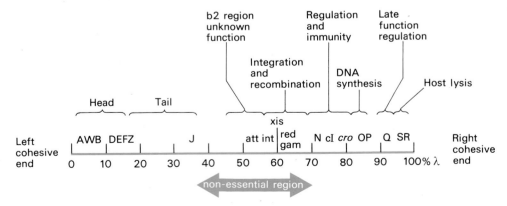

Fig. 4.1 Map of the λ chromosome, showing the physical position of some genes on the full-length DNA of wild-type bacteriophage λ. Clusters of functionally related genes are indicated.

concerned with gene regulation and prophage immunity to superinfection (*N*, *cro*, *cI*), followed by DNA synthesis (*O*, *P*), late function regulation (*Q*) and host cell lysis (*S*, *R*). Figure 4.2 illustrates the λ life-cycle.

Promoters and control circuits

As we shall see, it is possible to insert foreign DNA into the chromosome of phage λ derivatives, and in some cases foreign genes can be expressed efficiently via λ promoters. We must therefore briefly consider the promoters and control circuits affecting λ gene expression.

In the lytic cycle, λ transcription occurs in three temporal stages; early, middle and late. Basically, early gene transcription establishes the lytic cycle (in competition with lysogeny); middle gene products replicate and recombine the DNA, and late gene products package this DNA into mature phage particles. Following infection of a sensitive host, early transcription proceeds from major promoters situated immediately to the left (P_L) and right (P_R) of the repressor gene (*cI*) (Fig. 4.3). This transcription is subject to repression by the product of the *cI* gene and in a lysogen this repression is the basis of immunity to superinfecting λ. Early in infection transcripts

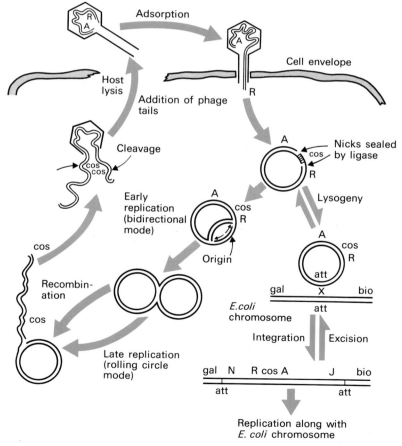

Fig. 4.2 Replication of phage λ DNA in lytic and lysogenic cycles.

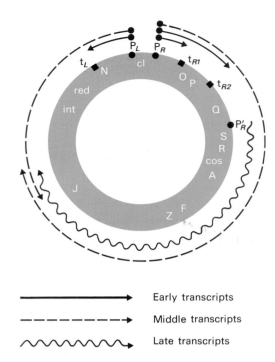

Fig. 4.3 Major promoters and transcriptional termination sites of phage λ. (See text for details.)

from P_L and P_R stop at termination sites t_L and t_{R_1}. The site t_{R_2} stops any transcripts that escape beyond t_{R_1}. Lambda switches from early- to middle-stage transcription by anti-termination. The N gene product, expressed from P_L, directs this switch. It interacts with RNA polymerase and, antagonizing the action of host termination protein ρ, permits it to ignore the stop signals so that P_L and P_R transcripts extend into genes such as *red* and O and P necessary for the middle stage. The early and middle transcripts and patterns of expression therefore overlap. The *cro* product, when sufficient has accumulated, prevents transcription from P_L and P_R. The gene Q is expressed from the distal portion of the extended P_R transcript and is responsible for the middle to late switch. This also operates by anti-termination. The Q product specifically anti-terminates the short P_R' transcript, extending it into the late genes, across the cohered *cos* region, so that many mature phage particles are ultimately produced.

Both N and Q play positive regulatory roles essential for phage growth and plaque formation; but an N^- phage *can* produce a small plaque if the termination site t_{R_2} is removed by a small deletion termed *nin* (N independent) as in λN^- *nin*.

Vector DNA

Wild-type λ DNA contains several target sites for most of the commonly used restriction endonucleases and so is not itself suitable as a vector. Derivatives of the wild-type phage have therefore been produced that either have a single target site at which foreign DNA can be inserted (*insertional* vectors), or have a pair of sites defining a fragment that can be removed and replaced by foreign DNA (*replacement* vectors). Since phage λ can accommodate only about 5% more than its normal complement of DNA, vector derivatives are constructed with deletions to increase the space within the genome. The shortest λ DNA molecules that produce plaques of nearly normal size are 25% deleted. Apparently if too much non-essential DNA is deleted from the genome it cannot be packaged into phage particles efficiently. This can be turned to advantage, for if the replaceable fragment of a replacement type vector either is removed by physical separation, or is effectively destroyed by treatment with a second restriction endonuclease that cuts it alone, then the deleted vector genome can give rise to plaques only if a new DNA segment is inserted into it. This amounts to positive selection for recombinant phage carrying foreign DNA.

Many vector derivatives of both the insertional and replacement types were produced by several groups of researchers (e.g. Thomas *et al.* 1974, Murray & Murray 1975, Blattner *et al.* 1977, Leder *et al.* 1977). Most of these vectors were constructed for use with *Eco* RI, *Bam* HI or *Hind* III, but their application can be extended to other endonucleases by the use of linker molecules. These early vectors have been largely superseded by improved vectors for rapid and efficient genomic library or cDNA library construction (see below and Chapter 7). However, we shall discuss them briefly here because they illustrate some basic principles.

The λWES.λB′ phage was an early vector, widely used at one time, which illustrates several important points. For details of the construction of this phage the reader is referred to papers of Thomas *et al.* (1974) and Leder *et al.* (1977). The DNA map of this vector is shown in Fig. 4.4. We can see that the phage has been constructed with three amber mutations in genes *W*, *E* and *S*. These reduce the likelihood of recombinants escaping from the laboratory environment, since appropriate amber suppressor strains are very uncommon in nature. The fragment designated C in wild-type λ has been deleted by restriction and religation *in vitro*. In addition, the two most righthand *Eco* RI sites have been eliminated and a *nin* deletion introduced. The deletions create space for insertion of foreign DNA. The B′ fragment is the replaceable fragment. (The B fragment has inadvertently been inverted during construction of the vector and is designated B′.) In use, the vector DNA is digested with *Eco* RI, then the B′ fragment may be removed by preparative gel electrophoresis or other physical methods. Alternatively, this fragment can be destroyed by treatment of the *Eco* RI digest with *Sst* I. The *Eco* RI treated foreign DNA is then added to a mixture of vector arms, the mixture is ligated and used to transfect an appropriate amber suppressor strain of *E. coli* so that viable recombinant phage are recovered. Joining of the two DNA arms without insertion of foreign DNA results in a molecule that is too short (9.8% (B′ fragment) + 11.3% (C fragment) + 6.1% (*nin* deletion) = 27.2% less than λ$^+$) to produce viable phage, even though it contains all of the genes necessary for lytic growth.

Figure 4.4 also shows the map of one of a set of 16 vector phages

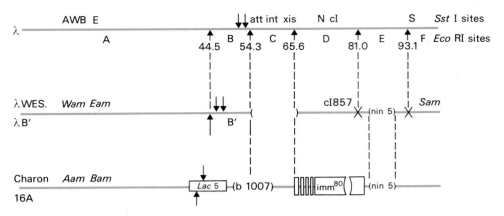

Fig. 4.4 Physical map of λDNA and two vector derivatives, λWES.λB' and charon 16A. Boxes indicate substitutions, but the lengths of substituted DNA are not exactly to scale. The λ regions are aligned. *Lac* 5 is a substitution from the *lac* region of *E. coli*. The box labelled *imm* 80 is a portion of phage φ80 DNA containing its immunity region. The region including four small boxes is derived from φ80 and is partially homologous to λ. Parentheses indicate deletions. Downward and upward arrows are *Sst* I and *Eco* RI restriction sites respectively. Numbers under *Eco* RI sites indicate the positions of the sites as percentages of the wild-type genome length.

constructed by Blattner *et al.* (1977). These have been aptly called Charon phages by their originators, after the old ferryman of Greek mythology who conveyed the spirits of the dead across the River Styx. Some of the Charon phages, like the one illustrated, have had amber mutations introduced in genes *A* and *B* in order to enhance biological containment. Charon 16A is an insertional vector with a single *Eco* RI site located in the gene for β-galactosidase (*lacZ*) which is included in the *lac5* DNA substitution. This is useful because there is a convenient colour test for the production of β-galactosidase. When the chromogenic substrate Xgal (see p. 78) is included in the plating medium, phage carrying *lac5* give dark blue plaques on Lac⁻ indicator bacteria. Potential success with Charon 16A cloning is detected by insertional inactivation of the Lac function which results in colourless plaques.

Another useful screening method employing insertional inactivation has been exploited by Murray *et al.* (1977). Insertion of foreign DNA at the single target within the immunity region of one of their vector molecules destroys the ability of the phage to produce a functional repressor so that recombinants give clear plaques which are readily distinguished from the turbid plaques of parental phages formed by simple rejoining of the two fragments of the vector DNA. This insertional inactivation of the *cI* gene also forms the basis of a powerful system for *selecting* recombinant formation in such phages as λgt10 (Nathans & Hogness 1983, Young & Davis 1983) and λNM1149 (Murray 1983). When the non-recombinant phages are plated on the *hflA* (*high frequency of lysogeny*) mutant of *E. coli* no plaques are formed. This is because lysogens, which are of course immune to superinfection, are created at such a high frequency in this host. However, recombinant phage have an inactive *cI* repressor gene, cannot form lysogens, and therefore do form plaques.

Improved phage λ replacement vectors

As with plasmid vectors, improved phage and vector derivatives have been developed in many laboratories. There have been several aims, among which are the following:

1 To increase the capacity for foreign DNA fragments, preferably for fragments generated by any one of several restriction enzymes (reviewed by Murray 1983).

2 To devise methods for positively selecting recombinant formation.

3 To allow RNA probes to be conveniently prepared by transcription of the foreign DNA insert. This facilitates the screening of libraries in chromosome walking procedures. An example of a vector with this property is λZAP.

4 To develop vectors for the insertion of eukaryotic cDNA such that expression of the cDNA, in the form of a fusion polypeptides with β-galactosidase, is driven in *E. coli*. This form of expression vector is useful in antibody screening. An example of such a vector is λgtll.

The first two points will be discussed here. The discussion of improved vectors in library construction and screening is deferred until Chapter 7.

The maximum capacity of phage λ derivatives can only be attained with vectors of the replacement type, so that there has also been an accompanying incentive to devise methods for positively selecting recombinant formation without the need for prior removal of the stuffer fragment. Even when steps are taken to remove the stuffer fragment by physical purification of vector arms, small contaminating amounts may remain, so that genetic selection for recombinant formation remains desirable. Two innovative means of achieving this are exploitation of the *A3* gene of phage T5 and the Spi⁻ phenotype.

Davison *et al.* (1979) have devised a vector in which the λB stuffer fragment of λWES.λB′ has been replaced by two identical 1.8 kb fragments of phage T5. These fragments carry the *A3* gene, which is known to prevent growth of T5 itself, or this adapted λ vector on *E. coli* carrying the plasmid Col Ib. Two fragments were found to be necessary in this new vector, called λgtWES.T5622, because a single 1.8 kb replacement of the λB fragment was too short to give viable phage. Positive selection for recombinant formation is imposed by plating on an *E. coli* host carrying plasmid Col Ib.

Wild-type λ cannot grow on *E. coli* strains lysogenic for phage P2; in other words the λ phage is Spi⁺ (sensitive to P2 inhibition). It has been shown that the products of λ genes *red* and *gam*, which lie in the region 64–69% on the physical map, are responsible for the inhibition of growth in a P2 lysogen (Herskowitz 1974, Sprague *et al.* 1978, Murray 1983). Hence vectors have been derived (e.g. λL47 and λ1059) in which the stuffer fragment includes the region 64–69%, so that recombinants in which this has been replaced by foreign DNA are phenotypically Spi⁻ and can be positively selected by plating on a P2 lysogen (Karn *et al.* 1980, Loenen & Brammar 1980).

Deletion of the *gam* gene has other consequences. The *gam* product is necessary for the normal switch in λDNA replication from the bidirectional mode to the rolling circle mode (see Fig. 4.2). Gam⁻ phage cannot generate the concatemeric linear DNA which is normally the substrate for packaging into phage heads. However, *gam⁻* phage do form plaques because the *rec* and *red* recombination systems act on circular DNA molecules to form multimers which can be packaged. Gam⁻ red⁻ phage are totally dependent upon *rec*-mediated exchange for plaque formation on *rec⁺* bacteria. λ DNA is a poor substrate for this *rec* mediated exchange. Therefore, such phage make vanishingly small plaques unless they contain one or more short DNA sequences called *chi* (cross-over *h*ot-spot *i*nstigator) sites, which stimulate *rec*-mediated exchange. The early replacement vector λWES.λB' generates *red⁻ gam⁺* clones. But many of the new replacement vectors with a large capacity (e.g. λL47, λ1059) generate *red⁻ gam⁻* clones. These vectors have, therefore, been constructed so as to include a *chi* site within the non-replaceable part of the phage.

The most recent generation of λ vectors combine a large capacity for foreign DNA, close to the theoretical limit of 23 kb, together with features that allow simple and efficient library construction (see Chapter 7). The replacement vectors EMBL3 and EMBL4 (Frischauf *et al.* 1983) have convenient polylinker (see p. 113) sequences flanking the replaceable fragment. Phages with inserts can be selected by their Spi⁻ phenotype, and are *chi⁺*. Derivatives of EMBL3 containing amber mutations are available (EMBL3 *Sam*, EMBL3 *Aam Bam*, EMBL3 *Aam Sam*). The inclusion of amber mutations in phage λ vectors not only increases biological containment, but can be used in a selective system for isolating DNA sequences linked to suppressor genes (see Chapter 13), and in recombinational screening (see Chapter 8).

High level expression of genes cloned in λ vectors

It is sometimes the aim of a gene manipulator to promote the expression of a gene which has been cloned so as to amplify the synthesis of a desirable gene product. There is much interest in improving the production of bacterial enzymes that are useful reagents in nucleic acid biochemistry itself, e.g. DNA ligase, DNA polymerase and restriction endonucleases. Panasenko *et al.* (1977) have described a recombinant phage, constructed *in vitro*, carrying the E. coli DNA ligase gene which, after induction of the recombinant lysogen, results in a 500-fold over-production of the enzyme so that it represents 5% of the total cellular protein of E. coli. Dramatic amplification depends *inter alia* upon both increasing the gene dosage and ensuring efficient transcription. The gene dosage is increased as a result of phage DNA replication within the host; the level of transcription may be improved by suitable choice of vector and subsequent manipulation of the recombinant phage.

A great deal of our knowledge about expression of genes cloned in phage λ comes from the studies of N. E. Murray, W. J. Brammar and their colleagues on a model system in which genes from the *trp* operon of *E. coli* are inserted in the phage genome either by manipulation *in vitro* or by genetic methods *in vivo* (Hopkins *et al.* 1976, Moir & Brammar 1976). The following discussion is based on their work.

First, we must distinguish between cases where the inserted DNA does or does not include the bacterial *trp* promoter. If the insert *does* include its own promoter, the yield of *trp* enzymes can be enhanced simply by delaying cell lysis so that the number of gene copies is increased and the time available for expression is extended. This was originally achieved by making the vector S^-. Moir and Brammar obtained better amplification of gene products by including mutations in gene Q or N. In Q^- phage all the late functions, including that of S, are blocked, and in addition, packaging of the replicated DNA is prevented which even further extends the availability of the DNA. An N^- phage is also defective in late functions and although it replicates more slowly than $N^+ Q^-$ phage, yields of enzyme achieved were at least as great. In such infected cells anthranilate synthetase, the product of the *trp E* gene, comprised more than 25% of the total soluble protein.

λ *trp* phage *lacking* the *trp* promoter have been constructed so that *trp* expression is initiated at the promoter P_L of the leftward operon of phage λ. This operon has two useful features:
1 P_L is a powerful promoter,
2 the anti-termination effect of gene N expression permits transcription through sequences which might otherwise prevent expression of a distant inserted gene.

Once again, cell lysis and DNA packaging were prevented by mutations in Q and S and additionally the *cro*⁻ mutation was introduced so as to derepress transcription from P_L. Cells infected with such a phage may contain as much as 10% of the soluble protein as anthranilate synthetase. However, these phage were difficult to construct and propagate so an alternative approach was adopted. The *cro* gene lies within the immunity region and its product is immune-specific. The *cro* product of the hetero-immune phage 434 will not interact with P_L of λ. Hybrid phage containing P_L from λ but *cro* and P_R from 434 are therefore phenotypically Cro⁻ as far as leftward transcription is concerned. Infection with such a λ*trp* derivative, which also carried the S^- mutation, gave cells in which 25% of the soluble protein was anthranilate synthetase. Derivatives of this type which are *Nam* can be grown on a non-suppressing host providing they also carry the *nin* deletion of t_{R_2}. Thus amplification can be modulated by controlling the suppression of the *Nam* gene. This is a useful property since extreme overproduction of a product relaxes the selection that can be imposed on a recombinant clone and may lead to problems of instability.

This elegant exploitation of the genetics of phage λ demonstrates the advantages of a well-characterized genetic system and shows that useful amplification of a variety of gene products may be achieved with λ vectors, even in cases where a strong promoter recognized by the host RNA polymerase does not accompany the inserted gene. Lessons learnt from this model system have been applied to the amplification of *E. coli* DNA polymerase I and T4 DNA ligase in induced lysogens of λ recombinants carrying these genes (Kelley *et al.* 1977, Murray *et al.* 1979, Wilson & Murray 1979).

Packaging phage λ DNA *in vitro*

So far, we have considered only one way of introducing manipulated phage DNA into the host bacterium, i.e. by transfection of competent bacteria (see Chapter 1). Using freshly prepared λ DNA that has not been subjected to any gene manipulation procedures, transfection will result in typically about 10^5 plaques per µg of DNA. In a gene manipulation experiment in which the vector DNA is restricted etc., and then ligated with foreign DNA, this figure is reduced to about 10^4-10^3 plaques per µg of vector DNA. Even with perfectly efficient nucleic acid biochemistry some of this reduction is inevitable. It is a consequence of the random association of fragments in the ligation reaction, which produces molecules with a variety of fragment combinations, many of which are inviable. Yet, in some contexts, 10^6 or more recombinants are required. The scale of such experiments can be kept within a reasonable limit (less than 100 µg vector DNA) by packaging the recombinant DNA into mature phage particles *in vitro*.

Placing the recombinant DNA in a phage coat allows it to be introduced into the host bacteria by the normal processes of phage infection, i.e. phage adsorption followed by DNA injection. Depending upon the details of the experimental design, packaging *in vitro* yields about 10^6 plaques per µg of vector DNA after the ligation reaction.

Figure 4.5 shows some of the events occurring during the packaging process that take place within the host during normal phage growth and which we now require to perform *in vitro*. Phage DNA in concatemeric form, produced by a rolling circle replication mechanism (see Fig. 4.2), is the substrate for the packaging reaction. In the presence of phage head precursor (the product of gene *E* is the major capsid protein) and the product of gene *A*, the concatemeric DNA is cleaved into monomers and encapsidated. Nicks are introduced in opposite strands of the DNA, 12 nucleotide pairs apart at each *cos* site, to produce the linear monomer with its cohesive termini. The product of gene *D* is then incorporated into what now becomes a completed phage head. The products of genes *W* and *FII*, among others, then unite the head with a separately assembled tail structure to form the mature particle.

The principle of packaging *in vitro* is to supply the ligated recombinant DNA with high concentrations of phage head precursor, packaging proteins and phage tails. Practically this is most efficiently performed in a very concentrated mixed lysate of two induced lysogens, one of which is blocked at the pre-head stage by an amber mutation in gene *D* and therefore accumulates this precursor while the other is prevented from forming any head structure by an amber mutation in gene *E* (Hohn & Murray 1977). In the mixed lysate, genetic complementation occurs and exogenous DNA is packaged (Fig. 4.6). Although concatemeric DNA is the substrate for packaging (covalently joined concatemers are of course produced in the ligation reaction by association of the natural cohesive ends of

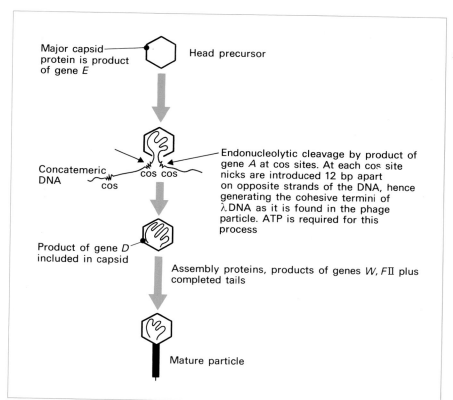

Fig. 4.5 Simplified scheme showing packaging of phage λDNA into phage particles.

λ), the *in-vitro* system *will* package added monomeric DNA, which presumably first concatemerizes non-covalently.

There are two potential problems associated with packaging *in vitro*. First, endogenous DNA derived from the induced prophages of the lysogens used to prepare the packaging lysate can itself be packaged. This can be overcome by choosing the appropriate genotype for these prophages, i.e. excision upon induction is inhibited by the *b2* deletion (Gottesman & Yarmolinsky 1968) and *imm 434* immunity will prevent plaque formation if an *imm 434* lysogenic bacterium is used for plating the complex reaction mixture. Additionally, if the vector does not contain any amber mutation an Su⁻ indicator can be used so that endogenous DNA will not give rise to plaques. The second potential problem arises from recombination in the lysate between exogenous DNA and induced prophage markers. If troublesome, this can be overcome by using recombination-deficient (i.e. *red⁻*, *rec⁻*) lysogens and by UV-irradiating the cells used to prepare the lysate, so eliminating the biological activity of the endogenous DNA (Hohn & Murray 1977).

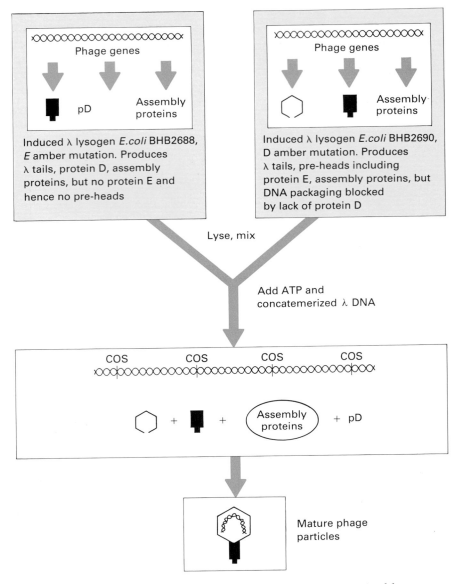

Fig. 4.6 *In-vitro* packaging of concatemerized phage λDNA in a mixed lysate.

Cosmid vectors

As we have seen, concatemers of unit-length λDNA molecules can be efficiently packaged if the *cos* sites — substrates for the packaging dependent cleavage — are 37–52 kb apart (75–105% the size of λ⁺DNA). In fact only a small region in the proximity of the *cos* site is required for recognition by the packaging system (Hohn 1975).

Plasmids have been constructed which contain a fragment of λDNA including the *cos* site (Collins & Brüning 1978, Collins & Hohn 1979). These plasmids have been termed *cosmids* and can be used as gene cloning vectors in conjunction with the *in-vitro* packaging system. Figure 4.7 shows a gene cloning scheme employing a cosmid. Packaging the cosmid recombinants into phage coats imposes a desirable selection upon their size. With a cosmid vector of 5 kb we demand the insertion of 32–47 kb of foreign DNA—much more than a phage λ vector can accommodate. Note that after packaging *in vitro*, the particle is used to infect a suitable host. The recombinant cosmid DNA is injected and circularizes like phage DNA but replicates as a normal plasmid without the expression of any phage functions. Transformed cells are selected on the basis of a vector drug-resistance marker.

Used in this way, cosmids provide an efficient means of cloning large pieces of foreign DNA. In addition, the system produces a very low

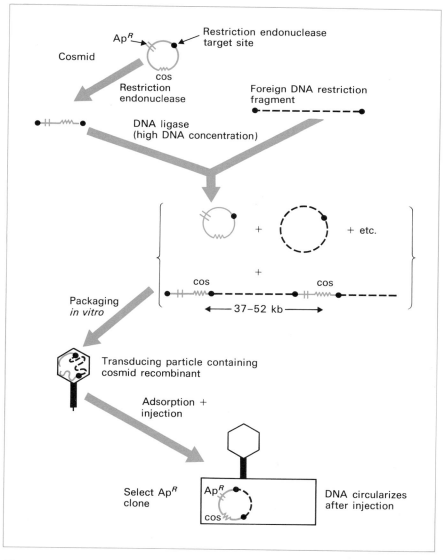

Fig. 4.7 Simple scheme for cloning in a cosmid vector. (See text for details.)

background of non-recombinant clones without recourse to selection by insertional inactivation or pretreatment of the linearized vector DNA with alkaline phosphatase (see p. 28). Because of their capacity for large fragments of DNA, cosmids are particularly attractive vectors for constructing libraries of eukaryotic genome fragments. Partial digestion with restriction endonuclease provides suitably large fragments. However, there is a potential problem associated with the use of partial digests in this way. This is due to the possibility of two or more genome fragments joining together in the ligation reaction, hence creating a clone containing fragments that were not initially adjacent in the genome. This would give an incorrect picture of their chromosomal organization. The problem can be overcome by size-fractionation of the partial digest.

Even with sized foreign DNA, in practice cosmid clones may be produced that contain non-contiguous DNA fragments ligated to form a single insert. The problem can be solved by dephosphorylating the foreign DNA fragments so as to prevent their ligation together. This method is very sensitive to the exact ratio of target-to-vector DNAs (Collins & Brüning 1978) because vector-to-vector ligation can occur. Furthermore, recombinants with a duplicated vector are unstable and break down in the host by recombination, resulting in the propagation of non-recombinant cosmid vector.

Such difficulties have been overcome in a cosmid cloning procedure devised by Ish-Horowicz & Burke (1981). By appropriate treatment of the cosmid vector pJB8 (Fig. 4.8) lefthand and righthand vector ends are purified which are incapable of self-ligation but which accept dephosphorylated foreign DNA. Thus the method eliminates the need to size the foreign DNA fragments, and prevents formation of clones containing short foreign DNA or multiple vector sequences.

An alternative solution to these problems has been devised by Bates & Swift (1983), who have constructed cosmid c2XB. This plasmid carries a *Bam* HI insertion site and two *cos* sites separated by a blunt-end restriction site (Fig. 4.9). The creation of these blunt ends, which ligate only very inefficiently under the conditions used, effectively prevents vector self-ligation in the ligation reaction.

Phasmid vectors

A second combination of plasmid and phage λ sequences has been devised to exploit the virtues of each type of vector. This combination consists of a plasmid vector carrying a λ attachment (λ*att*) site. The plasmid may insert into a phage λ genome by means of the site-specific recombination mechanism of the phage that is normally responsible for recombinational insertion of the phage into the bacterial chromosome during lysogen formation. This reversible recombinational insertion of plasmid into the phage is referred to as 'lifting' the plasmid and generates a phage genome containing one or more plasmid molecules (depending upon the length of the plasmid). These novel genetic combinations are called *phasmids* (Brenner *et al.* 1982). They contain functional origins of replication of the plasmids and of λ, and may be propagated as a plasmid

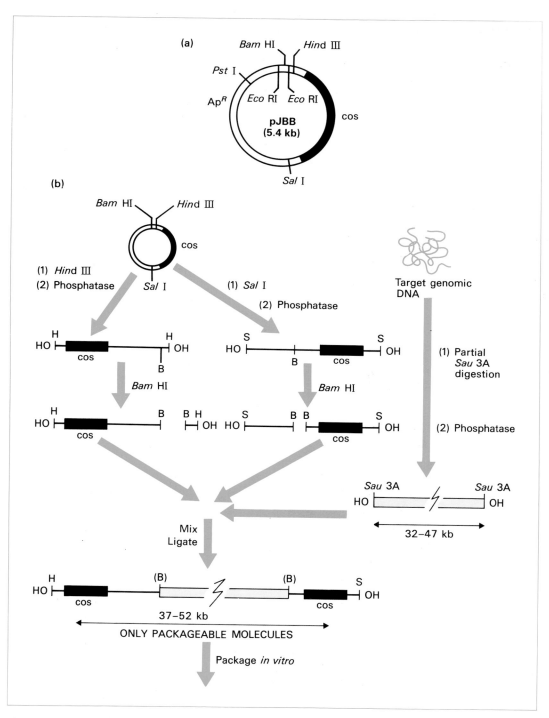

Fig. 4.8 Cosmid cloning scheme of Ish-Horowicz & Burke. (a) Map of cosmid pJB8. (b) Application to the construction of a genomic library of fragments obtained by partial digestion with *Sau* 3A. This restriction endonuclease has a tetranucleotide recognition site and generates fragments with the same cohesive termini as *Bam* HI (see p. 20).

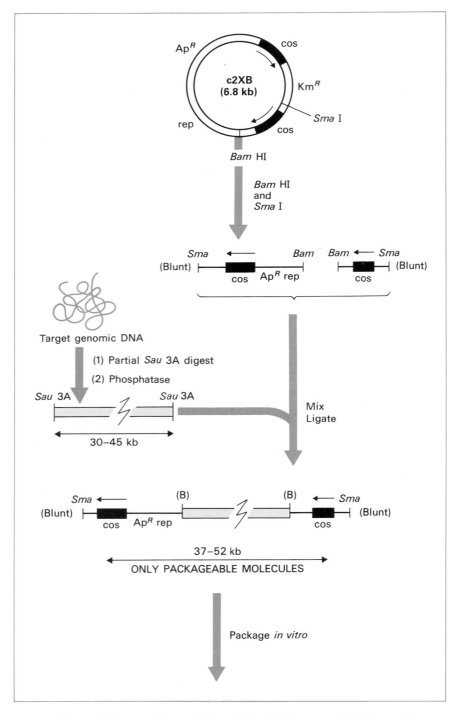

Fig. 4.9 Cosmid cloning scheme of Bates & Swift. The cosmid c2XB contains two *cos* sites, separated by a site for the restriction endonuclease *Sma* I which creates blunt ends. These blunt ends ligate only very inefficiently under the conditions used and effectively prevent the formation of recombinants containing multiple copies of the vector.

or as a phage in appropriate *E. coli* strains. Reversal of the lifting process releases the plasmid vector.

Phasmids may be used in a variety of ways; for instance, DNA may be cloned in the plasmid vector in a conventional way and then the recombinant plasmid can be lifted onto the phage. Phage particles are easy to store, they have an effectively infinite shelf-life, and screening phage plaques by molecular hybridization often gives cleaner results than screening bacterial colonies (see Chapter 8). Alternatively, a phasmid may be used as a phage-cloning vector, from which subsequently a recombinant plasmid may be released. A highly developed and novel phasmid vector, λZAP, with components of λ, M13, T3, and T7 phages, is described at the end of the next section, where the development of M13 vector derivatives is explained.

DNA CLONING WITH SINGLE-STRANDED DNA VECTORS

M13, f1 and fd are filamentous coliphages containing a circular single-stranded DNA molecule. These coliphages have been developed as cloning vectors for they have a number of advantages over other vectors, including the other two classes of vector for *E. coli*, plasmids and phage λ. However, in order to appreciate their advantages it is essential to have a basic understanding of the biology of filamentous phages.

The biology of the filamentous coliphages

The phage particles have dimensions 900 × 9 nm and contain a single-stranded circular DNA molecule which is 6407 (M13) or 6408 (fd) nucleotides long. The complete nucleotide sequences of fd and M13 are available and they are 97% homologous. The differences consist mainly of isolated nucleotides here and there, mostly affecting the redundant bases of codons, with no blocks of sequence divergence. Sequencing of f1 DNA indicates that it is very similar to M13 DNA.

The filamentous phages only infect strains of enteric bacteria harbouring F pili. The adsorption site appears to be the end of the F pilus, but exactly how the phage genome gets from the end of F pilus to the inside of the cell is not known. Replication of phage DNA does not result in host cell lysis. Rather, infected cells continue to grow and divide, albeit at a rate slower than uninfected cells, and extrude virus particles. Up to 1000 phage particles may be released into the medium per cell per generation (Fig. 4.10).

The single-stranded phage DNA enters the cell by a process in which decapsidation and replication are tightly coupled. The capsid proteins enter the cytoplasmic membrane as the viral DNA passes into the cell while being converted to a double-stranded replicative form (RF). The RF multiplies rapidly until about 100 RF molecules are formed inside the cell. Replication of the RF then becomes asymmetric, due to the accumulation of a viral-encoded single-stranded specific DNA binding protein. This

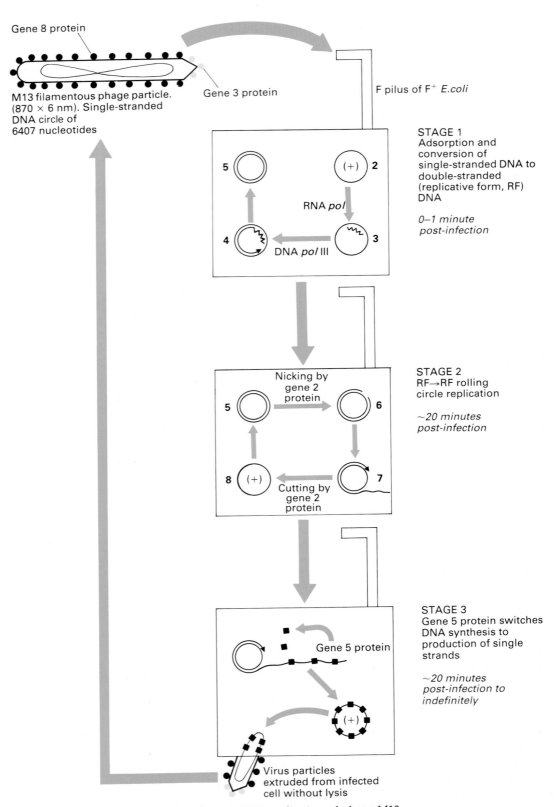

Fig. 4.10 Life cycle and DNA replication of phage M13.

protein binds to the viral strand and prevents synthesis of the complementary strand. From this point on only viral single strands are synthesized.
These progeny single strands are released from the cell as filamentous
particles following morphogenesis at the cell membrane. As the DNA
passes through the membrane the DNA binding protein is stripped off
and replaced with capsid protein.

Why use single-stranded vectors?

For several important applications of cloned DNA, single-stranded DNA
is required. Sequencing by the dideoxy method requires single-stranded
DNA, as do commonly used techniques for oligonucleotide-directed
mutagenesis, and certain methods of probe preparation (see p. 325). The
use of vectors that occur in single-stranded form is an attractive means of
combining cloning, amplification, and strand separation of an originally
double-stranded DNA fragment.

As single-stranded vectors the filamentous phages have a number of
advantages. First, the phage DNA is replicated via a double-stranded
circular DNA (RF) intermediate. This RF can be purified and manipulated
in vitro just like a plasmid. Second, both RF and single-stranded DNA
will transfect competent *E. coli* cells to yield either plaques or infected
colonies, depending on the assay method. Third, the size of the phage
particle is governed by the size of the viral DNA and therefore there are
no packaging constraints. Indeed, viral DNA up to six times the length of
M13 DNA has been packaged (Messing *et al.* 1981). Finally, with these
phages it is very easy to determine the orientation of an insert. Although
the relative orientation can be determined from restriction analysis of RF,
there is an easier method (Barnes 1980). If two clones carry the insert in
opposite directions, the single-stranded DNA from them will hybridize
and this can be detected by agarose gel electrophoresis. Phage from as
little as 0.1 ml of culture supernate can be used in assays of this sort,
making mass screening of cultures very easy.

In summary, as vectors, filamentous phages possess all the advantages
of plasmids while producing particles containing single-stranded DNA in
an easily obtainable form.

Development of filamentous phage vectors

Unlike λ, the filamentous coliphages do not have any non-essential genes
which can be used as cloning sites. However, in M13 there is a 507 base
pair intergenic region, from position 5498−6005 of the DNA sequence,
which contains the origins of DNA replication for both the viral and the
complementary strands. In most of the vectors developed so far foreign
DNA has been inserted at this site although it is possible to clone at the
carboxy-terminal end of gene IV (Boeke *et al.* 1979). The wild-type phages
are not very promising as vectors because they contain very few unique
sites within the intergenic region: *Asu* I in the case of fd, and *Asu* I and
Ava I in the case of M13. However, a site does not have to be unique to be
useful, as the example below shows.

The first example of M13 cloning made use of one of ten *Bsu* I sites in the genome, two of which are in the intergenic region (Messing *et al.* 1977). For cloning, M13 RF was partially digested with *Bsu* I and linear full-length molecules isolated by agarose gel electrophoresis. These linear monomers were blunt-end ligated to a *Hind* II restriction fragment comprising the *E. coli lac* regulatory region and the genetic information for the α-peptide of β-galactosidase. The complete ligation mixture was used to transform a strain of *E. coli* with a deletion of the β-galactosidase α-fragment and recombinant phage detected by intragenic complementation on media containing IPTG and Xgal. The IPTG is a gratuitous inducer of β-galactosidase, and Xgal a chromogenic substrate; where complementation occurs a blue colour is produced. One of the blue plaques was selected and the virus in it designated M13 mp1.

Insertion of DNA fragments into the *lac* region of M13 mp1 destroys its ability to form blue plaques, making detection of recombinants easy. However, the *lac* region only contains unique sites for *Ava* II, *Bgl* I and *Pvu* I and three sites for *Pvu* II, and there are no sites anywhere on the complete genome for the commonly used enzymes such as *Eco* RI of *Hind* III. To remedy this defect Gronenborn & Messing (1978) introduced an *Eco* RI site into the *lac* region of mp1 and the way they did so is particularly interesting. From DNA sequence data and restriction mapping (Fig. 4.11) it was known that a single base change (guanine residue 13 → adenine) in the codon for the 5th amino acid residue of the β-galactosidase α-fragment would create an *Eco* RI site. This in turn would lead to an aspartate residue being replaced by an asparagine residue that would have no significant effect on the complementation properties of the α-peptide.

Methylation of guanine has been shown to cause it to mispair with uracil. Therefore single-stranded DNA from M13 mp1 particles was treated with the methylating agent *N*-methyl-*N*-nitrosourea and then transformed into cells and allowed to undergo several cycles of replication. The CCC RF DNA was isolated from these cells and digested with *Eco* RI. Linear molecules of genome length were separated from undigested molecules by agarose gel electrophoresis, excised from the gel, and recircularized by way of their cohesive ends. The ligated molecules were transformed into cells and the resulting virus particles isolated. Following this procedure three individual clones with unique *Eco* RI restriction sites at different positions in the phage genome were isolated. Two of these, M13 mp2 and M13 mp3, were the result of the conversion of *Eco* RI* sequences in the *lac* fragment of M13 mp1 corresponding to the positions of amino acids 5 and 119 respectively. The *Eco* RI site of the third mutant was elsewhere in the genome.

The introduction of an *Eco* RI site into the *lac* region of M13 mp1, which is itself resistant to cleavage by *Eco* RI, creates a unique site to clone DNA fragments. Gronenborn & Messing (1978) have shown that insertion of *Eco* RI-derived fragments into M13 mp2 leads to inactivation or reduction of the β-galactosidase activity. Furthermore, expression of functions coded by the inserted DNA can be controlled by the *lac* regulatory region. However, to improve the versatility of M13 as a vector, Messing *et*

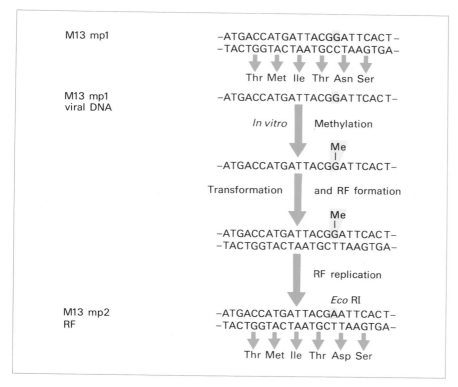

Fig. 4.11 *In-vitro* mutagenesis of the *lac* region of M13 mp1 to produce M13 mp2. Note that the methylguanine pairs with thymine but during replication this thymine will pair with adenine resulting in a G:C base pair being replaced by an A:T base pair.

al. (1981) constructed M13 mp7 which has a multipurpose cloning site in the *lac* region.

The actual construction of M13 mp7 is too complex to describe in detail here but a summary is provided in Fig. 4.12. Starting with M13 mp2, Rothstein *et al.* (1979) removed the single *Bam* HI site in gene III. Then a synthetic linker containing a *Bam* HI site was inserted at the *Eco* RI site to generate mWJ43. This phage still gives blue plaques on media containing IPTG and Xgal. Into this *Bam* HI site Messing *et al.* (1981) introduced yet another oligonucleotide linker, this time one containing *Pst* I and *Sal* I sites, to create M13 mp71. Since the reading frame of the *lac* region of M13 mp71 is still unaltered, a functional α-peptide of β-galactosidase is still produced. Although in this phage only the *Pst* I site is unique, the sites for *Sal* I, *Bam* HI and *Eco* RI are also usable for cloning because the ensuing loss of small inserts still results in a functional *lac* sequence. The sequence 5′-GTCGAC-3′ is cleaved by endonucleases *Sal* I, *Acc* I and *Hinc* II. Unfortunately, due to ambiguities in their recognition sequence the latter two endonucleases each cleave one additional site, both located in gene II. Consequently, these two sites were removed by chemical muta-genesis to generate M13 mp7.

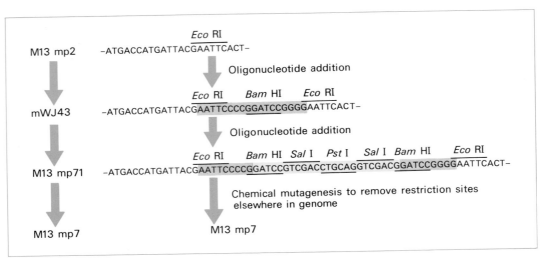

Fig. 4.12 The derivation of M13 mp7. Only the base sequence at the beginning of the *lac* region is shown.

The M13 mp7 vector has a symmetrical multiple restriction site, or *polylinker* region. This has the limitation that DNA fragments with dissimilar ends cannot be inserted because treatment of the polylinker site with a pair of restriction enzymes (e.g. *Sal* I and *Eco* RI) will generate a vector with ends derived from the outer pair of restriction sites (*Eco* RI). In order to overcome this problem Messing and his co-workers have constructed new derivatives, M13 mp8, mp10, mp11 (Messing & Vieira 1982), mp18 and mp19 (Norrander *et al.* 1983), which have unpaired restriction sites in non-symmetrical polylinker regions (see p. 58). These vectors have the advantage that DNA fragments with dissimilar ends can be cloned, and the orientation of the insert is fixed. M13 mp8 and mp9 have similar polylinker regions in opposite orientations, so that a foreign DNA fragment can be inserted either way round. The M13 mp10/mp11 pair, and the mp18/mp19 pair, also have their polylinker sites in opposite orientations. In practice, after such vectors have been digested with a pair of restriction enzymes it is often convenient to isolate the large vector band from an agarose gel, hence discarding the small restriction fragment. When ligation reactions are performed in the presence of the foreign DNA fragment, simple reclosure of the non-recombinant vector cannot occur, so that only white, recombinant plaques are obtained.

In this section we have concentrated on the vectors developed by Messing and his co-workers. There are several reasons for this. First, the clever use of mutagenesis to insert and remove restriction sites. Second, the construction of vectors with polylinkers. The convenience afforded by such polylinkers in gene manipulation experiments has been widely appreciated. Plasmid vectors have been constructed which also incorporate such polylinker sites. These are the pUC plasmids (Vieira & Messing 1982, Norrander *et al.* 1983). This principle has also been extended to the specialized vector πVX (see Chapter 7) to the phage λ vectors EMBL3 and EMBL4 and to many other recently developed vectors such as λZAP. Finally, the

M13 mp series of vectors have revolutionized site-directed mutagenesis and large-scale DNA sequencing. These topics are discussed in the following two chapters.

Exploitation of f1 (M13) biology to create a new family of single-stranded/double-stranded DNA vectors: pEMBL

The pUC series (see p. 56) of plasmid vectors was developed after the demonstration of the versatility of polylinker, or multiple, cloning sites in the M13 mp series of phage vectors, combined with blue/white screening of insertional inactivation of the α-peptide of β-galactosidase using Xgal plates. The pUC series have the additional advantage of being very small vectors. The important property of the M13 mp series of vectors is to provide cloned DNA in single-stranded form. However, one problem encountered with the single-stranded phage vectors is the instability of inserts when the insert exceeds a few kb in size (Zinder & Boeke 1982). The pEMBL series of vectors has been constructed so as to add to the features of single-stranded vectors the further advantages of a small vector size and stability of large inserts (Dente *et al.* 1983).

The basic principle that permitted this development was discovered by Dotto *et al.* (1981) and Dotto & Horiuchi (1981). They inserted into the *Eco* RI site of pBR322, a region of the f1 genome that contains all the *cis*-acting elements required for DNA replication and phage morphogenesis. They showed that, when F$^+$ *E. coli* containing the recombinant pBR322 were superinfected with f1 phage ('helper' phage), virion capsids were secreted that contained either f1 single-stranded DNA or the recombinant single-stranded pBR322 DNA, in about equal frequency.

In constructing pEMBL8, Dente *et al.* (1983) inserted a 1.3 kb fragment of the f1 genome containing all *cis*-acting elements required for DNA replication and phage morphogenesis into the unique *Nar* I site of pUC8 in both orientations, hence generating pEMBL8(+) and pEMBL8(−). The resulting plasmids (Fig. 4.13) conserved all the pUC8 features, including origin of plasmid replication, ampicillin resistance and blue colony phenotype on Xgal plates. Upon superinfection with f1 helper phage the intact origin of f1 replication is activated so that single-stranded plasmid DNA is

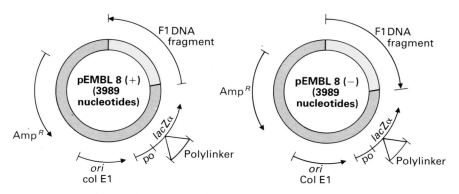

Fig. 4.13 Structures of pEMBL8(+) and pEMBL8(−).

produced, packaged, and secreted into the culture medium in virion-like particles. The orientation of the f1 DNA determines which of the two strands is found in the virion: pEMBL8 (+) and pEMBL8 (−) contain the non-coding and coding strands of the β-galactosidase gene, respectively. The plasmids pEMBL9 were constructed by replacing the polylinker of pEMBL8 with the polylinker of pUC9.

In practice, recombinant single-stranded DNA can be isolated from the mixture of particles in the culture medium and will contain a proportion of the helper phage DNA. For most purposes, such as DNA sequencing and oligonucleotide-directed mutagenesis, it is not necessary to perform purification steps to remove the helper DNA; its presence need not interfere with those applications. Thus, for example, the primer conventionally used in dideoxy sequencing is complementary to a region of the β-galactosidase gene which is present in the M13 mp series of vectors and also present in the pUC and pEMBL plasmid vectors. The 'wild-type' helper phage does not contain β-galactosidase sequences and so does not interfere with sequencing reactions because it does not hybridize with sequencing primer.

Double-stranded recombinant DNA can be prepared from cells harbouring the plasmid, in the absence of helper phage, where it replicates as a pUC-derived plasmid, and typically gives higher yields than RF preparations of the M13 mp series.

λZAP: exploitation of M13 biology and the specifity of phage T3, T7 RNA polymerases for their promoters

λZAP is a sophisticated vector that is particularly suitable for cloning cDNAs. This vector, developed and marketed commercially by Stratagene Cloning Systems (Short *et al.* 1988), illustrates several modern features, one of which is the exploitation of M13 biology so as to allow the cloned DNA insert to be automatically excised from the phage vector into a plasmid vector. In this regard λZAP is a form of phasmid vector. λZAP incorporates the following features:

1 Multiple unique cloning sites which can hold inserts up to 10 kb in length.
2 Insertional inactivation of β-galactosidase, giving blue/white screening on Xgal plates.
3 Expression of hybrid or fusion polypeptides in a manner analogous to λgt11 (see p. 133).
4 Automatic excision *in vivo* of the cloned DNA from the phage vector into a plasmid vector, Bluescript SK(−). This excision is brought about by M13 or f1 helper phage, and places the cloned DNA in a small plasmid vector convenient for restriction mapping and sequencing. This process eliminates the need to subclone DNA inserts from the λ phage into a plasmid by restriction and ligation.
5 The ability to prepare RNA transcripts of the inserted foreign DNA. Such transcripts are synthesized from either strand by using either the T3 or T7 phage RNA polymerases.

The structure of the lambda ZAP genome is shown in Fig. 4.14. Basically it is a lambda insertional vector. The foreign DNA—the figure illustrates an application with cDNA—is inserted at the multiple cloning site region, thus inactivating *lacZ* (and incidentally providing the opportunity for fusion polypeptide synthesis if the reading frame is correct). When an F$^+$ (or more commonly an F$'$) strain is infected with the recombinant phage and then superinfected with M13 helper phage, the helper phage supplies *trans*-acting proteins which recognize two DNA sequences incorporated in the λZAP arms. These two DNA sequences were derived from the f1 (M13) origin of replication, and signal the initiation and termination of DNA synthesis.

In normal f1 (M13) DNA replication these two distinct but overlapping DNA sequences act as follows. One site—the initiator—is recognized by the gene 2 protein which nicks the DNA(+) strand in RF DNA. This nick is then the site at which undirectional, rolling circle DNA synthesis is initiated, with displacement of the (+) strand. This (+) strand is then cleaved at the terminator site, which contains the same sequence that was nicked for initiation. Following cutting at the terminator by gene 2 protein, the two ends of the (+) strand are ligated to form a circular single (+) strand genome. This is subsequently converted to double-stranded RF DNA (refer to Fig. 4.10).

Short *et al.* (1988) exploited this knowledge by positioning f1 initiator and terminator sequences at separate places in the lambda vector. The DNA lying between these two sequences is that of a phagemid, Bluescript SK (−), which is analogous to the pEMBL vectors described above in that it can replicate as a plasmid conferring ampicillin resistance and in that it can be packaged into an M13 virion-like particle. This vector sequence has multiple cloning sites.

In use as a vector, the λZAP is cleaved into arms by cutting at one of the unique sites in the multiple cloning site. Commonly the unique *Eco* RI site is used. Foreign cDNA is inserted using *Eco* RI linkers. Once the required recombinant λZAP phage has been isolated, the automatic excision is accomplished by coinfecting F$'$ *E. coli* with the recombinant λZAP and f1 (M13) helper phage. This leads to the synthesis of a DNA strand from between the two signals, with strand displacement. The displaced strand is then automatically circularized in the helper phage-infected bacteria to form recombinant Bluescript SK(−), containing the cloned insert. This recombinant vector is packaged as a filamentous M13-like phage and secreted from the cell. Bluescript plasmids can be obtained by infecting an F$'$ strain and plating on ampicillin plates where the Col E1 origin of replication provides the replicon function and the *amp* gene provides ampicillin resistance. Colonies containing recombinant plasmid are obtained.

As shown in Fig. 4.14 the DNA insert of the Bluescript SK(−) is flanked by promoters for the RNA polymerases of phages T3 and T7. This allows RNA copies of the inserted DNA to be made conveniently *in vitro*. For example, if the recombinant plasmid is first linearized at the unique *Not*I site and then placed in a suitable reaction mixture containing T7 RNA polymerase and the four ribonucleoside triphosphates, a single-

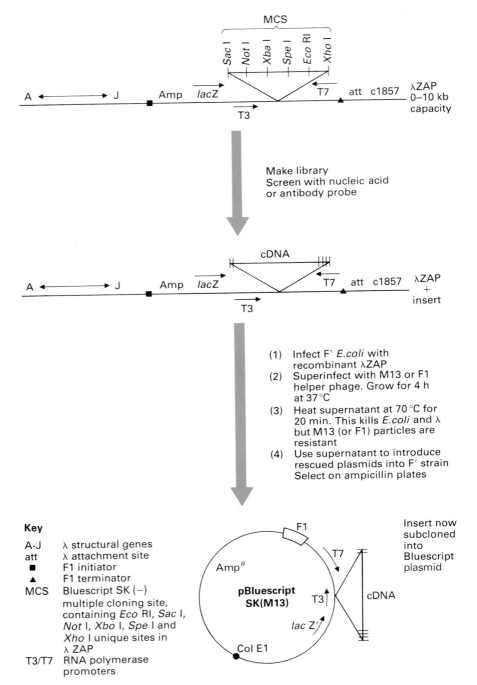

Fig. 4.14 The phage λ insertional vector, λZAP. Automatic excision.

stranded RNA molecule is synthesized initiating from the T7 promoter and terminating at the *Not*I cleaved end of the DNA. This facility is very useful for a number of applications. The RNA can be synthesized at high specific radioactivity by including an α-^{32}P- nucleoside triphosphate in

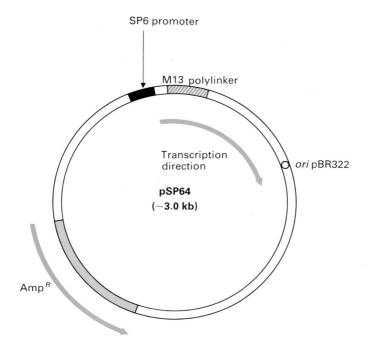

pSP64 promoter/polylinker region sequence:

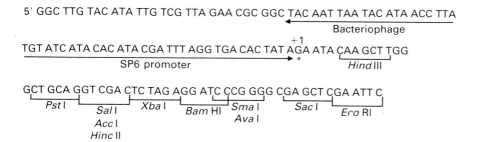

Fig. 4.15 Structure of pSP64, a vector for transcription of inserted DNA by phage SP6 RNA polymerase *in vitro*. The related plasmid pSP65 is similar, with an inverted polylinker. Transcription initiation occurs at the G marked with an asterisk (+1).

the reaction. Such a radioactive RNA copy can be used effectively to probe Northern blots (RNA–RNA hybrids have a melting temperature at about 15° higher than DNA–RNA hybrids and give 'clean' Northern blot results). These probes can also be used for RNase mapping (Zinn *et al.* 1983).

Clearly in such applications it is important that the correct strand of the DNA insert is transcribed into RNA. The positioning of T7 and T3 promoters at opposite ends of the inserts allows either strand to be transcribed, as desired, by appropriate choice of RNA polymerase, combined with linearization at a distal site with a suitable restriction endonuclease.

The purified RNA polymerases from coliphages T7, T3, T5 and the *Salmonella* phage SP6 are available commercially for applications such as these.

These polymerases are single-subunit enzymes, much simpler than the *E. coli* RNA polymerase. They exhibit high specificity for their individual promoter sequences. The T3 RNA polymerase, for example, is highly specific for a 23-base pair promoter sequence which differs by 3 base pairs from the T7 RNA polymerase promoter. Several simple plasmid vectors have been produced that incorporate promoters for phage RNA polymerase, situated adjacent to multiple cloning sites. The prototype on which these are based is the pSP64 and pSP65 pair (Fig. 4.13). This pair of vectors first established the usefulness of the phage polymerases, employing the phage SP6 RNA polymerase. In addition to the applications mentioned above, Krieg & Melton (1984) showed that transcripts of suitable inserts could function as synthetic mRNAs for translation in rabbit reticulocyte lysate or wheat-germ cell-free systems. For efficient translation in these systems the synthetic mRNAs must bear a 'cap' at the 5′ end. Fortunately this can conveniently be incorporated at the initiation of strand synthesis by the RNA polymerase if the cap dinucleotide GpppG is included in the RNA polymerase reaction mix. Capped RNAs have also been injected into the cytoplasm of *Xenopus* oöcytes for translation or, after injection into the oöcytes nucleus, for studies of RNA processing (Green *et al.* 1983).

5　Changing Genes:
Site-Directed Mutagenesis

Mutants are an essential prerequisite for any genetic study and no more so than in the study of gene structure and function relationships. Classically, mutants are generated by treating the test organism with chemical or physical agents that modify DNA (mutagens). This method of mutagenesis has been extremely successful, as witnessed by the growth of molecular biology, but suffers from a number of disadvantages. First, any gene in the organism can be mutated and the frequency with which mutants occur in the gene of interest can be very low. This means that selection strategies have to be developed. Second, even when mutants with the desired phenotype are isolated there is no guarantee that the mutation has occurred in the gene of interest. Third, prior to the development of gene cloning and sequencing techniques there was no way of knowing where in the gene the mutation had occurred and whether it arose by a single base change, an addition of DNA, or a deletion.

As techniques in molecular biology have developed, so that the isolation and study of a single gene is not just possible but routine, so mutagenesis has also been refined. Instead of crudely mutagenizing many cells or organisms and then analysing many thousands or millions of offspring to isolate a desired mutant, it is now possible to specifically change any given base in a cloned DNA sequence. This technique is known as *in-vitro* or site-directed mutagenesis. It has become a basic tool of gene manipulation for it simplifies DNA manipulations that in the past required a great deal of ingenuity and hard work, e.g. the creation or elimination of cleavage sites for restriction endonucleases. The importance of *in-vitro* mutagenesis goes beyond gene structure–function relationships for the technique enables mutant proteins to be generated with very specific changes in particular amino acids (protein engineering). Such mutants facilitate the study of the mechanisms of catalysis, substrate specificity, stability, etc. (see p. 353).

The single-primer method

The simplest method of site-directed mutagenesis is the single primer method (Gillam *et al.* 1980, Zoller & Smith 1983). The method involves priming *in-vitro* DNA synthesis with a chemically synthesized oligonucleotide (seven to twenty nucleotides long) that carries a base mismatch with the complementary sequence. As shown in Fig. 5.1 the method requires that the DNA to be mutated is available in single-stranded form, and cloning the gene in M13-based vectors makes this easy. However, DNA cloned in a plasmid and obtained in duplex form can also be converted to a partially single-stranded molecule that is suitable (Dalbadie-McFarland *et al.* 1982).

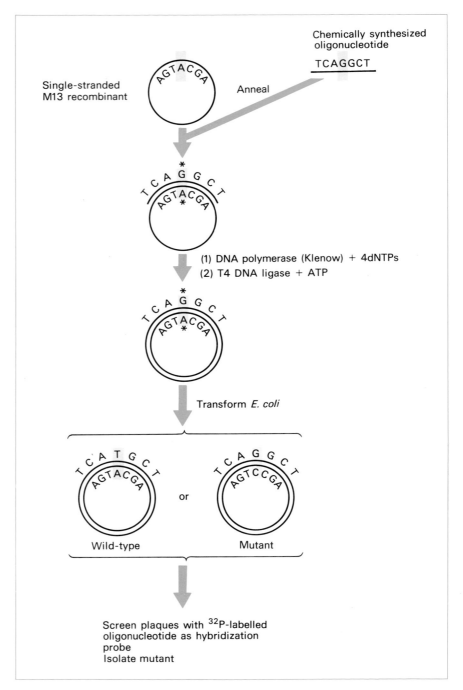

Fig. 5.1 Oligonucleotide-directed mutagenesis. Asterisks indicate mismatched bases.

The synthetic oligonucleotide primes DNA synthesis and is itself incorporated into the resulting heteroduplex molecule. After transformation of the host *E. coli*, this heteroduplex gives rise to homoduplexes whose sequences are either that of the original wild-type DNA or that containing the mutated base. The frequency with which mutated clones arise, compared with wild-type clones, may be low. In order to pick out mutants, the clones can be screened by nucleic acid hybridization (see Chapter 7) with ^{32}P-labelled oligonucleotide as probe. Under suitable conditions of stringency, i.e. temperature and cation concentrations, a positive signal will be obtained only with mutant clones. This allows ready detection of the desired mutant (Wallace *et al.* 1981, Traboni *et al.* 1983). In order to check that the procedure has not introduced other, adventitious, changes in the sequence it is prudent to check the sequence of the mutant directly by DNA sequencing.

A variation of the procedure (Fig. 5.2) outlined above involves oligonucleotides containing inserted or deleted sequences. As long as stable hybrids are formed with single-stranded wild-type DNA, priming of *in-vitro* DNA synthesis can occur giving rise ultimately to clones corresponding to the inserted or deleted sequence (Wallace *et al.* 1980, Norrander *et al.* 1983).

Deficiencies of the single-primer method

The efficiency with which the single-primer method yields mutants is dependent upon several factors. The double-stranded heteroduplex molecules that are generated will be contaminated both by any single-stranded non-mutant template DNA that has remained uncopied, and by partially double-stranded molecules. The presence of these species considerably reduces the proportion of mutant progeny. They can be removed by sucrose gradient centrifugation, or by agarose gel electrophoresis, but this is time-consuming and inconvenient.

Following transformation and *in-vivo* DNA synthesis, segregation of the two strands of the heteroduplex molecules can occur, yielding a mixed

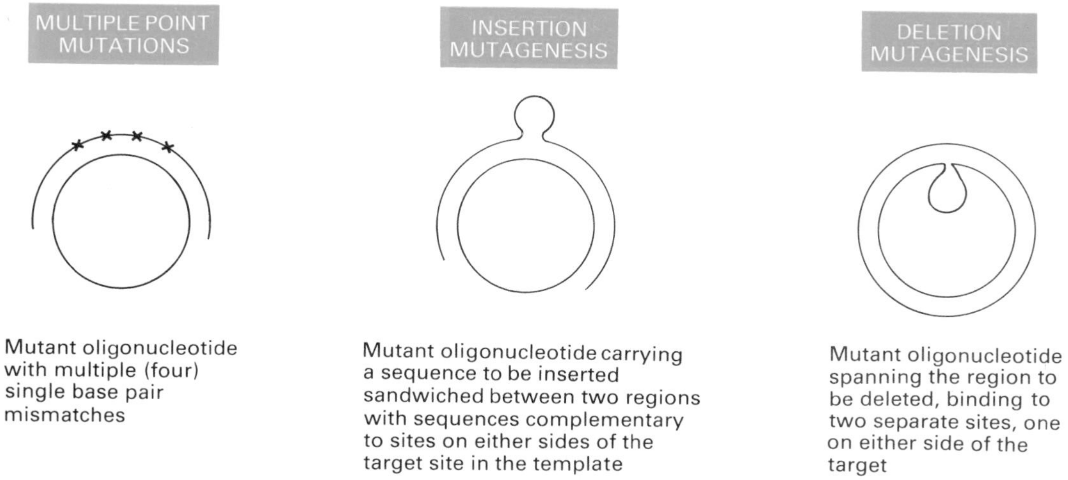

MULTIPLE POINT
MUTATIONS

INSERTION
MUTAGENESIS

DELETION
MUTAGENESIS

Mutant oligonucleotide
with multiple (four)
single base pair
mismatches

Mutant oligonucleotide carrying
a sequence to be inserted
sandwiched between two regions
with sequences complementary
to sites on either sides of the
target site in the template

Mutant oligonucleotide
spanning the region to
be deleted, binding to
two separate sites, one
on either side of the
target

Fig. 5.2 Oligonucleotide-directed mutagenesis used for multiple point mutation, insertion mutagenesis, and deletion mutagenesis.

population of mutant and non-mutant progeny. Mutant progeny have to be purified away from parental molecules, and this process is complicated by the cell's mismatch repair system. In theory, the mismatch repair system should yield equal numbers of mutant and non-mutant progeny, but in practice mutants are counter-selected. The major reason for this low yield of mutant progeny is that the methyl-directed mismatch repair system of *E. coli* favours the repair of non-methylated DNA. In the cell, newly synthesized DNA strands that have not yet been methylated are preferentially repaired at the position of the mismatch, thus preventing a mutation. In a similar way, the non-methylated *in-vitro*-generated mutant strand is repaired by the cell so that the majority of progeny are wild-type (Kramer *et al*. 1984a). The problems associated with the mismatch repair system can be overcome by using host strains carrying the *mut*L, *mut*S or *mut*H mutations which prevent the methyl-directed repair of mismatches.

Another problem with the single-primer method is that the mutant oligonucleotide is susceptible to digestion by 5′ → 3′ exonucleases. If these should degrade beyond the position of the single-base mismatch before the oligonucleotide is fully extended and ligated to give the double-stranded form, then a wild-type molecule would be generated. The solution is to use a high-quality preparation of the Klenow fragment of DNA polymerase. This should lack the 5′ → 3′ exonuclease activity, but if any is present the efficiency of mutation will drop markedly.

If mutant screening is to be avoided it is essential to maintain a high level of mutagenesis. Figure 5.3 shows the number of putative mutants which must be screened to give a 90% probability of isolating the desired mutant. The relationship is logarithmic. Consequently, if the efficiency of mutation drops below 10%, which can happen easily for the reasons cited above, the number of progeny to be screened will be very high.

Strand selection methods

A heteroduplex molecule with one mutant and one non-mutant strand must inevitably give rise to both mutant and non-mutant progeny upon replication. It would be desirable to suppress the growth of non-mutants, and various strategies have been adopted with this in mind.

The gapped duplex method. This method is illustrated in Fig. 5.4. The DNA to be mutagenized is cloned in a phage M13 vector and recovered in single-stranded form. The vector contains an amber (nonsense) mutation and therefore can be grown only on an amber suppressing host. A 'gapped duplex' is formed by annealing the single-stranded recombinant template with denatured linear double-stranded (RF) DNA derived from wild-type phage. As before, a synthetic oligonucleotide is used for mutagenesis and this is hybridized to the single-stranded region of the gapped duplex. The heteroduplex is converted to a fully double-stranded form with DNA polymerase and DNA ligase and transformed into *E. coli*. If a strain lacking an amber suppressor is used as the host, then only mutant progeny should be recovered. Kramer *et al*. (1984b) claim a maximum efficiency of 70% for the method. In practice the efficiency is much lower and the

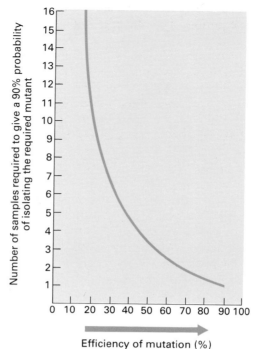

Fig. 5.3 Relationship between efficiency of mutation and the number of clones required to give a 90% probability of recovery of the required mutant.

mutation frequencies observed vary considerably depending on the target sequence and the mutant oligonucleotide. A disadvantage with this method is that should further rounds of mutation be necessary, the mutant sequence will need to be recloned in an amber vector.

Coupled priming and cyclic selection. Carter et al. (1985) have devised a system of strand selection which permits repeated rounds of mutagenesis without recloning. It depends on the sequence similarity between the recognition sites for the *Eco* K and *Eco* B restriction endonucleases (see p. 15). Two primers are used in this method, one to mutagenize the site of interest, and one to mutagenize the *Eco* K/*Eco* B strand selection site. To make full use of this approach a specialized vector has been constructed that carries a single *Eco* K restriction site. Thus, by mutation the selection site is altered from an *Eco* K to an *Eco* B site (Fig. 5.5). This means that the mutant progeny can be selected by growth in an *Eco* K-restricting host because progeny with an *Eco* K site do not survive. If required, the new *Eco* B-containing mutant clone can be subjected to a second mutagenesis step. A second selection primer is used to change the *Eco* B site back to an *Eco* K site, and an *Eco* B-restricting host is used to select for growth of the new mutant. Using this method, mutation efficiencies of up to 60% are possible.

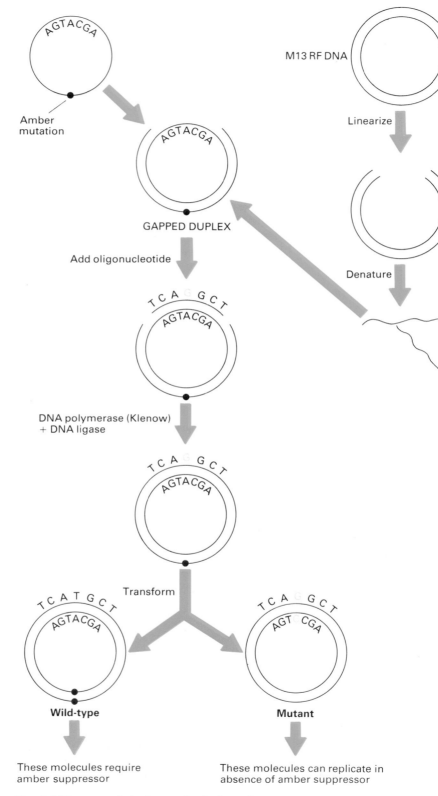

Fig. 5.4 The gapped duplex method of site-directed mutagenesis.

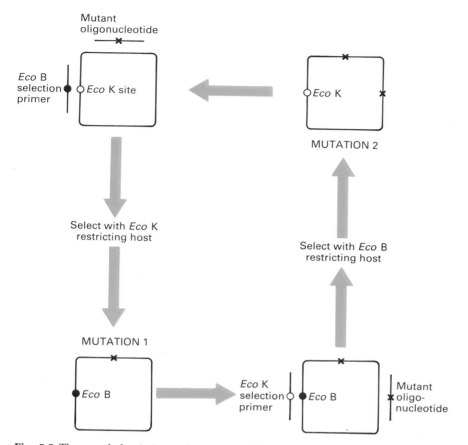

Fig. 5.5 The coupled priming technique used for cyclic selection mutagenesis.

The method of Kunkel (1985). In this method phage are grown on a specialized host *before* mutagenesis. The host used is deficient in dUTPase (*dut*) and uracil glycosylase (*ung*). The *dut* mutation results in increased intracellular dUTP levels and the *ung* mutation permits the incorporation of deoxyuridine into the DNA in place of thymidine at some positions. Phage M13 grown in a *dut ung* host contain 20–30 uracil residues per genome and are unable to grow in an *ung*⁺ host. Thus a heteroduplex composed of a uracil-containing parental strand and a mutant strand synthesized *in vitro* in the presence of dTTP will give rise to mostly mutant progeny when plated on a wild-type (*ung*⁺) host. Sequential mutations can be made using this system by growing the first mutant on a *dut ung* host before subjecting it to another round of mutagenesis.

Strand selection *in vitro*

The methods described above generate heteroduplex molecules, each con-sisting of a mutant and a non-mutant strand, and then attempt to select for the mutant strand *in vivo*. This can result in considerable loss of efficiency and often requires mutant vectors and/or host strains. In addition the heteroduplex mutant molecules give rise to mixed plaques containing both mutant and wild-type progeny. Clones identified as mutants require further plaque purification and identification.

The gapped duplex and primer cycling methods have another disadvantage. A background of homoduplex non-mutant molecules is often generated because template priming is able to occur in the absence of the mutant oligonucleotide. This background of non-mutant progeny considerably lowers the efficiency of the mutagenesis reaction, particularly when the mutant template binds poorly to the template.

An alternative method which has none of these disadvantages makes use of the observation that certain restriction enzymes (e.g. *Ava* I, *Ava* II, *Ban* II, *Hind* II, *Nci* I, *Pst* I and *Pvu* I) cannot cleave phosporothioate DNA (Taylor *et al.* 1985). The mutant oligonucleotide is annealed to the single-stranded DNA template in the usual manner, but is then extended by DNA polymerase in the presence of a thionucleotide (Fig. 5.6). This generates a heteroduplex in which the mutant strand is phosphorothioated. After sealing the gap in the mutant strand with DNA ligase the heteroduplex is treated with *Nci* I, which cleaves *only* the parental strand. The parental strand is partially digested with exonuclease III and then repolymerized (Fig. 5.7). This method permits strand selection *in vitro* and gives very high efficiencies of mutation, not just for point mutants, but for making deletions and insertions as well. Another advantage is that it does not require specialized hosts or phage vectors and thus can be used for repeated rounds of mutagenesis.

Transformation with oligonucleotides

Moerschell *et al.* (1988) have described a method whereby site-directed mutagenesis of a gene was carried out solely by transforming yeast with a synthetic oligonucleotide. The yeast strain in question was unable to grow on fermentable carbohydrates because of a mutation in the *CYC-1* gene.

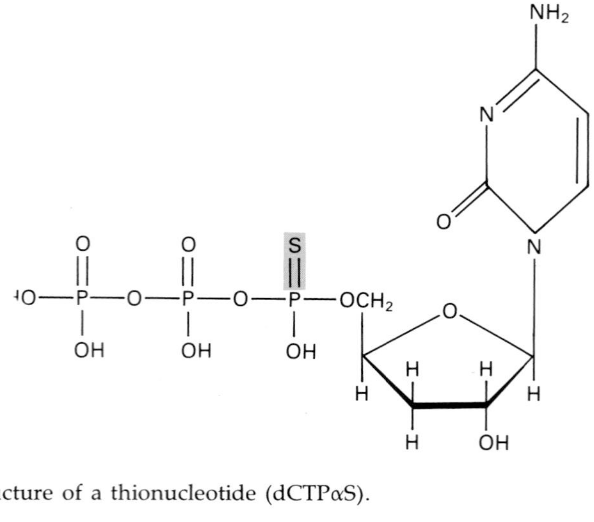

Fig. 5.6 Structure of a thionucleotide (dCTPαS).

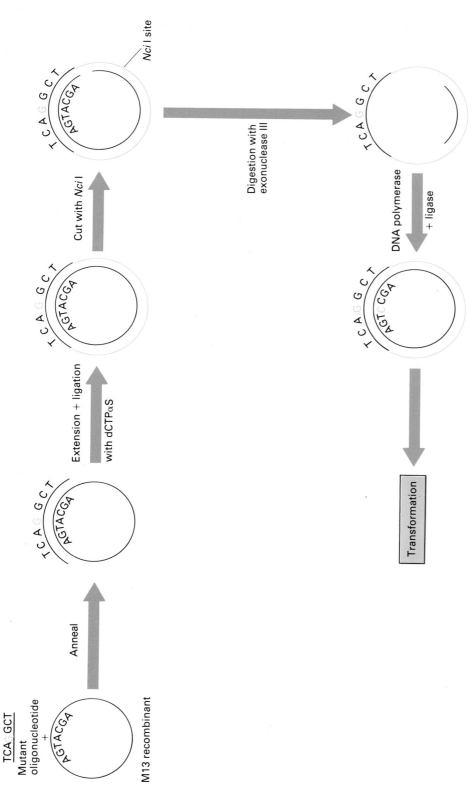

Fig. 5.7 *In-vitro* strand selection of mutants. The DNA-containing sulphur nucleotides is shown in colour.

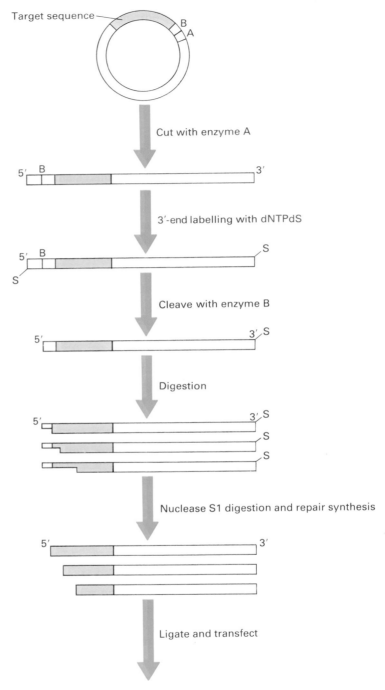

Fig. 5.8 Procedure for making unidirectional deletions in a DNA molecule. A and B represent recognition sequences for two different restriction endonucleases.

The mutation was created by changing a TC dinucleotide in the coding sequence to a single residue: in effect a base substitution coupled with a frame-shift. The yeast strain was transformed with oligonucleotides that

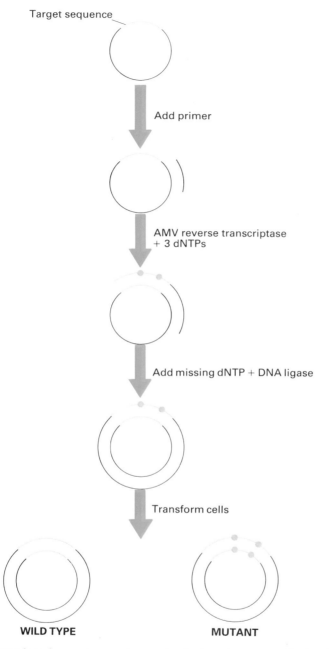

Fig. 5.9 Procedure for random mutagenesis of a target sequence. In the absence of one of the dNTPs the AMV reverse transciptase copies the template DNA slowly and inserts any of the dNTPs present for the missing dNTP. On adding back the missing dNTP, replication occurs at its usual high rate. Dots represent mutated bases.

could correct the *CYC-1* mutation, and revertants were obtained that had the expected base sequence. For this method to succeed it is necessary to have stable mutants and a positive selection method for revertants. High DNA concentrations (100 μg) also are required.

Making unidirectional deletions

All the methods described above can be used to make both additions and deletions. They are particularly suitable if one wants to make only a few addition or deletion mutants. However, where a whole family of mutants is desired then a corresponding family of oligonucleotide primers would be required. An alternative method is available for making a series of deletions of varying length (Barcack & Wolf 1986, Yanisch-Perron *et al.* 1985). The method is shown in Fig. 5.8 and makes use of the fact that α-thiophosphate-containing phosphodiester bonds are resistant to hydrolysis by the $3' \rightarrow 5'$ exonucleolytic activity of phage T4 DNA polymerase. Linear duplex DNA molecules blocked at one 3'-terminus with a thiophosphate are then degraded from the other end with the exonuclease. Digestion for different lengths of time followed by treatment with nuclease S1 and ligation allows the preparation and recovery of a nested set of deletion mutants.

Random mutagenesis

If one wants to isolate a whole series of point mutants in one particular DNA sequence, then the use of oligonucleotide mismatching is too laborious. One of the most effective methods of random introduction of point mutations at specifically defined sites is the enzymatic misincorporation of non-complementary nucleotides. Misincorporation can be forced during *in-vitro* enzymatic template-dependent DNA synthesis by omission of one of the dNTPs. Avian myeloblastosis virus (AMV) reverse transcriptase is particularly error-prone in this respect, and the lack of an associated $3' \rightarrow 5'$ exonuclease activity ensures that errors are not repaired *in vitro*. Singh *et al.* (1986) have described a method of carrying out mutagenesis in this way (Fig. 5.9).

Recombinant DNA techniques have permitted a refinement of the classical method of chemical mutagenesis. The segment of DNA to be mutated is cloned in an M13 vector, single-stranded DNA obtained, and treated with limiting concentrations of chemical mutagens. The mutated DNA then is used as a template for copying by reverse transcriptase which, when copying depurinated DNA, inserts any of the four bases at the site. The double-stranded target DNA is subcloned into a fresh vector for screening or selecting for mutants (Myers *et al.* 1985a). A method for identifying these mutants by denaturing gradient gel electrophoresis has been described by Myers *et al.* 1985b).

6 Analysing DNA Sequences

DNA sequencing is a fundamental capability for modern gene manipulation. Knowledge of the sequence of a DNA region may be an end in its own right, perhaps in understanding an inherited human disorder. In any event, sequence information is a prerequisite for planning any substantial manipulation of the DNA; for example, a computer search of the sequence for all known restriction endonuclease target sites will provide a complete and precise restriction map.

DNA sequencing by the Maxam and Gilbert method

This method for DNA sequencing makes use of chemical reagents to bring about base-specific cleavage of the DNA. For large-scale sequencing it is now less favoured than the enzymatic, 'dideoxy', method. However, it still finds application and illustrates principles of polyacrylamide gel electrophoresis as applied to sequence determination.

In the Maxam and Gilbert method (Maxam & Gilbert 1977), the starting point is a defined DNA restriction fragment. The DNA strand to be sequenced must be radioactively labelled at one end with a ^{32}P-phosphate group. [A detailed practical account of the entire sequencing procedure, including end-labelling methods, is available (Maxam & Gilbert 1980).] The base-specific cleavages depend upon the following points.

1 Chemical reagents have been characterized which alter one or two bases in DNA (Table 6.1). These are base-specific reactions; for example, dimethylsulphate methylates guanine (at the N7 position).
2 An altered base can then be removed from the sugar-phosphate backbone of DNA (Table 6.1).
3 The strand is cleaved with piperidine at the sugar residue lacking the base. This cleavage is dependent upon the previous step.

When each of the base-specific reagents is used in a limited reaction

Table 6.1 Reagents for Maxam and Gilbert DNA sequencing.

Base specificity	Base reaction	Altered base removal	Strand cleavage
G	dimethylsulphate	piperidine	piperidine
G + A	acid	acid-catalysed depurination	piperidine
T + C	hydrazine	piperidine	piperidine
C	hydrazine + NaCl	piperidine	piperidine
A > C	NaOH	piperidine	piperidine

with end-labelled DNA, a nested set of end-labelled fragments of different lengths is generated. It is important to emphasize that the base-specific reactions are deliberately limited to give about one, or a few, cleavages per molecule. This is illustrated in Fig. 6.1 where the nested set of fragments produced by the G-specific reaction is given as an example. Sets of fragments are produced by reacting the DNA with each of the reagents separately (Table 6.1). All five reactions (1−5, Table 6.1) may be performed; the fifth reaction gives redundant information but is confirmatory. These sequencing reactions are analysed by running the four or five samples side by side on a sequencing gel.

A sequencing gel is a high-resolution gel designed to fractionate single-stranded (denatured) DNA fragments on the basis of their length. Such routinely contain 6−20% polyacrylamide and 7 M urea. The urea is a denaturant whose function is to minimize DNA secondary structure which affects electrophoretic mobility. The gel is run at sufficient power to heat up to about 70°C. This also minimizes DNA secondary structure. The labelled DNA bands obtained after such electrophoresis are revealed by autoradiography on large sheets of X-ray film. The sequence can then be read directly from the sequencing ladders in the adjacent base-specific tracks (Fig. 6.2).

Sequencing by the chain-terminator, or dideoxy procedure

This method is now favoured for large-scale DNA sequence determination. It is used in conjunction with cloning the DNA to be sequenced in the M13 mp series of single-stranded vectors (or alternatively pEMBL or Bluescript vectors may be used: see Chapter 4). This technology is very powerful.

In order to appreciate the elegance of the combined technology it is necessary first to understand the chain-terminator DNA sequencing procedure (Sanger *et al*. 1977). This procedure capitalizes on two properties of DNA polymerase. First, its ability to synthesize faithfully a complementary copy of a single-stranded DNA template. Second, its ability to use

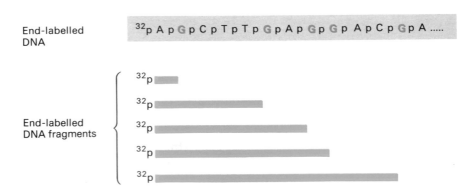

Fig. 6.1 Chemical cleavage of hypothetical DNA at G residues. A nested set of end-labelled DNA fragments is produced by limited reaction of an end-labelled DNA with G-specific reagents. Other fragments are produced, but only the terminal fragments bear the label.

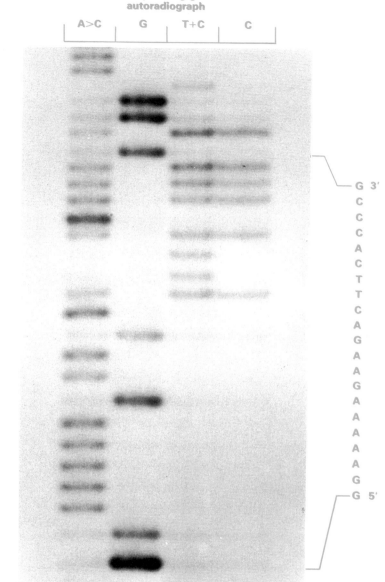

Fig. 6.2 Photograph of the Maxam and Gilbert autoradiograph. (Photograph courtesy of N. Warburton.)

2'3'-dideoxynucleoside triphosphates as substrates (Fig. 6.3). Once the analogue is incorporated at the growing point of the DNA chain, the 3' end lacks a hydroxyl group and no longer is a substrate for chain elongation; thus the growing DNA chain is terminated.

In practice, the Klenow fragment of DNA polymerase I, which lacks the 5'→3' exonuclease activity of the intact enzyme, is used to synthesize a complementary copy of the single-stranded target sequence. Initiation of

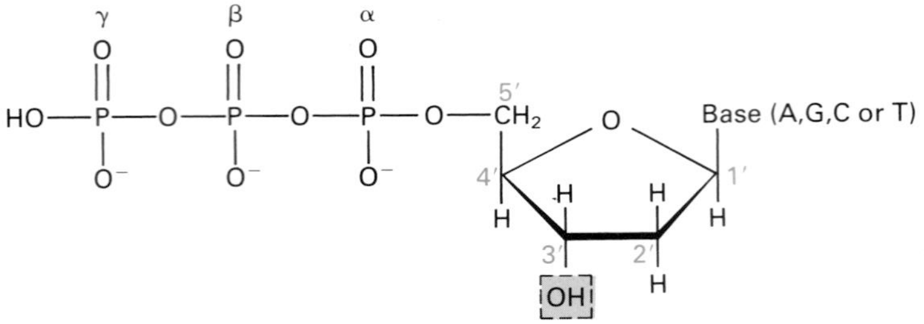

Normal deoxynucleoside triphosphate (i.e. 2'-deoxynucleotide)

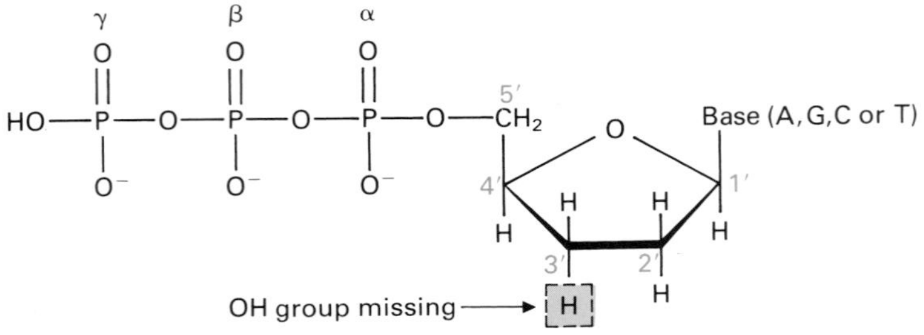

Dideoxynucleoside triphosphate (i.e. 2',3' dideoxynucleotide)

Fig. 6.3 Dideoxynucleoside triphosphates act as chain terminators because they lack a 3'-OH group. Numbering of the carbon atoms of the pentose is shown (primes distinguish these from atoms in the bases). The α, β, and γ phosphorus atoms are indicated.

DNA synthesis requires a primer and usually this is a chemically synthesized oligonucleotide which is annealed close by.

DNA synthesis is carried out in the presence of the four deoxynucleoside triphosphates, one or more of which is labelled with ^{32}P, and in four separate incubation mixes containing a low concentration of one each of the four dideoxynucleoside triphosphate analogues. Therefore, in each reaction there is a population of partially synthesized radioactive DNA molecules, each having a common 5' end, but each varying in length to a base-specific 3' end. After a suitable incubation period the DNA in each mixture is denatured, electrophoresed side by side, and the radioactive bands of single-stranded DNA detected by autoradiography. The sequence can then be read off directly from the autoradiograph as shown in Figs 6.4 and 6.5.

The sharpness of the autoradiographic images can be improved by replacing the ^{32}P-radiolabel with the much weaker β-emitter ^{32}S. This improvement is achieved by including an α-^{35}S-deoxynucleoside triphosphate (Fig. 6.6) as a supplement to the normal deoxynucleoside triphosphates in the sequencing reaction. The α-^{35}S-deoxynucleoside triphosphate

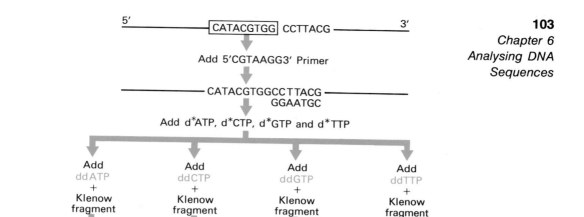

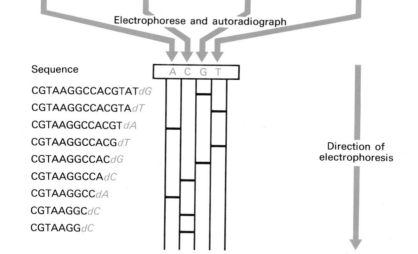

Fig. 6.4 DNA sequencing with dideoxynucleoside triphosphates as chain terminators. In this figure asterisks indicate the presence of ^{32}P and the prefix 'd' indicates the presence of a dideoxynucleoside. At the top of the figure the DNA to be sequenced is enclosed within the box. Note that unless the primer is also labelled with a radioisotope the smallest band with the sequence CGTAAGGdC will not be detected by autoradiography as no labelled bases were incorporated.

is accepted and incorporated by the DNA polymerase. Other technical improvements to Sanger's original method have been made. Some workers prefer to use DNA polymerases other than the Klenow fragment of *E. coli* DNA polymerase I. Natural or modified forms of the phage T7 DNA polymerase ('sequenase') have found favour, as has the DNA polymerase

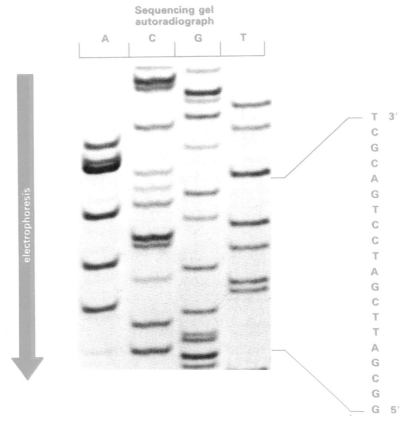

Fig. 6.5 Autoradiograph of a sequencing gel obtained with the chain terminator DNA sequencing method.

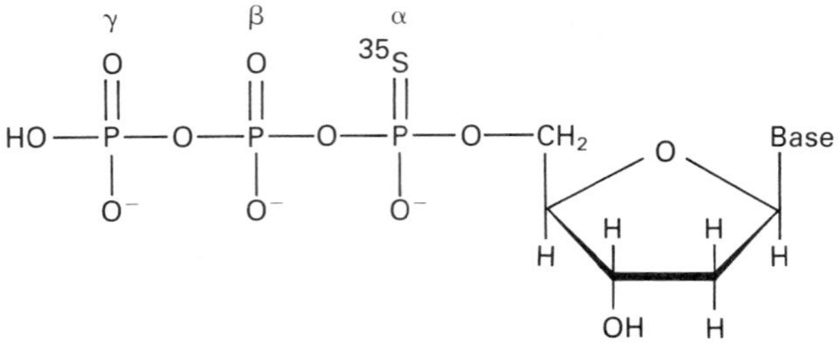

Fig. 6.6 Structure of $\alpha\text{-}^{35}\text{S}$-deoxynucleoside triphosphate.

of the thermophilic bacterium *Thermus aquatus* (Taq DNA polymerase). The Taq DNA polymerase can be used in a chain termination reaction carried out at high temperature (65–70°C). This minimizes chain termination artefacts caused by secondary structure in the DNA.

The DNA to be sequenced is first cloned into one of the clustered cloning sites in the *lac* region of M13 mp series of vectors. Recombinants are detected by the formation of white plaques on media containing IPTG and Xgal (Fig. 6.7). Virus is isolated from these white plaques, stocks

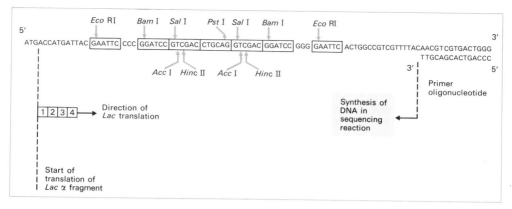

Fig. 6.7 Sequence of M13 mp7 DNA in the vicinity of the multipurpose cloning region. The upper sequence is that of M13 mp7 from the ATG start codon of the β-galactosidase α-fragment, through the multipurpose cloning region, and back into the β-galactosidase gene. The horizontal bars indicate the recognition sites for the enzymes shown. The short sequence at the righthand is that of the primer used to initiate DNA synthesis across the cloned insert. The numbered boxes correspond to the amino acids of the β-galactosidase fragment.

prepared and the single-stranded viral DNA extracted for sequencing. The real beauty of M13 mp series is that cloning into the same, specific region of the genome obviates the need for isolation of many different primers, since a single primer can be used for all inserts. The original primer was a short restriction fragment which was cloned in pBR322 and which was complementary to a region of the *lac* Z gene immediately adjacent to the righthand *Eco* RI insertion site (Anderson *et al.* 1980). Messing *et al.* (1981) have developed a more suitable 15-base synthetic oligonucleotide which primes just to the right of the polylinker. This has subsequently been found to have low homology with a second site in M13, so that an improved 15-base oligonucleotide primer has been synthesized. This has no homology at any secondary site (Norrander *et al.* 1983).

Analysis of DNA sequence data

Any substantial DNA sequence undertaking will necessitate computer analysis of the data. The analysis will often fall into two stages. In the first stage, compiling the sequence data, computer files will be more or less directly inputted with sequence information read from the autoradiograph of the gel or other sequencing system. In the most direct approach, an automated laser scanner will actually read the gel tracks and enter the data. Software is being developed that can interpret ambiguities in the autoradiograph at least as expertly as the experienced human gel reader. Various less direct systems are available that combine manual positioning of a computerised 'pen' probe over the autoradiographic bands with automatic entry of the data into a file. These automated systems have the obvious advantage of speed and, just as importantly, they help to minimize the ever-present difficulty of clerical errors accumulating in sequence data. The computer will then be used to compile readings from several sequencing runs, to search for and identify overlaps and hence *contiguities* between runs, and to compile data from sequencing complementary strands of the DNA. The software will also identify inconsistencies, indicating sequencing errors, between sequences determined more than once on the same or complementary strands.

In the second stage, the analysis proper, the deduced sequence may be processed in a variety of ways.

1 The sequence can be searched for all known restriction endonuclease target sites, thus producing a comprehensive and precise map.

2 The sequence can be searched for features such as tandem repeats or inverted repeats. Inverted repeats lead to the possibility of hairpin-loop formation within one strand of the DNA or in RNA transcribed from it.

3 The sequence can be conceptually translated into protein in all three possible reading frames on both strands—six frames in all. An 'open' reading frame, ORF, is a frame that does not include a termination codon. A long open reading frame may indicate that a previously unknown polypeptide coding region exists (Doolittle 1986).

4 The DNA sequence itself, or the deduced polypeptide sequence, may be compared with a data bank of other sequences. Often, because of the degeneracy of the genetic code, similarities are found between two poly-peptide sequences, which would not have been apparent had the comparison been carried out at the DNA level.

Similarities between two sequences can, in principle, arise by two routes: either by convergent evolution, or through their being related by descent from a common ancestral sequence. Convergent evolution implies that the two sequences have not arisen in evolutionary time through having had a common ancestral sequence, but that selection for a particular function in two lineages has converged on a particular structure or two related structures. By contrast, two sequences may remain similar in evol-utionary time because of selection pressure limiting the scope for divergence.

This selection pressure need not necessarily act on whole proteins. It is possible for protein coding regions to be assembled *piecemeal*, with con-served domains fulfilling a particular function, e.g. DNA binding or nucleoside triphosphate binding, in polypeptides that are otherwise dis-tinct in function. The fact that certain polypeptide sequence motifs or domains are often found as exon units is consistent with a piecemeal gene assembly model. So-called exon-shuffling is evident.

Whatever the evolutionary route sequence to similarity has been, any sequence similarity is taken to be an indication of similar function. It has been a feature of modern molecular biology that striking and unexpected sequence similarities have been discovered in a range of situations. The message seems to be that evolution is reluctant to discard a good idea for building up functional polypeptide domains. The sequence data banks are of sufficient size even at our present state of knowledge (and they are expanding very rapidly) to make a search of them for similarities with any newly discovered sequence an undertaking that has a reasonable prospect of turning up something of interest. Two final points should be made in this context. First, it is important that development continues in international arrangements for gaining access to data banks that are as absolutely up-to-date and as accurate as possible. The current rate of expansion is so great that a significant proportion of known sequences is, at any time, not accessible. Second, it has been the authors' experience that, in the cases where one has been in a position to check particular sequences in data banks or scientific journals, a disturbingly high proportion of entries contain errors. Errors are especially common in non-coding regions and appear to arise from a variety of causes ranging from outright sequencing errors to clerical errors.

7 Cloning Strategies, Gene Libraries and cDNA Cloning

Cloning strategies

Any DNA cloning procedure has four essential parts: a method for generating DNA fragments, reactions which join foreign DNA to the vector, a means of introducing the artificial recombinant into a host cell in which it can replicate, and a method of selecting or screening for a clone of recipient cells that has acquired the recombinant (Fig. 7.1). In previous chapters DNA cutting and joining reactions have been described, and the properties of several phage and plasmid vectors have been discussed together with the factors governing the choice between the various cutting and joining methods and different vector molecules. These choices will depend upon what type of clones are wanted, e.g. cDNA or genomic DNA clones. We also have to consider whether we wish to increase the frequency of the desired sequence in our starting material by a prior enrichment procedure. Alternatively we can compare gene libraries in which the desired sequence has not been enriched. The latter is sometimes called a *shotgun* approach.

Genomic DNA libraries

As an example, let us suppose that we wish to clone a single-copy gene from the human genome. We might simply digest total human DNA with a restriction endonuclease such as *Eco* RI, insert the fragments into a suitable phage λ vector and then attempt to isolate the desired clone. How many recombinants would we have to screen in order to isolate the right one? Assuming *Eco* RI gives, on average, fragments about 4 kb long, and given that the human haploid genome is 2.8×10^6 kb, we can see that over 7×10^5 independent recombinants must be prepared and screened in order to have a reasonable chance of including the desired sequence. In other words we have to obtain a very large number of recombinants, which together contain a complete collection of all (or nearly all) of the DNA sequences in the entire human genome. Such a collection from which we withdraw the desired clone is called a *gene library* or *gene bank*.

There are two problems with the above approach. First, the gene may be cut internally one or more times by *Eco* RI so that it is not obtained as a single fragment. This is likely if the gene is large. Also, it may be desired to obtain extensive regions flanking the gene or whole gene clusters. Fragments averaging about 4 kb are likely to be inconveniently short. Alternatively, the gene may be contained on an *Eco* RI fragment that is larger than the vector can accept. In this case the appropriate gene would not be cloned at all.

These problems can be overcome by cloning *random* DNA fragments of

109

*Chapter 7
Cloning
Strategies,
Gene Libraries
and cDNA
Cloning*

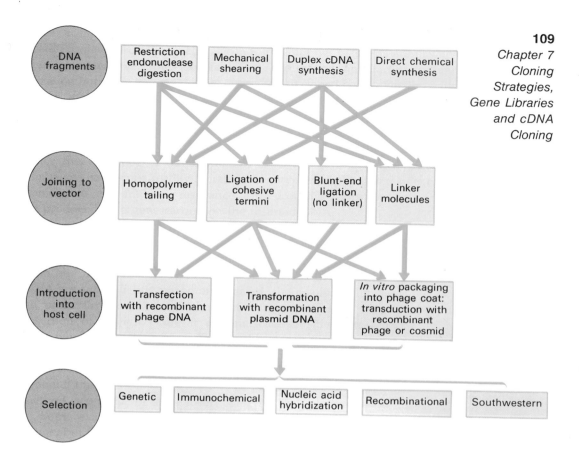

Fig. 7.1 Generalized scheme for DNA cloning in *E. coli*. Favoured routes are shown by arrows.

a large size (~20 kb). Since the DNA is randomly fragmented, there will be no systematic exclusion of any sequence. Furthermore, clones will overlap one another, giving an opportunity to 'walk' from one clone to an adjacent one (see p. 115). Because of the larger size of each cloned DNA, fewer clones are required for a complete or nearly complete library. How many clones are required? Let n be the size of the genome relative to a single cloned fragment. Thus for the human genome, 2.8×10^6 kb* and for a cloned fragment size of 20 kb, $n = 1.4 \times 10^5$. The number of independent recombinants required in the library must be greater than n, because sampling variation will lead to the inclusion of some sequences several times, and the exclusion of other sequences in a library of just n recombinants. Clarke & Carbon (1976) have derived a formula which relates the probability (P) of including any DNA sequence in a random library of N independent recombinants:

* See Table 7.1 for the genome size of various organisms.

$$N = \frac{ln(1 - P)}{ln\left(1 - \frac{1}{n}\right)}.$$

Therefore, to achieve a 95% probability ($P = 0.95$) of including any particular sequence in a random human genomic DNA library of 20 kb fragment size:

$$N = \frac{ln(1 - 0.95)}{ln\left(1 - \frac{1}{1.4 \times 10^5}\right)}$$

$$= 4.2 \times 10^5.$$

Notice that a considerably higher number of recombinants is required to achieve a 99% probability, for here $N = 6.5 \times 10^5$.

How can appropriately sized random fragments be produced? A variety of methods are available. Random breakage by mechanical shearing is appropriate, but a much more commonly used procedure involves restriction endonucleases. In the strategy devised by Maniatis *et al.* (1978) (Fig. 7.2) the target DNA is restricted with a mixture of *two* restriction enzymes. These enzymes have tetranucleotide recognition sites, which therefore occur frequently in the target DNA and in a *limit* double-digest would produce fragments averaging less than 1 kb. The restriction is carried out only to a partial extent, so that the bulk of fragments are relatively large, in the range 10–30 kb. These are effectively a random set of overlapping fragments. These can be fractionated by velocity centrifugation on a sucrose gradient or by preparative gel electrophoresis, so as to give a random population of fragments of about 20 kb, which are suitable for insertion into the phage lambda vector. Packaging *in vitro* ensures that an appropriately large number of independent recombinants can be recovered, which will give an almost completely representative library.

In Maniatis' strategy, the use of two different restriction endonucleases with completely unrelated recognition sites, *Hae* III and *Alu* I, assists in

Table 7.1 Genome sizes of some organisms.

Organism	Genome size in kb (haploid where appropriate)
Escherichia coli	4.0×10^3
Yeast (*Saccharomyces cerevisiae*)	1.35×10^4
Arabidopsis thaliana (higher plant)	7.0×10^4
Tobacco	1.6×10^6
Wheat	5.9×10^6
Zea mays	15×10^6
Drosophila melanogaster	1.8×10^5
mouse	2.3×10^5
human	2.8×10^6
Xenopus laevis	3.0×10^6

111
*Chapter 7
Cloning
Strategies,
Gene Libraries
and cDNA
Cloning*

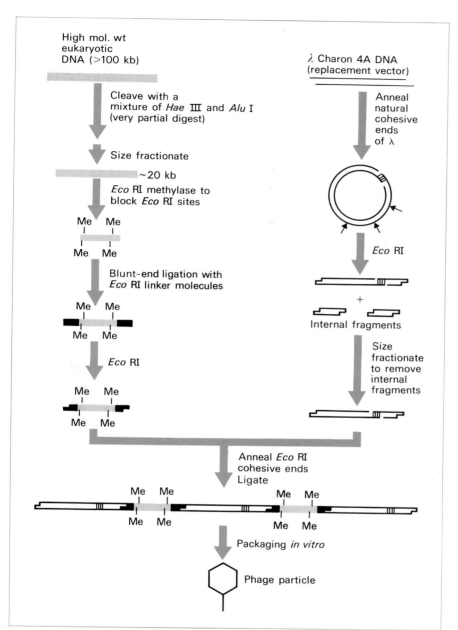

Fig. 7.2 Maniatis' strategy for producing a representative gene library.

obtaining fragmentation that is nearly random. These enzymes both produce blunt ends, and the cloning strategy requires linkers (see Fig. 7.2). A
convenient simplification can be achieved by using a *single* restriction
endonuclease which cuts frequently, such as *Sau* 3A. This will create a
partial digest that is slightly less close to random than that achieved with
a pair of enzymes. However, it has the great advantage that the *Sau* 3A
fragments can be readily inserted into high-capacity phage λ vectors, such

as λL47 or λEMBL3 (see Chapter 4), which have been digested with *Bam* HI (Fig. 7.3). This is because *Sau* 3A and *Bam* HI create the same cohesive ends (see p. 20). These partial digestion methods, coupled with packaging the phage λ recombinants, have been widely employed strategies for creating genomic DNA libraries.

In place of the phage λ vectors, cosmid vectors may be chosen. These also have the high efficiency afforded by packaging *in vitro* and have an even higher capacity than any phage lambda vector. However, there are two drawbacks in practice. First, most workers find that screening libraries of phage λ recombinants by plaque hybridization gives cleaner results than screening libraries of bacteria containing cosmid recombinants by colony hybridization (see Chapter 8). Plaques usually give less of a background hybridization than do colonies. Second, it may be desired to retain and store an amplified genomic library. With phage, the initial recombinant DNA population is packaged and plated out. It can be screened at this stage. Alternatively, the plates containing the recombinant plaques can be washed to give an *amplified* library of recombinant phage. The amplified library can then be stored almost indefinitely; phage have a long shelf-life. The amplification is so great that samples of this amplified library could be plated out and screened with different probes on hundreds of occasions. With bacterial colonies containing cosmids it is also possible to store an amplified library (Hanahan & Meselson 1980), but bacterial populations cannot be stored as readily as phage populations. There is often an unacceptable loss of viability when the bacteria are stored.

A word of caution is necessary when considering the use of any amplified library. This is the possibility of *distortion*. Not all recombinants in a population will propagate equally well, e.g. variations in target DNA size or sequence may affect replication of a recombinant phage, plasmid

Fig. 7.3 (*Facing page*) Creation of a genomic DNA library using the phage λ vector EMBL3A.

High mol. wt genomic DNA is partially digested with *Sau* 3A. The fragments are treated with phosphatase to remove their 5'-phosphate groups. The vector is digested with *Bam* HI and *Eco* RI which cut within the polylinker sites. The tiny *Bam* HI/*Eco* RI polylinker fragments are discarded in the isopropanol precipitation, or alternatively the vector arms may be purified by preparative agarose gel electrophoresis. The vector arms are then ligated with the partially digested genomic DNA. The phosphatase treatment prevents the genomic DNA fragments from ligating together. Non-recombinant vector cannot reform because the small polylinker fragments have been discarded. The only packageable molecules are recombinant phages. These are obtained as plaques on a P2 lysogen of sup^+ *E. coli*. The Spi$^-$ selection ensures recovery of phage lacking *red* and *gam* genes. A sup^+ host is necessary because, in this example, the vector carries amber mutations in genes A and B. These mutations increase biological containment, and can be applied to selection procedures such as recombinational selection (see Chapter 8), or tagging DNA with a sup^+ gene (see Chapter 13). Ultimately, the foreign DNA can be excised from the vector by virtue of the *Sal* I sites in the polylinker. (N.B. Rogers *et al.* 1988 have shown that the EMBL3 polylinker sequence is not exactly as originally described. It contains an extra sequence with a previously unreported *Pst* I site. This does not affect most applications as a vector.)

113

Chapter 7
Cloning
Strategies,
Gene Libraries
and cDNA
Cloning

or cosmid. Therefore, when a library is put through an amplification step particular recombinants may be increased in frequency, decreased in frequency, or lost altogether. Development of modern vectors and cloning strategies has simplified library construction to the point where many workers now prefer to create a new library for each screening, rather than risk using a previously amplified one.

The ease with which random libraries can be created and screened, and the possibility of chromosome walking, means that the shotgun approach has now become very widely adopted. However, as an alternative

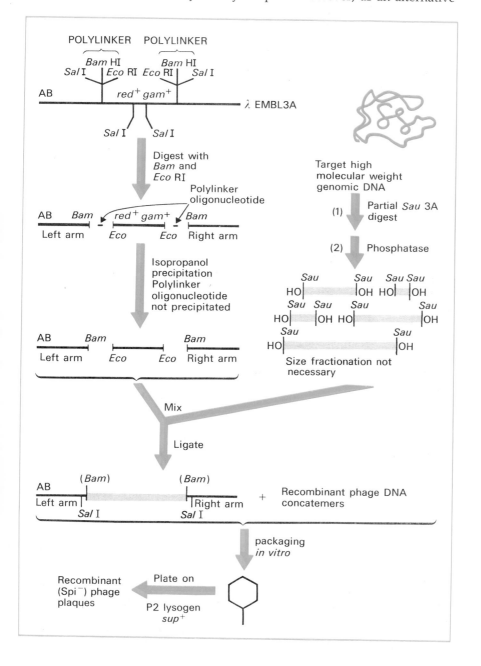

approach we could obtain a partially purified DNA fraction which is enriched in the desired sequence. The task of screening would then be diminished correspondingly. This approach is now rather outdated but it was necessary at a time before packaging *in vitro* had been developed, because then it was not possible readily to produce enough independent clones for a complete library. One method that was employed very successfully for such enrichment is chromatography upon the medium RPC-5, which consists of a quaternary ammonium salt, the extractant, supported upon a matrix of plastic beads. Originally developed for high-resolution reversed-phase chromatography of tRNAs, it will fractionate milligram quantities of DNA fragments generated by restriction endonucle- ᵔ (Hardies & Wells 1976). Leder and his co-workers (Tilghman *et al.* ᵓᵧᵎ loaded an *Eco* RI digest of total mouse genomic DNA onto an RPC-⊙ column, and upon elution with a concentration gradient of sodium acetate, obtained a series of DNA fractions which were assayed for their ability to hybridize with mouse globin cDNA. A fraction was identified as being substantially enriched in a DNA fragment bearing β-globin sequences. The discrimination on the RPC-5 column chromatography is only slightly dependent upon fragment size, so that additional enrichment (to about 500-fold) could be obtained by combining it with preparative gel electrophoresis. The final enriched fraction was ligated into the λWES.λB vector, and out of about 4300 plaques obtained by transfection, three positive clones were detected.

E. coli hosts for library construction in phage vectors

Genomic DNA libraries in phage λ vectors are expected to contain most of the sequences of the genome from which they have been derived. However, anecdotal reports of sequences that cannot be found in libraries of eukaryotic genomes are not uncommon. There is also the related observation that deletions can occur during the cloning of mammalian DNA. Only some of these deletions are preventable by growth on an E. coli recA recombination-deficient host.

Leach & Stahl (1983) have shown that inverted repeat sequences (large palindromes) cannot be cloned in phage λ vectors unless special host strains are used. Wyman et al. (1985) have extended these observations by showing that a host with mutant recB, recC and sbcB recombination-function genes will propagate many phage λ recombinants containing human DNA fragments that would not be propagated on other host strains.

Chromosome walking

Walking along the chromosome is a term used to describe an approach which allows the isolation of gene sequences whose function is quite unknown but whose genetic location is known. A particularly dramatic application of this approach involves homeotic mutations in *Drosophila*. Homeotic mutations are an especially interesting class of genetic defects which characteristically cause one imaginal disk of the larva to develop in the form of a region of the body derived from an entirely different imaginal disk (for review see Ingham 1988). *Antennapedia*, for example, is a mutation which can result in adult flies whose antennae are homeotically transformed into leg-like structures. Although the developmental basis of this transformation is unknown, the genetic locus of the mutation has been accurately mapped by classical genetic methods and assigned to a region comprising just a few bands in the polytene chromosomes of *Drosophila* salivary glands. This allows DNA from this region to be isolated by molecular cloning procedures, which in turn enables the mutation to be defined precisely at the DNA sequence level. This provides reagents for analysing the molecular biology of development in normal and mutant imaginal disks. The principle is as follows (see also Fig. 7.4).

115

Chapter 7
Cloning
Strategies,
Gene Libraries
and cDNA
Cloning

1   Cloned genomic sequences can be localized in the genome by radio-labelling the cloned fragment and using it as a probe in a hybridization experiment *in situ* with cytological preparations of polytene chromosomes.

2   A random set of cloned genomic DNA is localized in this way, and one is chosen whose location on the chromosome in question is closest to the map position of the mutation under investigation.

3   The genomic library is then screened with this chosen clone as probe to identify other clones containing DNA with which it reacts and which represent clones overlapping with it. The overlap can be to the left or to the right.

4   Repetition of this single walking step along the chromosome.

Such walking will necessarily occur in both directions along the chromosome unless there is some means of distinguishing the direction. In *Drosophila* this can be achieved by hybridization *in situ* to polytene chromosomes.

A possibility that has to be recognized arises from the existence of repeated DNA sequences. These may occur dispersed at several places in the genome and could disrupt the orderly progress of the walk. For this reason the probe used for stepping from one genomic clone to the next

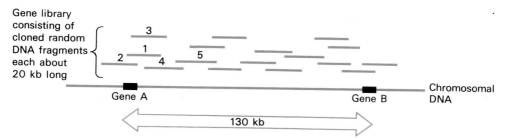

Fig. 7.4 Chromosome walking. It is desired to clone DNA sequences of gene *B*, which has been identified genetically but for which no probe is available. Sequences of a nearby gene *A* are available in cloned fragment 1. Alternatively, a sequence close to gene *B* could be identified by *in-situ* hybridization to *Drosophila* polytene chromosomes. In a large, random genomic DNA library many overlapping cloned fragments are present. Clone 1 sequences can be used as a probe to identify overlapping clones 2, 3 and 4. Clone 4 can, in turn, be used as a probe to identify clone 5, and so on. It is, therefore, possible to walk along the chromosome until gene *B* is reached. (See text for details.)

must be a unique sequence clone, or a subclone which has been shown to contain only a unique sequence. Once the investigator has arrived at, or very near, the locus in question it should be possible to compare mutant and wild-type DNA sequences directly or by restriction mapping with the Southern blot technique. Cloned probes could also be applied to analysing transcripts from the region, hence making inroads into the analysis of the previously unknown molecular biology of the locus.

As outlined here the process of walking is simple in principle. On a small scale it can be applied to almost any eukaryote. If starting with a cloned fragment of a gene, or a cDNA clone, the entire genomic sequence together with substantial flanking sequences may be obtained with a single step in each direction. Walking on a larger scale from a clone chosen to be near another gene of interest in order to walk to the other gene, is very demanding. However, here the advanced genetics of *Drosophila* comes to the rescue, for combined with the inherent usefulness of polytene chromosome hybridization are the numerous inversions and translocations that may be exploited to reduce the distance to be walked. In one of the first applications of this now common technology, Hogness and his co-workers (Bender *et al.* 1983) cloned DNA from the *Ace* and *rosy* loci and the homeotic *Bithorax* gene complex in *D. melanogaster*.

The existence of the polytene chromosomes in *Drosophila* permits a different, more direct, approach. It has, by a tremendous technical *tour de force*, been found possible to physically excise a region of such a salivary gland chromosome by micromanipulation, and thence to extract its DNA, restrict it and ligate it to a phage λ vector, all within a microdrop under oil, and thereafter obtain clones with reasonable efficiency (Scalenghe *et al.* 1981). With this technique it should be possible to isolate clones from any desired region of the genome of the fruit fly. The main interest in these micromanipulation experiments lies not in the ability to clone sequences from particular regions of the *Drosophila* genome—this can be achieved by other means such as chromosome walking—but rather in demonstrating that minute quantities of DNA can be cloned successfully.

Following the early demonstrations of the power of chromosome walking in *Drosophila*, the walking approach has been applied to important regions of the human genome; for example, the muscular dystrophy locus on the X-chromosome (Hoffman *et al.* 1987). Improved vectors and strategies have been developed in response to the need for easier chromosome walking. For example, two phage λ derivatives, λDASH and λFIX, have been designed by Stratege Cloning systems. The map of the λDASH genome is shown in Fig. 7.5. This replacement vector can accomodate inserts of 9–22 kb and includes multiple cloning sites. These sites define the stuffer fragment which bears *red*$^+$ *gam*$^+$ genes and therefore permits Spi selection. In these regards, λDASH resembles EMBL3 (p. 113). Immediately adjacent to the cloning sites are T3 and T7 promoters for the T3 and T7 RNA polymerases. These promoters allow RNA probes to be made from insert sequences (see p. 85). Importantly, from the point of view of chromosome walking, *end-specific* RNA probes will be produced if the recombinant DNA to be transcribed into RNA is first digested with a restriction endonuclease that cuts relatively frequently (e.g. a tetranucleotide target site), and which

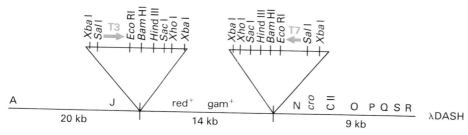

Fig. 7.5 The replacement vector λDASH can accommodate inserts of 9–22 kb. In the arms, immediately adjacent to the cloning sites, are promoters specific for T3 and T7 RNA polymerases. Thus RNA probes can be made from inserted sequences without subcloning. End-specific RNA probes are generated from recombinant DNA predigested with a restriction enzyme which has a tetranucleotide target site and which leaves only a short insert sequence attached to the promoter.

thus leaves only a short insert DNA fragment attached to the T3 and T7 promoters. These probes are ideal for probing a library for the next step in a chromosome walk and have the great advantage that they can be made conveniently without recourse to subcloning.

λFIX (Fig. 7.6) is similar to λDASH except that it incorporates XhoI sites situated so as to take advantage of a cloning strategy that prevents the ligation of vector arms without included foreign DNA, and that eliminates multiple inserts. The principle of this strategy is to digest the vector with XhoI and then partially fill in the sticky ends of the XhoI site to leave dinucleotide 5′ overhangs (Fig. 7.7). Such filled-in ends cannot reassociate. Similarly, partially filled-in Sau 3A ends cannot reassociate, but *can* combine with the vector bearing the partially filled-in XhoI ends, and hence can be joined with the vector arms.

Analysing very large DNA sequences and long-distance chromosome walking

Conventional cloning techniques cannot readily accept chromosome walks of thousands of kb, yet distances of this magnitude commonly separate linked marker mutations in mammals. As a rule of thumb, about 1000 kb corresponds to a recombination distance of 1 cm (1% recombination) in humans. The problems of analysing (mapping) and walking such large distances have been largely overcome by progress in three areas.

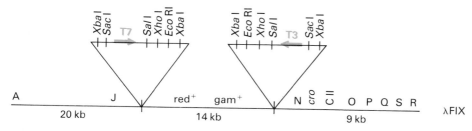

Fig. 7.6 The replacement vector λFIX is similar to λDASH.

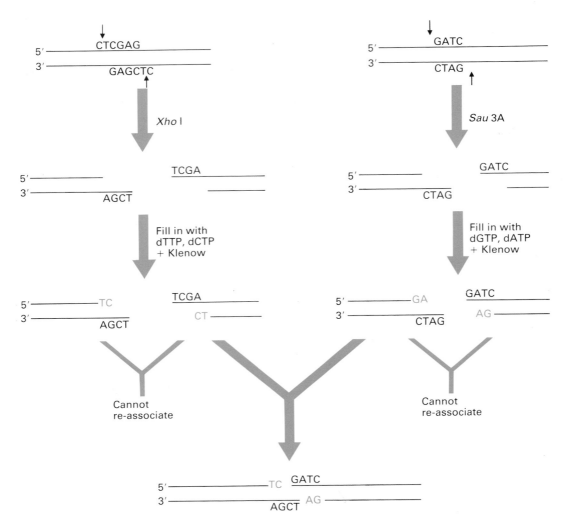

Partially filled-in *Xho* I and *Sau* 3A ends *can* combine

Fig. 7.7 Partially filled-in XhoI sites cannot self-ligate but can accommodate partially filled-in, non-self-ligatable *Sau* 3A ends.

1 The discovery of restriction endonucleases, or combinations of methylases and endonucleases, which cut DNA at very infrequent sites.
2 The development of *pulsed field gel electrophoresis* and *field inversion gel electrophoresis* to resolve very large DNA fragments.
3 The deployment of specialized *chromosome-jumping* strategies.

These developments are described in the following sections. The reader is also referred to the use of yeast artificial chromosomes as a means of cloning very large DNA fragments (Chapter 11).

Cutting DNA at very rare target sites

There are two recently discovered type II restriction endonucleases that have octanucleotide recognition sequences and that therefore cut DNA at

119

*Chapter 7
Cloning
Strategies,
Gene Libraries
and cDNA
Cloning*

very widely spaced sites. These are *Not* I (GCGGCCGC) and *Sfi* I (GGCCNNNNNGGCC). Other enzymes with similarly large recognition sequences will probably be discovered and extend this list. In addition, there are restriction endonucleases with shorter recognition sequences that have the property of being uncommon sequences in mammalian DNA. Examples of this category are *Nru* I (TCGCGA) and *Bss* H II (GCGCGC). Their target sequences include two CG dinucleotides. The CG dinucleotide is rare in mammalian DNA and the target sites for these enzymes are consequently very infrequent.

Additional specificities in target sites may be created by combining the specificity of certain methylases with that of restriction endonucleases. An example of this approach involves the restriction endonuclease, *Dpn*I, which cuts DNA at the sequence G—A—T$\overset{M}{-}$C to produce flush ends. This enzyme is unusual in that it requires the methylation of the adenines in *both* strands of the DNA in order for cleavage to occur. Thus DNA from *dam*$^+$ *E. coli* is cleaved by *Dpn* I, whereas DNA from *dam*$^-$ mutants is not (see Chapter 2). McClelland *et al.* (1984) exploited this property by combining it with the specificity of the modification methylases M.Taq I or M.Cla I. The enzyme M.Taq I methylates both strands of the sequence TCGA to create T—C—G$\overset{M}{-}$A. Thus in DNA that is not previously methylated at these sites, cleavage by *Dpn* I will only occur at the following octanucleotide sequence: T—C—G—A—T—C—G—A. Similarly M.Cla I methylates both strands of the sequence A—T—C—G—A—T to produce A—T—C—G$\overset{M}{-}$A—T, and so in combination with *Dpn* I cleavage will only occur at the decanucleotide sequence ATCGATCGAT.

Pulsed field gel electrophoresis and field inversion gel electrophoresis

A new type of gel electrophoresis was developed by Schwartz & Cantor (1984) which resolves DNA molecules up to 2000 kb in length. The technique uses conventional agarose and conventional buffers. The innovation is the use of alternately pulsed, perpendicularly (i.e. orthogonally) oriented electrical fields. In their demonstration of the power of this technique Schwartz and Cantor separated intact yeast, *Saccharomyces cerevisiae*, chromosomal DNA. Each chromosome was evidently a single piece of DNA and could be analysed further by Southern blotting the gel. The separation appears to depend upon the electrical perturbation of the orientation of the DNA, and on the degree of extension of long DNA molecules. The relaxation time of such molecules in free solution is a very sensitive function of the molecular weight. This orthogonal pulsed field gel electrophoresis, PFGE, technique has been further developed (Carle & Olson, 1984), but has the disadvantages that the DNA samples do not run in straight line tracks.

These difficulties have been overcome in field inversion gel electrophoresis, FIGE (Carle *et al.* 1986). This produces good resolution up to 2000 kb without the need for a complicated perpendicular-field gel apparatus. FIGE uses a conventional electrophoresis apparatus with an electrical field that pulses forward—reverse combined with a pause between each phase. The pulses are in the form of a time-ramp. The DNA tracks run in straight lines and are comparable across the gel, thus making calibration with markers and the interpretation of Southern blots much simpler than in orthogonal PFGE.

Chromosome jumping strategies

Chromosome jumping, as described by Collins & Weissman (1984) and Poustka & Lehrach (1986), depends upon the circularization of very large genomic DNA fragments, followed by cloning DNA from the region covering the closure site of these circles, thus bringing together DNA sequences that were originally located a considerable distance apart in the genome. These cloned DNAs from the closure sites make up a 'jumping library'. Such a jumping library greatly speeds up the process of long-range chromosome walking.

Figure 7.8 (Poustka *et al.* 1987) shows a strategy for creating a jumping library using *Not* I—digested human genomic DNA. This enzyme cuts only rarely in mammalian DNA, and so drastically reduces the size of the library required to cover the mammalian genome. This strategy has the additional advantage that analysis of jumps is easy by FIGE analysis and Southern blotting of *Not* I digested genomic DNA. However, with a complete digest such as this, overlaps between *Not* I fragments have to be obtained indirectly: one solution is to use 'linking clones' containing conventionally cloned fragments with an internal *Not* I site (Poustka *et al.* 1987). A variant of the strategy shown in Fig. 7.8 employs *partial* digestion with a restriction endonuclease to generate large genomic DNA fragments (Collins & Weissman, 1984). This technique has been applied to a jump of 100 kb in the cystic fibrosis locus (Collins *et al.* 1987) and to a jump of 200 kb in the region of the Huntington's disease gene (Richards *et al.* 1988).

cDNA cloning

Cloned eukaryotic cDNAs have their own special uses, which derive from the fact that they lack the intron sequences that are usually present in the corresponding genomic DNA. Introns are non-coding sequences that often occur within eukaryotic gene sequences. They can be situated within the coding sequence itself, where they then interrupt the co-linear relationship of the gene with the encoded polypeptide. They may also occur in the 5' or 3' untranslated regions of the gene, but in any event they are copied into RNA by RNA polymerase when it transcribes the gene. The initial, primary transcript is a precursor to mRNA. It goes through a series of processing events in the nucleus before appearing in the cytoplasm as mature mRNA. These events include the removal of intron sequences by a process called splicing. When cDNA is derived from mRNA it therefore lacks intron sequences. Since removal of eukaryotic intron transcripts by splicing does not occur in bacteria, eukaryotic cDNA clones find application where bacterial expression of the foreign DNA is necessary, either as a prerequisite for detecting the clone (see Chapter 8), or because the polypeptide product is the primary objective. Also, where the sequence of the genomic DNA is

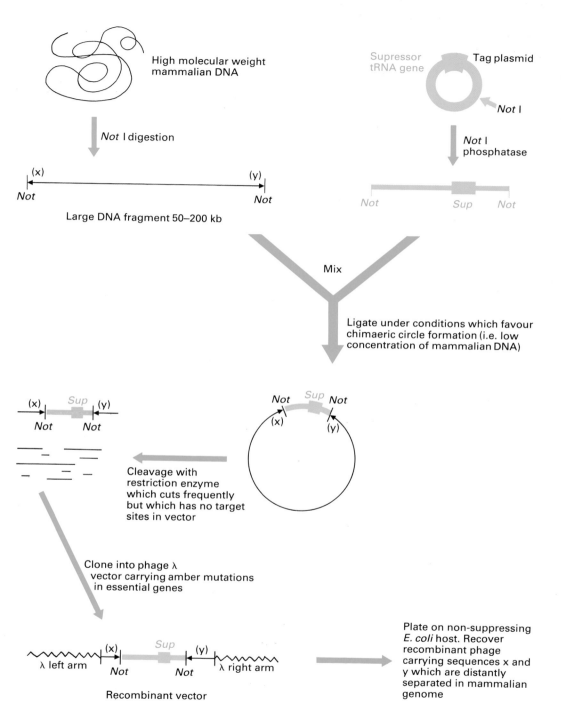

Fig. 7.8 Jumping library construction. This example illustrates the use of *Not* I-digested mammalian DNA. A variant of this procedure could use mammalian DNA which has been *partially* digested with an enzyme that cuts more frequently than *Not* I. Such a procedure would produce a jumping library containing overlapping DNA fragments.

known, the position of intron/exon boundaries can be assigned by comparison with the cDNA sequence.

A second situation where cDNA cloning is carried out involves the analysis of temporally regulated gene expression in development, or tissue-specific gene expression. By using the 'plus and minus' screening procedure (see Chapter 8) it is possible to screen a cDNA clone library to identify cDNA clones derived from mRNA molecules present in one cell type but absent in another cell type.

As with genomic DNA, it may rarely be appropriate to isolate cDNA from purified mRNA. Much more commonly a cDNA clone library may be prepared and screened for particular sequences. Before proceeding further it is necessary to consider the nature of mRNA populations in tissues. In many tissues and cultured cells mRNAs are present at widely different *abundances*, i.e. some mRNA types are present in large numbers per cell, others may be present at just a few copies per cell. Table 7.2 gives some representative examples. Notice that in the chick oviduct one mRNA type is superabundant. This is the mRNA encoding ovalbumin, the major egg-white protein. Therefore, this mRNA population is naturally so enriched in ovalbumin mRNA that cloning the ovalbumin cDNA presents no problem in screening. The clones could be identified by screening a small number of recombinants; the hybrid released translation procedure would be appropriate (see Chapter 8).

Another appropriate strategy for obtaining abundant cDNAs is to clone the cDNA directly in an M13 vector such as M13 mp8. A set of clones can then be sequenced immediately and identified on the basis of the polypeptide that each encodes. A successful demonstration of this *shotgun sequencing* strategy is given by Putney *et al.* (1983) who determined DNA sequences of 178 randomly chosen muscle cDNA recombinants. Complete amino acid sequences were available for 19 abundant muscle-related proteins. Altogether, they were able to identify clones corresponding to 13 of these 19 proteins, including interesting protein variants.

For cDNA clones in the low abundance class it is usual to construct a cDNA library. The formula of Clarke & Carbon (1976) (see p. 110 can be adapted to estimate the number of recombinants required for a reasonably complete library. Typically, 10^6–10^7 clones will be sufficient for low-abundance mRNAs from most cell types. Once again the high efficiency

Table 7.2 Abundance classes of typical mRNA populations.

Source	Number of different mRNAs	Abundance (molecules/cell)
mouse liver cytoplasmic poly(A)$^+$	9	12 000
	700	300
	11 500	15
chick oviduct polysomal poly(A)$^+$	1	100 000
	7	4000
	12 500	5

References: mouse (Young *et al.* 1976); chick oviduct (Axel *et al.* 1976).

123

*Chapter 7
Cloning
Strategies,
Gene Libraries
and cDNA
Cloning*

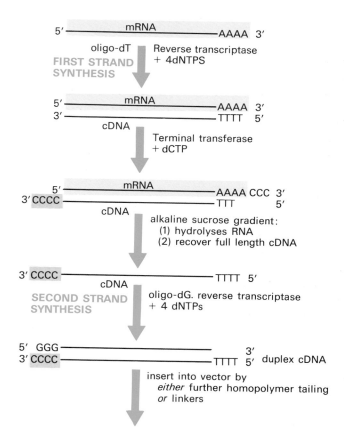

Fig. 7.9 Improved method for full-length duplex cDNA snythesis. The first strand is tailed with oligo-dC so as to allow priming of the second-strand synthesis by oligo-dG.

obtained by packaging *in vitro* makes phage λ vectors attractive for obtaining large numbers of cDNA clones. Insertional vectors such as λgt10, λNM1149, λZAP or λgt11 (see Chapter 8) are particularly well-suited for such cDNA cloning.

Is it worth enriching for a particular mRNA of cDNA before cloning? Only in special circumstances is a ready purification possible and attractive. In general, the most commonly used primary screening technique involves colony or plaque hybridization with radioactive or immunochemical probes. This is applicable to very large libraries and the effort involved is largely independent of the number of recombinants to be screened. There is, therefore, usually little to be gained by attempting to enrich the starting mRNA or cDNA in order to reduce the number of recombinants to be screened. The isolation of clones by such screening procedures has become a commonplace technical feat that effectively performs a purification impossible by any other means. However, the use of the polymerase chain reaction (see p. 331), to amplify particular cDNA sequences before cloning, may be useful in certain circumstances.

(a)

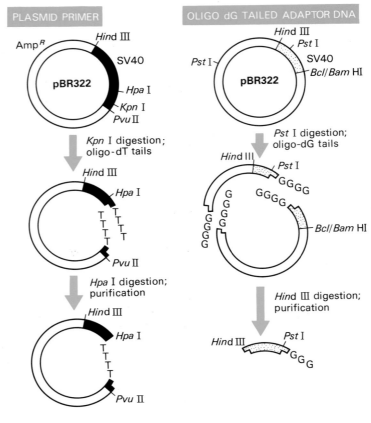

(b)

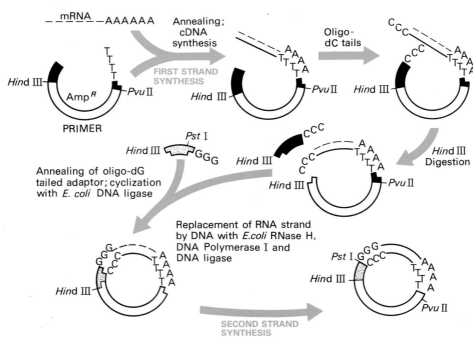

125

*Chapter 7
Cloning
Strategies,
Gene Libraries
and cDNA
Cloning*

Full-length cDNA cloning

In Chapter 4 some early strategies for synthesizing cDNA molecules for cloning were described. These involved self-priming of second-strand synthesis and therefore led to the loss of some sequence corresponding to the 5′ end of the mRNA.

Several strategies have been developed to overcome this difficulty. One of the simplest of these is shown in Fig. 7.9. The dC tailing of single-stranded cDNA followed by oligo-dG priming of second-strand synthesis does not lead to hairpin formation, nuclease S1 treatment is not required, and consequently this is an effective method for generating full-length cDNA clones (Land *et al.* 1981).

Two further methods, shown schematically in Figs 7.10 and 7.11, have been devised to eliminate the use of nuclease S1. Additionally, in both methods the oligo-dT sequence for priming the first-strand cDNA synthesis is linked to the vector DNA in a prior reaction. Both methods have been reported to promote full-length cDNA cloning with a very high efficiency (Okayama & Berg 1982, Heidecker & Messing 1983). It is thought that full-length reverse transcripts are obtained *preferentially* because in each case an RNA−DNA hybrid molecule, which is the result of first-strand synthesis, is the substrate for a terminal transferase reaction. A cDNA that does not extend to the end of the mRNA will present a shielded 3-hydroxyl group, which is a poor substrate for tailing.

It will also be noticed that the Okayama & Berg strategy employs a second-strand synthesis step in which the RNA strand is replaced in a DNA polymerase reaction that is primed by nicking the RNA with RNase H. The second-strand synthesis occurs at these nicks by a nick-translation type of reaction. This type of second-strand reaction has been exploited in efficient cDNA library cloning schemes that are simpler than that of Okayama & Berg; e.g. those developed by Gubler & Hoffman (1983), and Lapeyre & Amalric (1985). The Gubler & Hoffman protocol and modifications of it have been popular. A particular advantage of the Okayama & Berg strategy is that the cDNA is inserted into the vector in a defined orientation. This is useful in derivatives of the strategy in which the cDNA is inserted into a vector adjacent to a phage T3, T7 or SP6 promoter, because then an RNA copy can be synthesized *in vitro* from a defined cDNA strand with purified phage RNA polymerase (see Chapter 4).

The use of random primers in cDNA cloning

In the examples discussed so far, cDNA synthesis has been initiated on poly (A)$^+$ mRNA by priming with oligo-dT sequences. There are three limitations to this approach.

Fig. 7.10 (*Facing page*) The Okayama & Berg method of cDNA cloning. (a) Preparation of plasmid primer and adaptor DNA. The unshaded portion of each ring is pBR322 DNA, and the shaded or stippled segments are from SV40 DNA. (b) Steps in the construction of plasmid-cDNA recombinants. The designations for the DNA segments are as mentioned in 7.10 (a).

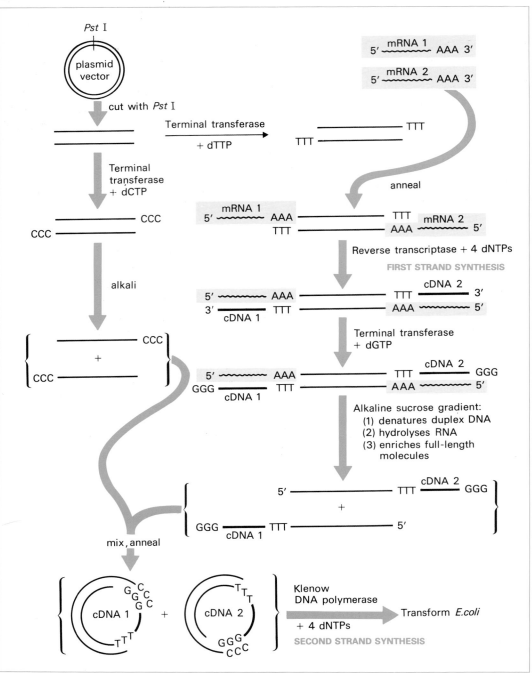

Fig. 7.11 Efficient full-length cDNA cloning (Heidecker & Messing 1983). The mRNA is annealed to linearized and oligo-dT tailed plasmid DNA, which then primes synthesis of the first cDNA strand. Oligo-dG tails are added to the cDNA-plasmid molecules, which are then centrifuged through an alkaline sucrose gradient. This step removes small molecules, hydrolyses the mRNA and separates the two cDNAs which were formerly attached to the same duplex plasmid. Denatured, oligo-dC tailed plasmid DNA is added (in excess) and conditions adjusted to favour circularization by the complementary homopolymer tails. The excess oligo-dC tailed plasmid may simply renature, but cannot circularize. The circular molecules have a free 3-hydroxyl on the oligo-dC tail which primes second-strand synthesis of the cDNA to create duplex recombinant plasmids which transform *E. coli*. Clones can be obtained with the cDNA inserted in both orientations.

127
Chapter 7
Cloning
Strategies,
Gene Libraries
and cDNA
Cloning

1 Not all RNAs bear a 3'-terminal poly (A) sequence. It is possible to add a poly(A) sequence *in vitro* with purified poly(A) polymerase, but this can be problematical.

2 Large mRNAs can be difficult to deal with because it is not reasonable to expect to synthesize and clone sequences which lie further than about 10 kb from the oligo-dT primer.

3 3' end bias. Because priming occurs at the 3' end of poly(A)$^+$ mRNAs and because cDNA synthesis is often incomplete, cDNA libraries are enriched in 3'-terminal sequences.

The first limitations are common in cloning genomic RNAs from RNA viruses. The third limitation may be important where cDNA libraries are to be made in a vector such as λgtll or λZAP and screened for expression of fusion polypeptides, either immunochemically or by the South-western approach (Chapter 8). These limitations can often be overcome by priming the first-strand cDNA synthesis not with oligo-dT, but rather with 'random' oligonucleotide primers. Usually the primer consists of a mixture of all possible chemically synthesized hexadeoxynucleotides. These hybridize at random sites along the RNA, prime cDNA synthesis, and generate cDNA sequences of a sufficient size to be useful when cloned. Commonly a Gubler & Hoffman-type second-strand step is employed.

8 Recombinant Selection and Screening

The task of isolating a desired recombinant from a population of bacteria or phage depends very much upon the cloning strategy that has been adopted; for instance, when a cDNA derived from an abundant mRNA is to be cloned, the task is relatively simple — only a small number of clones need to be screened. Isolating a particular single-copy gene sequence from a complete mammalian genomic library requires techniques in which hundreds of thousands of recombinants can be screened.

In the first part of this chapter we give an overview of the general principles employed in recombinant selection and screening procedures, under the headings of genetic, immunochemical, South-western, nucleic acid hybridization, and recombinational methods. As we shall see, nucleic acid hybridization with labelled probes is the most generally applicable method. The technology for labelling nucleic acids for such hybridization methods has become an interesting field in its own right, central to the achievements in recombinant DNA research. Therefore Chapter 15 is devoted to this technology.

Genetic methods

Selection for presence of vector. When combined with microbiological techniques, genetic selection is a very powerful tool since it can be applied to large populations. All useful vector molecules carry a selectable genetic marker or property. Plasmid and cosmid vectors carry drug-resistance or nutritional markers, and in the case of phage vectors plaque formation is itself the selected property. Genetic selection for presence of the vector is a prerequisite stage in obtaining the recombinant population. As we have seen, this can be refined to distinguish recombinant molecules and non-recombinant, parental vector. Insertional inactivation of a drug-resistance marker, or of a gene such as β-galactosidase for which there is a colour test, are examples of this (see p. 78). With certain replacement type lambda vectors, and with cosmid vectors, size selection by the phage particle selects recombinant formation.

Selection of inserted sequences. If an inserted foreign gene in the desired recombinant is expressed, then genetic selection may provide the simplest method for isolating clones containing the gene. Cloned *E. coli* DNA fragments carrying biosynthetic genes can be identified by complementation of non-revertible auxotrophic mutations in the host *E. coli* strain. A related example comes from the work of Cameron *et al.* (1975) who have cloned the *E. coli* DNA ligase gene in a phage λgt.λB vector. They exploited the inability of λ*red*⁻ phage (the vector is *red*⁻ by deletion of the C fragment) to form plaques on *E. coli lig* ts at the permissive temperature, whereas λ*red*⁻ phage will form plaques on *E. coli* Lig⁺. Recombinant phage carrying the wild-type ligase function could therefore be selected simply by their ability to form plaques through complementation of the host deficiency when plated on *E. coli lig* ts.

It has been found that certain eukaryotic genes are expressed in *E. coli* and can complement auxotrophic mutations in the host bacterium. Ratzkin & Carbon (1977) inserted fragments of yeast DNA, obtained by mechanical shearing, into the plasmid Col E1 using a homopolymer tailing procedure. They transformed *E. coli his* B mutants with recombinant plasmid and, by selecting for complementation, isolated clones carrying an expressed yeast *his* gene.

A similar approach has even been applied successfully to cloned mouse sequences. Chang *et al.* (1978) constructed a population of recombinant plasmids containing cDNA that was derived from an unfractionated mouse cell mRNA preparation in which dihydrofolate reductase (DHFR) mRNA was present. Mouse DHFR is much less sensitive to inhibition by the drug trimethoprim than is *E. coli* DHFR, so that by selecting transformants in medium containing the drug, clones were isolated in which resistance was conferred by synthesis of the mouse enzyme. This was an early example of expression of a mammalian structural gene in *E. coli*. The factors affecting expression of heterologous genes are complex, and an efficient selection procedure was required in order to identify clones actually synthesizing mouse DHFR amongst those containing non-expressed DHFR cDNA.

Immunochemical methods

Immunochemical detection of clones synthesizing a foreign protein has also been successful in cases where the inserted gene sequence is expressed. A particular advantage of the method is that genes that do not confer any selectable property on the host can be detected, but it does of course require that specific antibody is available.

A number of laboratories have developed similar immunochemical detection methods (Skalka & Shapiro 1976, Ehrlich *et al.* 1978a). The method of Broome & Gilbert (1978) is illustrative. It depends upon three points:

1 an immune serum contains several IgG types that bind to different determinants on the antigen molecule;

2 antibody molecules absorb very strongly to plastics such as polyvinyl, from which they are not removed by washing;

3 IgG antibody can be readily radiolabelled with ¹²⁵I by iodination *in vitro*.

These properties are elegantly exploited in the following way. First, transformed cells are plated on agar in a conventional Petri dish. A replica plate must also be prepared because subsequent procedures kill these colonies. The bacterial colonies are then lysed in one of a number of ways — by exposure to chloroform vapour, by spraying with an aerosol of virulent phage, or by using a host bacterium that carries a thermo-inducible prophage. This releases the antigen from positive colonies. A sheet of polyvinyl that has been coated with the appropriate antibody (unlabelled) is applied to the surface of the plate, whereupon the antigen complexes with the bound IgG. The sheet is removed and exposed to ^{125}I-labelled IgG. The ^{125}I-IgG can react with the bound antigen via antigenic determinants at sites other than those involved in the initial binding of antigen to the IgG-coated sheet, as shown in Fig. 8.1. Positively reacting colonies are detected by washing the sheet and making an autoradiographic image. The required clones can then be recovered from the replica plate. The method can be applied with only minor modification to plates bearing phage plaques instead of transformed colonies.

Two further aspects of the immunochemical method deserve mention. First, detection of altered protein molecules is possible providing that the alteration does not prevent cross-reaction with antibody. Thus Villa-Komaroff *et al*. (1978) have isolated *E. coli* clones containing cDNA sequences from rat preproinsulin mRNA. The cDNA was inserted by dG-dC homopolymer tailing at the *Pst* I site of pBR322. Using anti-insulin antibody, they isolated a clone which expressed a fused protein composed of the N-terminal region of β-lactamase (from pBR322) and a region of the proinsulin protein linked through a stretch of six glycine residues encoded by $d(G)_{18}$ of the joint (see p. 147). Second, the two-site detection method employed by Broome & Gilbert is particularly suited to the detection of novel genetic constructions; for instance, by coating polyvinyl discs with IgG prepared from an immune serum directed against one protein, and detecting the immobilized antigen with ^{125}I-antibodies directed against another protein, only hybrid polypeptide molecules, synthesized as a result of DNA recombination, would produce an autoradiographic response.

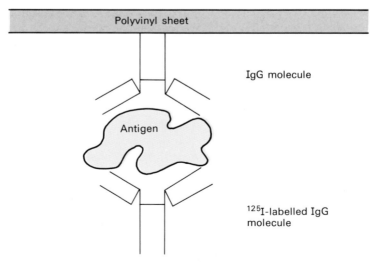

Fig. 8.1 Antigen−antibody complex formation in the immunochemical detection method of Broome & Gilbert. (See text for details.)

A very efficient exploitation of the immunochemical detection method
131

*Chapter 8
Recombinant
Selection and
Screening*

A very efficient exploitation of the immunochemical detection method
involves the phage λ expression vector λgt11 (Young & Davis 1983).
This vector carries the *E. coli lacZ* gene. A unique *Eco* RI site is located
within the β-galactosidase coding region. Recombinant libraries can be
constructed in which eukaryotic cDNA has been inserted, by means of
linkers, into the *Eco* RI sites. In such recombinants the β-galactosidase is
insertionally inactivated and, depending upon the translational phase at
the fusion junction, hybrid proteins are expressed. In a population of
cDNA recombinants in which duplex cDNA has been synthesized by any
of the common methods, we can expect a proportion of recombinants
containing any particular cDNA to be in phase. The vector can accept up
to 8.3 kb of insert DNA and complete cDNA libraries containing large
numbers of independent recombinants can be constructed readily because
of the efficiency endowed by packaging *in vitro*.

Immunochemical screening of the library can be carried out upon
colonies of induced lysogenic bacteria *or*, as is now more common, the
screening is carried out on plaques of the recombinant phage. The original
approach with induced lysogens is shown in Fig. 8.2. In this approach the
library of recombinant λgt11 is first used to lysogenize *E. coli*. This is
efficiently carried out with a *hfl*A (*h*igh *f*requency of *l*ysogeny) mutant of
E. coli. Lysogens produce detectable amounts of hybrid proteins upon induction. It has been claimed that up to 10^6 colonies can be screened on a
single 8.2 cm diameter filter. This immunochemical approach has largely
been replaced by a very similar but much more convenient method that is
carried out on phage plaques. In this procedure the library is plated out at
moderately high density (up to 5×10^4 plaques per 9 cm square plate),
with *E. coli* strain Y1090 as host (Fig. 8.3). This *E. coli* strain overproduces
the *lac* repressor and ensures that no expression of cloned sequences
(which may be deleterious to the host) takes place until the inducer IPTG
is presented to the infected cells. Y1090 is also deficient in the *lon* protease,
hence increasing the stability of recombinant fusion proteins. Fusion proteins expressed in plaques are absorbed onto a nitrocellulose membrane
overlay and this membrane is then processed for antibody screening.
When a positive signal is identified on the membrane, the positive plaque
can be picked from the agar plate (a replica is not necessary) and hence
recombinant phage can be isolated.

South-western screening for DNA binding proteins

In the previous section we have seen how fusion proteins may be detected
immunochemically when expressed in plaques produced by recombinant
λgt11 or λZAP. A closely related approach has been used very successfully
for screening and isolation of clones expressing fusion proteins where the
foreign sequence encodes a DNA-binding protein that binds specifically
to a particular DNA sequence. The method involves a nitrocellulose membrane 'plaque-lift' on which expressed fusion proteins are adsorbed. The
procedure for performing the plaque-lift is the same as that just described.
However, the screening is carried out by incubating the membrane with a
radiolabelled *duplex* DNA oligonucleotide containing the sequence for

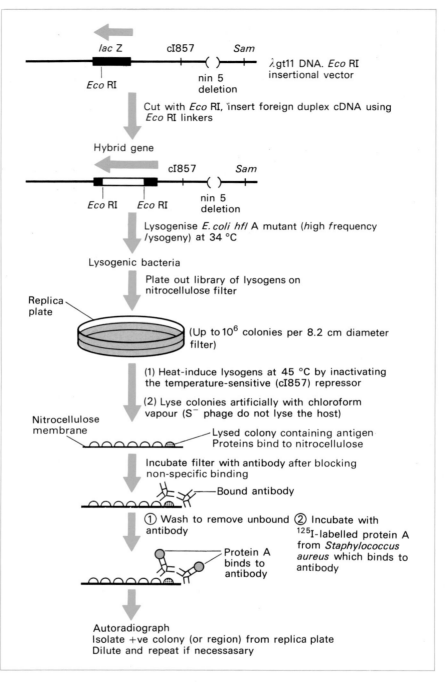

Fig. 8.2 Immunochemical screening applied to induced lysogens with the expression vector λgt11 and its derivatives. The duplex cDNA is inserted within the *lacZ* gene. In a proportion of recombinants the insertion will be in the correct translational reading phase so as to direct the synthesis of a hybrid protein that will be detected by reaction with antibody raised against the required protein. The functions of the *cI* and *S* genes are discussed in Chapter 4. The *hfl*A mutation of *E. coli* results in a very high frequency of lysogenization by phage λ. These lysogens express detectable amounts of hybrid protein when they are induced by

which the DNA-binding protein is specific. Conditions are chosen in which non-specific binding is minimized. Positively reacting plaques are identified by autoradiography of the filter. This technique therefore uses a radiolabelled DNA to detect polypeptide on the nitrocellulose, and has been called a 'South-western' procedure. It has been spectacularly successful in the isolation of clones expressing cDNA sequences corresponding to certain mammalian transcription factors (Vinson *et al.* 1987, Staudt *et al.* 1988). Clearly the procedure can only be successful: (a) where the binding activity is due to a single polypeptide chain, and (b) where it can be expressed in functional form when in a fusion polypeptide. (Note that several *different* fusions of any particular polypeptide with the β-galactosidase polypeptide are likely to be created in any random cloning of a particular cDNA in a single cDNA library.) It may also be significant that success has been obtained on some occasions where random priming has been used to generate the cDNA first strand. It is also clear that the affinity of the polypeptide for the specific DNA sequence must be high. The procedure has been found most efficient when the oligonucleotide containing the binding sequence has been ligated into multimeric form. This may mean that a single DNA multimer may be bound by more than one fusion polypeptide molecule on the filter, hence greatly increasing the average dissociation time.

Nucleic acid hybridization methods

Various recombinant detection methods employing hybridization with DNA isolated and purified from the transformed cells have been developed. However, these have been almost entirely superseded by the method of Grunstein & Hogness (1975) who have developed a screening procedure to detect DNA sequences in transformed colonies by hybridization *in situ* with radioactive 'probe' RNA. Their procedure can rapidly determine which colony amongst thousands contains the required sequence. A modification of the method allows screening of colonies plated at a very high density (Hanahan & Meselson 1980).

   The colonies to be screened are first replica plated onto a nitrocellulose filter disc that has been placed on the surface of an agar plate prior to inoculation (Fig. 8.4). A reference set of these colonies on the master plate is retained. The filter bearing the colonies is removed and treated with alkali so that the bacterial colonies are lysed and their DNAs are denatured. The filter is then treated with proteinase K to remove protein and leave denatured DNA bound to the nitrocellulose, for which it has a high affinity, in the form of a 'DNA-print' of the colonies. The DNA is fixed firmly by baking the filter at 80°C. The defining, labelled RNA is hybridized to this DNA and the result of this hybridization is monitored by autoradio-

raising the temperature so as to inactivate the temperature-sensitive *c*I repressor carrying the c1857 mutation. The frequency of lysogenization is high on the *hfl* A strain, but some non-lysogens will be present on the filter. The procedure can be improved by incorporating a drug-resistance marker, e.g. ampicillin resistance (Kemp *et al.* 1983) or kanamycin resistance, borne on transposon Tn5, into the vector. Lysogenic bacteria can then be selected in the presence of the drug.

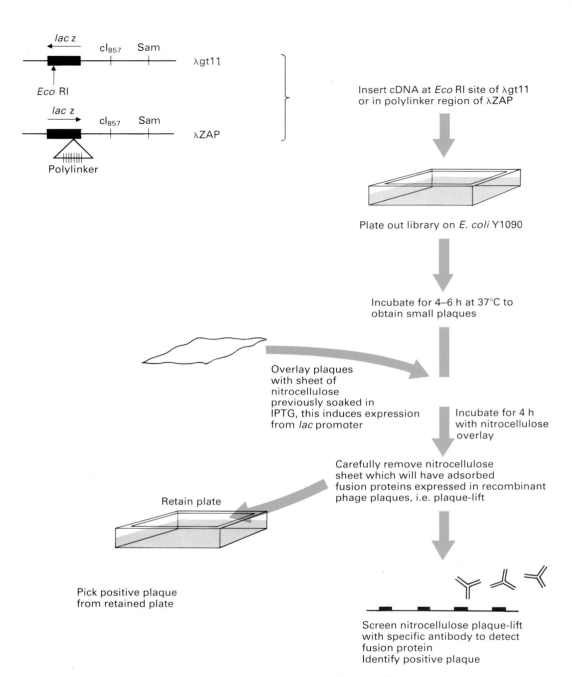

Fig. 8.3 Immunochemical screening of λgt11 or λZAP recombinant plaques.

graphy. A colony whose DNA-print gives a positive autoradiographic result can then be picked from the reference plate.

Variations of this procedure can be applied to phage plaques (Jones & Murray 1975, Kramer *et al.* 1976). Benton & Davis (1977) devised a method in which the nitrocellulose filter is applied to the upper surface of agar plates, making direct contact between plaques and filter. The plaques

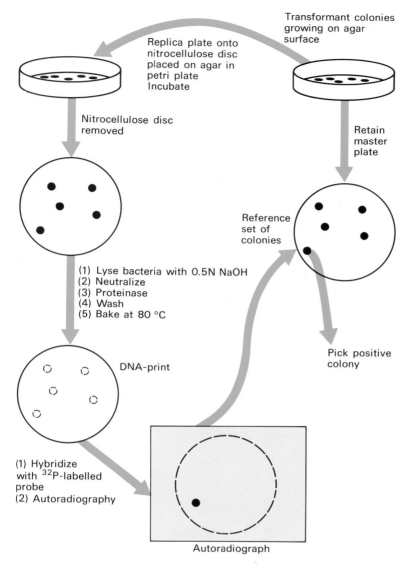

Fig. 8.4 Grunstein–Hogness method for detection of recombinant clones by colony hybridization.

contain phage particles as well as a considerable amount of unpackaged recombinant DNA. Both phage and unpackaged DNA bind to the filter and can be denatured, fixed, hybridized, etc. This method is the original 'plaque-lift' procedure and has the advantage that several identical DNA prints can easily be made from a single phage plate: this allows the screening to be performed in duplicate, and hence with increased reliability, and also allows a single set of recombinants to be screened with two or more probes. The Benton & Davis plaque-lift procedure must be the most widely applied method of library screening, successfully applied in thousands of laboratories to the isolation of recombinant phage by nucleic acid hybridization. Figure 8.5 shows a scheme of the procedure.

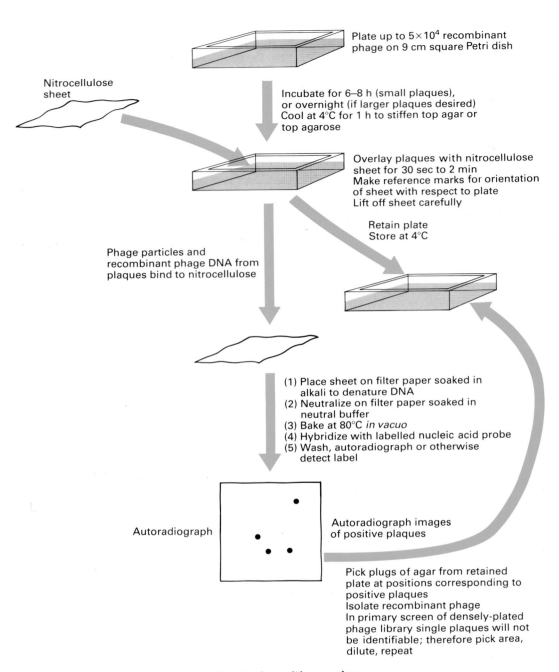

Plate up to 5×10^4 recombinant phage on 9 cm square Petri dish

Incubate for 6–8 h (small plaques),
or overnight (if larger plaques desired)
Cool at 4°C for 1 h to stiffen top agar or
top agarose

Nitrocellulose
sheet

Overlay plaques with nitrocellulose
sheet for 30 sec to 2 min
Make reference marks for orientation
of sheet with respect to plate
Lift off sheet carefully

Retain plate
Store at 4°C

Phage particles and
recombinant phage DNA from
plaques bind to nitrocellulose

(1) Place sheet on filter paper soaked in
    alkali to denature DNA
(2) Neutralize on filter paper soaked in
    neutral buffer
(3) Bake at 80°C *in vacuo*
(4) Hybridize with labelled nucleic acid probe
(5) Wash, autoradiograph or otherwise
    detect label

Autoradiograph

Autoradiograph images
of positive plaques

Pick plugs of agar from retained
plate at positions corresponding to
positive plaques
Isolate recombinant phage
In primary screen of densely-plated
phage library single plaques will not
be identifiable; therefore pick area,
dilute, repeat

Fig. 8.5 Benton & Davis' plaque-lift procedure.

This is the prototype for a variety of plaque-lift screening procedures previously illustrated in this chapter.

The great advantage of the hybridization methods is generality. They do not require expression of the inserted sequences and can be applied to any sequence, provided a suitable radioactive probe is available.

The probe problem

137

*Chapter 8
Recombinant
Selection and
Screening*

Screening procedures which rely on nucleic acid hybridization are, as we have seen, general in application and powerful. Using these procedures it is now possible easily to isolate any gene sequence from virtually any organism, *if a probe is available*. The problem of gene isolation, then, is a problem of obtaining a suitable probe. There are several solutions to this.

1 In certain specialized cell types or tissues, particular mRNAs are abundant or super-abundant. The corresponding cDNA clones can be isolated by screening small numbers of recombinants directly by such methods as shotgun sequencing or HRT (see below). These cDNA probes can then be used to isolate genomic sequences.

2 Nucleic acid sequences encoding certain proteins have been sufficiently conserved in evolution such that cross-species nucleic acid hybridization is possible. Examples of effective heterologous probes include histone sequences [sea urchin *vs. Xenopus* (Old *et al.* 1982b)], and actin sequences [*Dictyostelium vs. Xenopus* (Cross *et al.* 1984)]. This means that the particular biological advantages of one experimental system can be exploited to isolate a gene sequence, which may then provide a probe for corresponding genes in other organisms.

3 An oligonucleotide, which needs to be only about 15–20 nucleotides long, and which corresponds to a part of the sequence encoding the protein in question, can be synthesized chemically. This requires that short, oligopeptide, stretches of amino acid sequence must be known. The DNA sequence (or its complement) encoding the protein can then be deduced from the genetic code. However, a problem arises here because of the degeneracy of the code. Most amino acids are encoded by more than one codon. For this reason, oligopeptide sequences known to contain tryptophan or methionine residues are particularly valuable, because these two amino acids have single codons, and the number of possibilities is thereby reduced. Thus, for example, the oligopeptide His–Phe–Pro–Phe–Met may be identified and chosen to provide a probe sequence. This oligopeptide could be encoded by the following sequences:

$$5' \quad CA^{T}_{C}TT^{T}_{C}CCCTT^{T}_{C}ATG \quad 3'$$

$$\begin{matrix} & & & & T & & & \\ & & & & A & & \\ & & & & G & & \end{matrix}$$

These 32 different sequences do not have to be synthesized individually. It is possible to perform a mixed addition reaction for each polymerization step where the sequence is degenerate. Therefore only one, mixed probe is prepared and radiolabelled.

Plus and minus screening

This is a variant of the nucleic acid hybridization method that is particularly suitable for isolating tissue-specific or developmentally regulated cDNA sequences or clones derived from mRNAs that are induced by particular treatments.

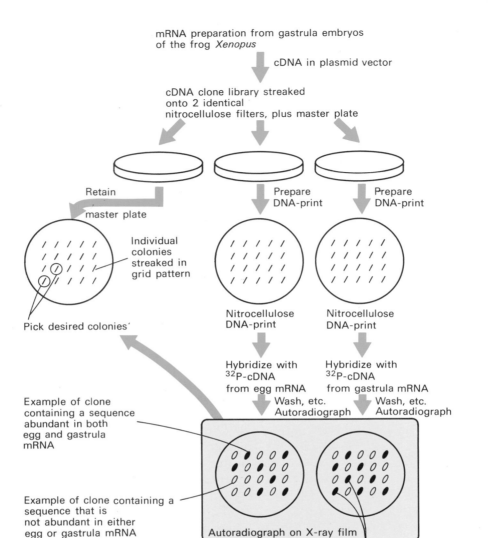

Fig. 8.6 Plus and minus screening. (See text for details.)

Let us consider, for example, the isolation of cDNAs derived from mRNAs which are abundant in the gastrula embryo of the frog *Xenopus* but which are absent, or present at low abundance, in the egg. A cDNA clone library is prepared from gastrula mRNA. Replica filters carrying identical sets of recombinant clones are then prepared (Fig. 8.6). One of these filters is then probed with ^{32}P-labelled mRNA (or cDNA) from gastrula embryos, and one with ^{32}P-labelled mRNA (or cDNA) from the egg. Some colonies will give a positive signal with both probes, these represent cDNAs derived from mRNA types that are abundant at both stages of development. Some colonies will not give a positive signal with either probe, these correspond to mRNA types present at undetectably low abundance in both tissues. This is a feature of using probes derived from mRNA populations: only abundant or moderately abundant sequences in the probe carry a significant proportion of the label and are effective in hybridization. Importantly, some colonies give a positive signal with the gastrula probe, but not with the egg probe. These should, therefore, correspond to the required sequences.

Such plus and minus screening has been applied to the analysis of the development of *Xenopus* by Dworkin & Dawid (1980) and to the slime mould *Dictyostelium* by Williams & Lloyd (1979). In both instances it was estimated that a fivefold difference in mRNA abundance could be detected in this procedure.

Recombinational probe

This ingenious and powerful method is based upon homologous recombination in the *E. coli* host (Seed 1983). The probe sequence here is inserted into a specially constructed plasmid vector, πVX. This is a very small plasmid of 902 base pairs, derived from the Col E1 replicon, which contains a convenient polylinker sequence and a suppressor tRNA gene, *sup* F (Fig. 8.7). Genomic phage λ libraries are propagated on recombination-proficient *E. coli* containing the probe-πVX recombinant plasmid. Phage carrying sequences homologous to the probe acquire an integrated copy of the plasmid by homologous recombination. Phage bearing integrated probe-πVX can then be recovered and isolated by growth under appropriate selective conditions. This is most easily achieved by using a phage λ vector carrying an amber mutation suppressible by *sup* F (e.g. EMBL3 derivatives, see Chapter 4). By finally plating on a non-suppressing *E. coli* only those phage that have integrated a *sup* F gene, by virtue of homology with the probe, can form plaques.

This method can be applied readily to very large numbers ($>5 \times 10^6$) of phage in a genomic library, and has the advantage of being very quick, providing that the probe-πVX recombinant plasmid has been constructed at a prior stage. The shortest probe segment giving high recombination has been found to be about 60 base pairs long. A certain amount of sequence divergence can be tolerated in the homologous recombination event, but it has been shown that the probe does discriminate between a perfectly homologous sequence and one with 8.2% sequence divergence (Seed 1983).

Hybrid released translation (HRT) and hybrid arrested translation (HART)

These methods enable a cloned DNA to be correlated with the protein(s) which it encodes. Hybrid *released* translation is a direct method in which cloned DNA is bound to a nitrocellulose filter and hybridized with an unfractionated preparation of mRNA or even total cellular RNA. The filter is then washed, and hybridized messenger is eluted by heating in low-salt buffer or in a buffer containing formamide. The recovered mRNA is then

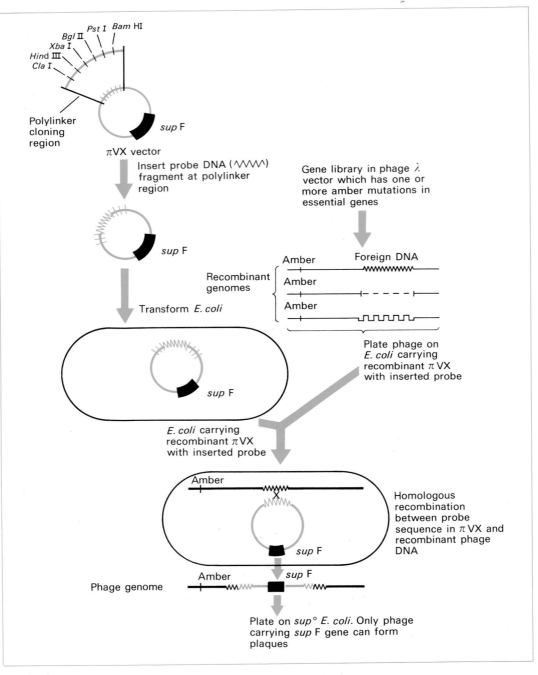

Fig. 8.7 Recombinational probing. (See text for details.)

translated in a cell-free translation system such as that derived from wheat germ or rabbit reticulocytes, and radiolabelled polypeptide products whose synthesis it directs are analysed on an appropriate gel electrophoresis system (Fig. 8.8).

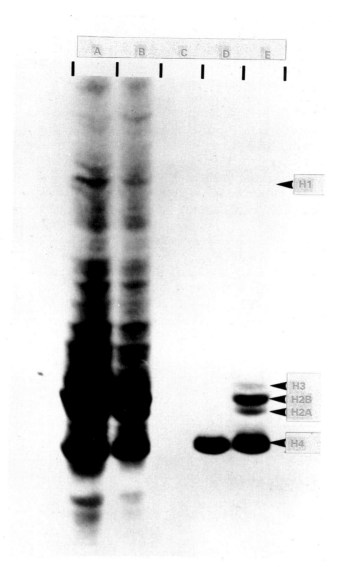

Fig. 8.8 *Hybrid released translation*. Application of HRT to two cloned DNAs containing histone sequences of X. *laevis*. The figure shows an autoradiograph of a polyacrylamide gel analysis of radiolabelled polypeptides synthesized in a wheat germ cell-free translation system with ^{3}H-lysine. The tracks correspond to the following RNA additions to the cell-free system: A and B, total RNA from X. *laevis* ovary; C, no added RNA; D, RNA released from a filter bearing histone H4 cDNA sequences cloned in pAT153; E, RNA released from a filter bearing H3, H2B, H2A and H4 sequences clustered in a genomic DNA fragment cloned in λWES. These filters had been hybridized with total ovary RNA. Positions of marker histones are indicated.

Hybrid *arrested* translation is now less widely used than HRT. It is based upon the fact that an mRNA will not direct the synthesis of protein in a cell-free translation system when it is in a hybrid form with its DNA complement (Paterson *et al.* 1977). In a model experiment, rabbit globin mRNA, which is a mixture of mRNA species encoding α- and β-globin polypeptides, was mixed with denatured DNA of the recombinant plasmid pβG1 which contains a rabbit β-globin cDNA sequence. Conditions were chosen so that DNA−RNA hybridization was favoured whilst suppressing re-association of the plasmid DNA (i.e. high formamide concentration). The nucleic acids were recovered from the hybridization mixture and added to a cell-free translation system prepared from wheat germ. ^{35}S-methionine was included so that synthesis of radioactive polypeptides could be analysed by polyacrylamide gel electrophoresis and autoradiography. It was apparent that the prehybridized mRNA was rendered inactive by hybridization with its DNA complement. Full translational activity of the β-globin mRNA was recovered when the prehybridized mRNA preparation was dissociated by a brief heating before addition to the cell-free system.

HRT and HART are powerful methods. Unfractionated mRNA preparations from most cell types direct the synthesis of several hundreds of distinct abundant polypeptides that can be resolved on 2-dimensional electrophoresis systems. It should be quite feasible to isolate recombinant cDNAs derived from such a mixed population of mRNAs, and correlate them with their corresponding polypeptides.

9 Expression in *E. coli* of Cloned DNA Molecules

Following the demonstration (see p. 46) that a gene (Ap^R) originating from *Staphylococcus aureus* can function in an unrelated bacterium, *Escherichia coli*, it was widely assumed that genes from any bacterium, could be expressed in any other. This belief was based on the expectation that, in parallel with the universality of the genetic code, the other parts of the gene-to-phenotype biochemical pathway would also be universal. This idea was strengthened by the observation that genes from two lower eukaryotes, *Saccharomyces cerevisiae* (Struhl *et al.* 1976) and *Neurospora crassa* (Vapnek *et al.* 1977) are also expressed in *E. coli*. For that reason the genes from higher organisms were also expected to be expressed in bacteria. However, these favourable reports were followed by a whole series which indicated that many cloned genes were not expressed in their new genetic background. An explanation of these failures is provided by consideration of the steps involved in the gene-to-phenotype pathway.

Synthesis of a functional protein depends upon transcription of the appropriate gene, efficient translation of the mRNA and, in many cases, post-translational processing and compartmentalization of the nascent polypeptide. A failure to perform correctly any one of these processes can result in the failure of a given gene to be expressed. Transcription of a cloned insert requires the presence of a promoter recognized by the host RNA polymerase. Efficient translation requires that the mRNA bears a ribosome binding site. In the case of an *E. coli* mRNA a ribosome binding site includes the translational start codon (AUG or GUG) and a sequence that is complementary to bases on the 3' end of 16s ribosomal RNA. Shine & Dalgarno (1975) first postulated the requirement for this homology, and various S-D sequences, as they are often called, have been found in almost all *E. coli* mRNAs examined. Identified S-D sequences vary in length from three to nine bases and preceded the translational start codon by three to twelve bases (Steitz 1979). A common post-translational modification of proteins involves cleavage of a *signal* sequence whose function is to direct the passage of the protein through the cell membrane (Blobel & Doberstein 1975, Inouye & Beckwith 1977). Another phenomenon, possibly related to the processing of proteins, is their degradation. It is known that the short polypeptides encoded by genes which have undergone nonsense mutations are rapidly degraded in *E. coli* while the wild-type proteins are stable. It can be envisaged that the foreign proteins would be rapidly degraded in the new host if their configuration or amino acid sequence did not protect them from intracellular proteases.

From the above discussion it is clear that the first requirement for expression in *E. coli* of a structural gene, inserted *in vitro* in a DNA molecule, is that the gene be placed under the control of an *E. coli* promoter.

When this is done the protein product may be synthesized from its own N-terminus, but more often than not a fusion protein is produced. A few examples are given in Table 9.1 and some of them are discussed in more detail below. It should be noted that in some cases it may be possible to cleave the fusion protein *in vitro* to yield native protein and one example already discussed is somatostatin (see p. 52).

Construction of vectors that give improved transcription of inserts

In order to place genes under the control of an *E. coli* promoter, Fuller (see Backmann *et al.* 1976) constructed a 'portable promoter' in the form of a DNA fragment known as *Eco* RI (UV5). This fragment carries the regulatory region of the *lac* operon and the UV5 mutation bounded by *Eco* RI generated cohesive ends. The UV5 mutation makes the system insensitive to catabolite repression. The fragment also carries the L8 mutation but this does not affect its use. The L8 mutation is only mentioned here for the sake of completeness and to avoid confusion when reading the literature. One of the *Eco* RI sites is located 65 base pairs downstream from the start point of transcription, a position corresponding to the ninth amino acid residue of the *lacZ* gene. Backmann *et al.* (1976) have constructed a plasmid in which this *Eco* RI (UV5) fragment is fused to the repressor gene (*cI*) of bacteriophage λ. Strains carrying this plasmid overproduced the repressor and transcription of the *cI* gene was largely under *lac* control. Many of the expression vectors in current use are based on either the *lac* UV5 promoter, as described above, the tryptophan (*trp*) promoter which controls the expression of the genes involved in tryptophan biosynthesis, or the λp$_2$ promoter (see Chapter 4). Hybrids of these promoters also have been constructed and these are discussed later (p. 155).

There are two tests that can be used to show that expression of a cloned gene is under the control of the selected *E. coli* promoter. First, expression of the foreign gene should be obtained when it is placed in the correct orientation relative to the promoter but not when it is in the wrong

Tables 9.1 Kinds of protein produced when cloned DNA fragments are linked to bacterial promoters.

Kind of protein	Selected examples	Reference
fusion protein	rat growth hormone ovalbumin fibroblast interferon	Seeburg *et al.* 1977 Mercerau-Puijalon *et al.* 1978 Houghton *et al.* 1980
cleavable fusion protein	somatostatin β-endorphin human insulin	Itakura *et al.* 1977 Shine *et al.* 1980 Goeddel *et al.* 1979b
native protein (protein synthesized from its own N- terminus	SV40 t antigen human growth hormone mouse dihydrofolate reductase	Roberts *et al.* 1979a Goeddel *et al.* 1979a Chang *et al.* 1978

orientation. Second, expression of the gene should be enhanced by those environmental conditions that normally lead to activation of the promoter, e.g. addition of an inducer such as IPTG for *lac*-based expression vectors or starvation for tryptophan with *trp*-based vectors.

Positioning cloned inserts in the correct translational reading frame: expression of fusion proteins

When expression of a foreign gene is dependent on its fusion to an *E. coli* gene it is important that the correct translational reading frame is maintained. There is only a one in three chance that two randomly selected fragments will be fused in phase. Thus for most gene fusions, adjustments would have to be made at the gene junction. To avoid this laborious procedure Charnay *et al.* (1978) constructed a set of vectors, each having a single *Eco* RI restriction site, in which the cloned gene can be placed in each of the three possible reading frames relative to the translation initiation site of the *lacZ* gene. They started with a λ*lac* derivative (γ△ZUV5) having a unique *Eco* RI site in the ninth codon of the β-galactosidase gene. The reading frame of this *Eco* RI site was defined arbitrarily as φ1. λ△Z vectors in which the inserted fragments can be positioned in phases φ2 and φ3 (λ△Z2 and λ△Z3) were constructed by two successive additions of 2G:C base pairs to each of the *Eco* RI (UV5) fragments (Fig. 9.1).

The procedure for adding two base pairs to each end of the 203 base pair *Eco* RI (UV5) fragment is shown in Fig. 9.2. The fragment was treated with S1 nuclease in order to produce flush ends with G:C as the final base pair. Using T4 DNA ligase a synthetic *Eco* RI linker was then attached to each end of the fragment and this was followed by digestion with *Eco* RI

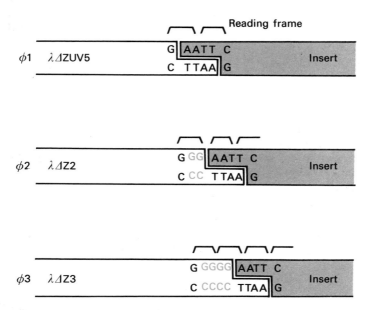

Fig. 9.1 The different reading frames with respect to the translation initiation site of the *lacZ* gene presented by the three λ vectors.

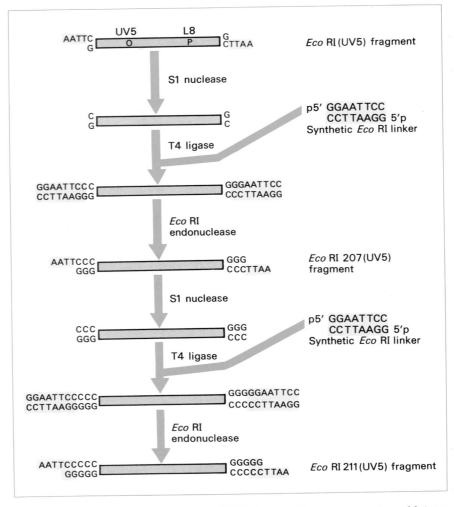

Fig. 9.2 Construction of the *Eco* RI 211 (UV5) fragment by two successive additions of two G:C base pairs to each end of the *Eco* RI (UV5) fragment.

endonuclease. This resulted in the production of a 207 base pair fragment called *Eco* RI 207 (UV5). Using the same procedures, *Eco* RI 207 (UV5) was converted to *Eco* RI 211 (UV5). Both the 207 and 211 base pair fragments were then inserted into a phage λ vector. Using methods which are irrelevant to this discussion the unwanted *Eco* RI sites upstream from the promoter on each fragment were removed to generate λ△Z2 and λ△Z3.

Tacon *et al.* (1980) using a slightly different methodology have constructed a similar set of plasmid vectors but in which the expression of the cloned inserts is under the control of the *trp* promoter. As with the *lac* vectors of Charnay *et al.* (1978) use of this set of vectors results in the production of a fused protein. These *trp*-based vectors are used widely and among the proteins expressed with them are influenza virus surface antigens (Emtage *et al.* 1980) and synthetic genes coding for poly (aspartyl-L-phenylalanine) (Doel *et al.* 1980) and urogastrone (Smith *et al.* 1982). A whole series of similar vectors has been developed subsequently whose other properties have been tailored to specific needs. Representative examples are described by Simons *et al.* (1987), Stark (1987) and Zaballos *et al.* (1987).

An alternative strategy for obtaining fusion proteins: cloning in the *Pst* I site of pBR322

An alternative method of putting an insert under the control of an *E. coli* promoter without the concomitant problem of translational reading frame adjustment has been used by Villa-Komaroff *et al.* (1978). They cloned cDNA transcripts of rat preproinsulin mRNA at the unique *Pst* I site of pBR322 (see Fig. 3.8). This site lies in the Ap^R gene at a position corresponding to amino acid residues 183 and 184. Consequently, an insert at the *Pst* I site should result in the production of a fused gene product. To effect this insertion, use was made of the homopolymer tailing procedure (Fig. 9.3). The advantage of this method is that in each recombinant molecule the lengths of the repeating G:C base-paired joints may be different and at least some of them will be in the correct reading frame.

Manipulation of cloned genes to achieve expression of native proteins

So far we have described methods for placing a cloned DNA fragment under the control of an *E. coli* promoter. In many instances these methods result in the production of a hybrid protein carrying amino-terminal amino acids from β-lactamase, β-galactosidase, or the *trp* E gene product. It would be more satisfactory if the gene-promoter fusion always produced a native protein.

Tacon *et al.* (1983) have described the construction of two plasmid vectors which facilitate the expression of native proteins. Both make use of the *trp* promoter and associated S-D sequence. In one of them, pWT551, the gene to be cloned is inserted at a *Hind* III site but expression requires the presence of an initiator ATG or GTG codon at the start of the coding region. The second vector, pWT571, has an *Eco* RI site overlapping the initiation codon (ATG AATTC). Cloning a gene in the correct reading frame in this site will result in expression of a protein having the sequence fMet—Asn at the N-terminus. Alternatively, digestion with endonuclease *Eco* RI followed by S1 nuclease to remove the 5'-AATT extensions permits blunt-end ligation to the initiation codon of a fragment encoding a protein lacking *N*-formylmethionine.

As with vectors for the production of fusion proteins, a whole series of similar vectors with specialist properties has been developed. Representative examples have been described by Stark (1987) and Zaballos *et al.* (1987).

Secretion of foreign proteins

β-lactamase, the product of the Ap^R gene of pBR322, is a periplasmic

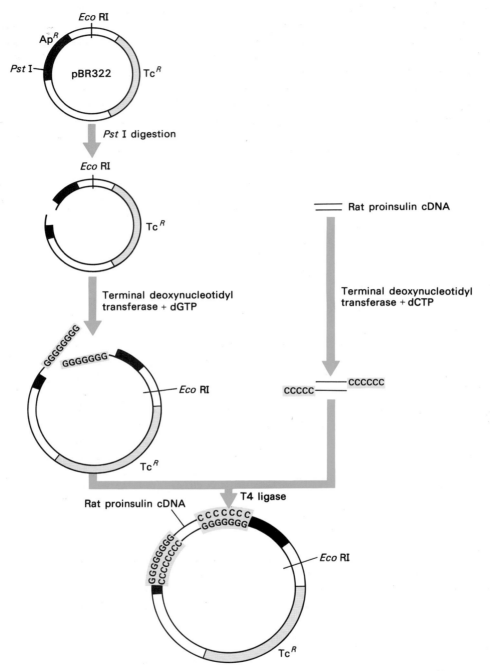

Fig. 9.3 Insertion of rat proinsulin cDNA at the *Pst* I site of pBR322.

protein. It is synthesized as a preprotein with a 23 amino acid leader sequence that serves as a signal to direct the secretion of the protein to the periplasmic space. This signal sequence is removed as the protein traverses the membrane. Insertion of a foreign gene into the Ap^R gene should permit expression of the foreign sequence as a fusion product that is transported to the periplasmic space. The insertion of the rat proinsulin gene at the *Pst* I site of pBR322 (see p. 147) provides a good test of this hypothesis. As expected, the β-lactamase-proinsulin fusion protein was found in the periplasmic space (Villa-Komaroff *et al.* 1978).

Talmadge and Gilbert (1980) have constructed a series of plasmids in which the unique *Pst* I site of pBR322 has been moved so that it lies within or near the pre-β-lactamase signal sequence. The genes for rat proinsulin and rat preproinsulin were cloned into the *Pst* I sites of these different plasmids resulting in the formation of hybrid β-lactamase (prokaryotic) and insulin (eukaryotic) signal sequences. By comparing the levels of insulin antigen in the *E. coli* periplasmic space, Talmadge *et al.* (1980) were able to deduce the structural requirements of a signal sequence and to show that a eukaryotic signal sequence can function in a prokaryote. Subsequent experiments have defined the factors governing the secretion of polypeptides into the periplasm of *E. coli* (for review see Hirst & Welch 1988). These are shown in Table 9.2.

Secretion of cytoplasmic proteins

The experiments described above show that a protein which normally is secreted by a eukaryote can also be secreted by *E. coli*. The question remains, can a protein which normally is cytoplasmic be engineered in such a way that it is secreted? Analysis of the signal sequences from a number of secreted proteins indicated a number of features common to all of them. At the N-terminus they have a run of polar amino acid residues, including one which is positively charged, followed by a 'core' of 10−15 hydrophobic residues. However, more than just the signal sequence may be required for secretion. Koshland & Botstein (1982) isolated chain-terminating mutants of the Ap^R gene. Cells carrying these mutant genes synthesized truncated β-lactamases that did not appear in the periplasmic space as usual but remained in the cytoplasmic membrane. Thus at least

Table 9.2 Factors required for secretion of proteins.

Factor	Comment
N-terminal signal sequence	not required with a few proteins (see text)
ATP and electrochemical potential	provides energy for secretion
leader peptidase or lipoprotein signal peptidase	removes signal sequence
export apparatus	*secA*, *secD* and *secY* (*priA*) have been identified as components, since mutations in relevant genes cause accumulation of precursors
cytoplasmic factors	

the carboxy-terminus of β-lactamase plays an important role in the secretory process. In the case of β-galactosidase, which normally is cytoplasmic, Itoh *et al.* (1981) were unable to obtain secretion.

Further evidence for a role of the structural sequence of the protein in determining its cellular location comes from the work of Tommassen *et al.* (1983). They constructed a hybrid gene which encoded the signal sequence as well as 158 amino acids of β-lactamase and the complete structural sequence of the outer membrane PhoE protein. The fusion protein expressed by this gene was found in the outer membrane rather than the periplasm.

Most of those proteins which are natural products of *E. coli* and destined for secretion are not released into the growth medium. Rather, they are located in the periplasmic space or embedded in the outer membrane. Those few *E. coli* proteins which are released into the growth medium do not obey all the criteria listed in Table 9.2. For example, with both haemolysin and the colicins the amino-termini of the molecules are identical to the initial translation product, i.e. no signal sequence is involved in secretion (Hirst & Welch 1988). However, other proteins are required for secretion, the HlyB and HlyD gene products in the case of haemolysin, and a lysis protein in the case of colicins. In *Klebsiella pneumoniae*, a close relative of *E. coli*, the enzyme pullulanase is first localized to the cell surface and subsequently is released into the medium as high mol. wt complexes (d' Enfert *et al.* 1987). Of critical importance to this dual localization are the products of two or more *K. pneumoniae* genes whose expression in *E. coli* allow this bacterium to secrete pullulanase in an identical way. D'Enfert & Pugsley (1987) constructed four different fusions between the pullulanase gene and the *phoA* gene. Although the hybrid proteins were apparently membrane-associated, they were not secreted into the medium either by *E. coli* carrying the pullulanase secretion genes or by *K. pneumoniae*. These results show that extragenic factors are widespread but usually highly specific for secretion of a particular protein. Furthermore, the results obtained with pullulanase fusion proteins indicates that export to the growth medium is not readily attainable.

An alternative way of exporting proteins to the medium, which does not rely on extragenic factors, is found with the gonococcal IgA protease. When the gene for IgA protease was cloned and expressed in *E. coli*, all of the protein was found in the growth medium (Pohlner *et al.* 1987). Apparently this protein has a carboxy-terminal 'helper' domain required for extracellular secretion, in addition to a signal sequence required for transport across the cytoplasmic membrane. It remains to be seen if this 'helper' domain is specific to IgA protease or if it can be attached to other proteins to facilitate export.

A different strategy has been adopted by Abrahmsen *et al.* (1985, 1986). They constructed vectors in which the staphylococcal gene for the IgG-binding protein A was placed under the control of a *lac* promoter. Subsequently the protein A gene was substantially truncated so that only two of five IgG-binding domains remain. Following induction, truncated protein A molecules were secreted into the medium. When a synthetic gene for human growth hormone was fused to the modified protein A gene, the fusion protein was again exported to the growth medium. The success of this strategy is indicated by the fact that titres as high as 1 g/litre were obtained under optimal conditions (Josephson & Bishop 1988).

Stability of foreign proteins in *E. coli*

There are a number of examples of the instability of cloned gene products. Details of the degradation of somatostatin were given earlier (see p. 53). Human fibroblast pre-interferon and interferon also are known to be unstable (Taniguchi *et al.* 1980). Talmadge & Gilbert (1982) have shown that cellular location can affect stability. They measured the half-life of rat proinsulin in *E. coli* and found it to be two minutes when located in the cytoplasm but over 10 times greater in the periplasm.

Various strategies have been developed to cope with the instability of foreign proteins in *E. coli*. In the case of somatostatin, degradation was prevented by producing a fused protein consisting of somatostatin and β-galactosidase and subsequently cleaving off the somatostatin with cyanogen bromide. Although there are a number of examples in the literature of the stabilization of proteins in this way, the disadvantage is that the subsequent production of a native protein *in vitro* is not always possible (see p. 144). An alternative approach is to use certain mutants of *E. coli* which have a reduced complement of intracellular proteases. One particularly useful mutant lacks the *lon* protease, a DNA-binding protein with ATP-dependent proteolytic activity. A quite different strategy has been developed by Simon *et al.* (1983). It is based on the observation that during bacteriophage T4 infection of *E. coli*, turnover of normal cellular proteins continues but there is a marked decrease in the degradation of abnormal proteins (e.g. fragments produced as a result of nonsense mutations). They cloned the T4 gene responsible for this effect and called it *pin* for *p*rotease *in*hibition. They found that labile eukaryotic proteins, e.g. fibroblast interferon, encoded by genes cloned in *E. coli* were stabilized in cells in which the T4 *pin* gene was expressed.

For any given protein a variety of factors singly or in combination may modulate half-life *in vivo*. Among such factors are the flexibility, accessibility and sequence of the N- and C-termini, the presence of chemically blocking amino-terminal groups such as the acetyl group, and the exposure on the surface of the folded protein of protease cleavage sites. Bachmair *et al.* (1986) have shown that in yeast the half-life of proteins derived from β-galactosidase varies from 3 minutes to 20 hours, depending on the amino acid at the N-terminus. It is not known if a similar phenomenon occurs in *E. coli*, but if it does it could explain the great variation in stability of different heterologous proteins made from similar gene constructs.

Detecting expression of cloned genes

When it is known that a cloned DNA fragment codes for a particular protein or RNA species, then suitable tests can be devised to determine if that gene is expressed in its new host. Thus somatostatin or insulin can be detected by sensitive radioimmunoassays, 5*s* RNA by hybridization to suitable probes, and enzymes either by appropriate assay procedures or by complementation of auxotrophic mutations in a suitable host (see p. 128). But how does one detect expression of cloned genes when the function of these genes is not known? The answer is to look for the synthesis of novel proteins or RNA species in cells carrying the recombinant molecules. However, with wild-type cells the detection of new proteins or RNA species is almost impossible because of the concurrent expression of the host genome. Three approaches have been devised to circumvent this problem. Two of them are *in-vivo* methods: expression of plasmid-borne genes can be detected by the use of *mini-cells* or by UV irradiation of the host cell. The latter method is known as the *maxi-cell* method and is also applicable to genes cloned in λ vectors. The third method is a coupled *in-vitro* transcription and translation system.

Mini-cells are small, spherical, anucleate cells produced continuously during the growth of certain mutant strains of bacteria. Because of their size difference, mini-cells and normal-sized cells can be separated easily on sucrose gradients. Mini-cells purified in this way from plasmid-free parents can be shown to contain normal amounts of protein and RNA but to lack DNA. *In vivo*, these mini-cells do not incorporate radioactive precursors into RNA or protein. By contrast, mini-cells produced from plasmid-carrying parents contain significant amounts of plasmid DNA. Plasmid-containing mini-cells are capable of RNA and protein synthesis and would appear to be ideal for detecting the expression of genes carried by recombinant plasmids.

In general, there is a correlation between the genotype of a plasmid and the polypeptides synthesized by mini-cells containing that plasmid. However, there are complications, since the introduction of deletions or DNA insertions into plasmids can have unpredictable results; for example, a 1 Mdal. deletion in a Col E1-derived plasmid prevented the synthesis in mini-cells of polypeptides of 56 000, 42 000, 30 000 and 28 000 daltons. The explanation of this result was apparent after the demonstration that the latter three polypeptides were degradation products of Col E1 (Meagher *et al.* 1977b). The insertion into a kanamycin-resistance gene of a DNA fragment containing the *Eco* RI methylase gene caused the synthesis in mini-cells of *Eco* RI methylase and two other polypeptides. Again, this result could be explained since it was known that the kanamycin-resistance protein was inactivated by the insertion and that there was a portion of another gene on the inserted DNA fragment. These facts permit the prediction that two new polypeptides would be produced. When the DNA insertion is of unknown function, e.g. randomly cleaved eukaryotic DNA sequences, interpretation of the expression of the DNA fragments in mini-cells could still be complicated.

The maxi-cell method (Sancar *et al.* 1979) is based on two observations. First, when irradiated with UV light, *E. coli recA uvrA* cells stop DNA synthesis and chromosomal DNA is extensively degraded so that only a small amount remains several hours later. Second, if these cells contain a Col E1-like multicopy plasmid (e.g. pBR322), the plasmid molecules that do not receive a UV hit continue to replicate with plasmid DNA levels increasing about 10-fold by 6 hours post-irradiation. If a radioactive amino acid is added a few hours after irradiation, the vast majority of the proteins which are labelled are plasmid-encoded.

153
Chapter 9
Expression of
Cloned DNA
Molecules

The expression of genes cloned in λ vectors can be studied by infection of bacteria previously irradiated so severely with UV light that their own gene expression is effectively eliminated by DNA damage. Under these conditions the products of the cloned genes are seen against a background of λ-specified proteins. If the expression of the DNA insert is independent of λ promoters then even this background can be eliminated by infecting UV-irradiated cells lysogenized with a non-inducible mutant of λ (Newman *et al.* 1979). Under these conditions sufficient λ repressor is present in the cells to prevent transcription of phage genes.

Bacterial cell-free systems for coupled transcription and translation of DNA templates were first described by Zubay and colleagues as early as 1967. More recently these have been modified (Pratt *et al.* 1981) to permit expression *in vitro* of genes contained on bacterial plasmids or bacteriophage genomes. Since DNA fragments generated by restriction endonuclease cleavage can be used to programme polypeptide synthesis, it is possible to readily assign polypeptides to small regions of the coding template. There are two other advantages to the system. First, incorporation of radioactive label into protein is far more efficient than is possible using *in-vivo* methods. This makes the system very sensitive and ^{35}S-labelled polypeptides can be identified by polyacrylamide gel electrophoresis after a few hours autoradiography (Fig. 9.4). Second, DNA from other prokaryotes can be expressed efficiently. This is particularly useful when cloning in organisms other than *E. coli* in which the *mini-cell* and *maxi-cell* techniques have not been developed.

MAXIMIZING THE EXPRESSION OF CLONED GENES

In the previous section we discussed those features of gene expression in *E. coli* which need to be considered if detectable synthesis of a cloned gene product is to be obtained. However, much of the interest surrounding *in-vitro* gene manipulation concerns the commercial applications, e.g. production of vaccines or human hormones in *E. coli*. Here detectable expression is not sufficient; rather it must be maximized. The key factors affecting the level of expression of a cloned gene are shown in Table 9.3, and, as will be seen in the following section, most of them exert their effect at the level of translation.

Constructing the optimal promoter

A large number of promoters for *E. coli* have been analysed (Hawley & McLure 1983, Harley & Reynolds 1987) and the most recent compilation gives the sequence of 263 of them. Clearly we would like to use this information to develop the most efficient promoter possible. Comparison of many promoters has led to the formulation of a consensus sequence which consists of the −35 region (5′-TTGACA-) and the −10 region or

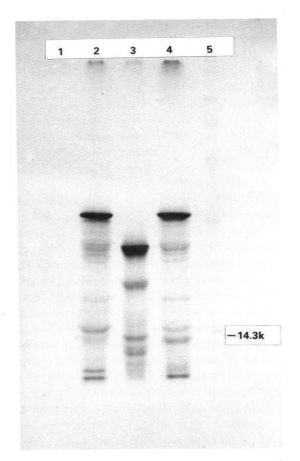

Fig. 9.4 Autoradioagraph produced after transcription and translation of plasmid DNA *in vitro*. Plasmid DNA was transcribed and translated *in vitro* in the presence of ^{35}S methionine, the protein products separated by SDS-polyacrylamide electrophoresis and the individual proteins detected by autoradiography. Track 1, DNA-free control; track 2, plasmid pAT153; track 3, plasmid pACYC184; track 4, plasmid pWT111; track 5, molecular-weight marker protein. Tracks 2–4 were loaded with equivalent amounts of radioactivity (500 000 c.p.m.) and the gel was exposed to the photographic film for 3.5 h. The major band in tracks 2 and 4 is β-lactamase and in track 3 chloramphenicol acetyl transferase. Note that many truncated polypeptides are produced but that these are minor constituents.

Pribnow box (5'-TATAAT), the transcription start point being assigned position +1. Of the four promoters used most widely in expression vectors none shows absolute identity with the consensus sequence (Fig. 9.5). In trying to identify the strongest promoter it is essential to have a measure of the relative efficiencies of the different candidates and a suitable system has been devised (Russell & Bennett 1982, de Boer *et al.* 1983a). The promoter to be tested is placed in front of a promoter-less galactokinase (*gal*K) gene carried on a plasmid and the level of galactokinase synthesized in a GalK⁻ host used as a measure of promoter strength. Using the galacto-kinase system it was shown that promoter strength is directly proportional to the degree of similarity with the consensus sequence. Thus it is not

Table 9.3 Factors affecting the expression of cloned genes.

promoters strength
translational initiation sequences
codon choice
secondary structure of mRNA
transcriptional termination
plasmid copy number
plasmid stability
host cell physiology

surprising that compared with the *lac* promoter the two synthetic hybrid promoters *tac* I and *tac* II (Fig. 9.5) were 11 and 8 times stronger respectively. While the −35 and −10 regions show the greatest conservation across promoters and also are the sites of nearly all mutations which affect transcriptional strength, other bases flanking the −35 and −10 regions occur at greater than random frequency and can affect promoter activity (Hawley & McLure 1983, Dueschle *et al.* 1986, Keilty & Rosenberg 1987). Furthermore, the distance between the −35 and −10 regions is important. Hawley & McClure analysed a number of mutations which affect this spacing. In all cases the promoter was stronger if the spacing was moved closer to 17 base pairs and weaker if moved further away from 17 base pairs.

Because of the increased strength of the *tac* promoters relative to the *lac* and *trp* promoters, Amann *et al.* (1983) have cloned a *tac* promoter on a series of plasmid vectors that facilitates the expression of cloned genes. These vectors contain various cloning sites followed by transcription signals. In addition, Amann *et al.* (1983) have described plasmids that facilitate the conversion of the *lac* promoter to the *tac* promoter. Maximal transcription of a cloned gene may require more than just −35 and −10 regions that are close to the consensus sequence. With some promoters, DNA sequences 10−100 base pairs upstream from the −35 site can act as 'upstream activators'. The deletion and/or insertion of DNA sequences can distort these activators and decrease transcription (Lamond & Travers 1983, Bossi & Smith 1984, Gourse *et al.* 1986). Upstream activation is associated with sequences rich in A and T, which result in DNA curvature (Plaskon & Wartell 1987).

```
               -35 REGION                                        -10 REGION

                          1 2 3 4 5 6 7 8 9  10 11 12 13 14 15 16 17
CONSENSUS   • • • T T G A C A • • • • • • • • •  • • • • • • • • • T A T A A T • •

lac           G G C T T T A C A C T T T A T G C T T C C G G C T C G T A T A T T G T

trp           C T G T T G A C A A T T A A T C A T  C G A A C T A G T T A A C T A G

λPL           G T G T T G A C A T A A A T A C C A  C T G G C G G T G A T A C T G A

rec A         C A C T T G A T A C T G T A T G A A  G C A T A C A G T A T A A T T G

tac I         C T G T T G A C A A T T A A T C A T    C G G C T C G T A T A A T G T

tac II        C T G T T G A C A A T T A A T C A T  C G A A C T A G T T T A A T G T
```

Fig. 9.5 The base sequence of the −10 and −35 regions of four natural promoters and two hybrid promoters.

Optimizing translation initiation

As well as giving clues to the essential structural features of promoters, DNA sequence analysis has yielded considerable information about those factors which affect translational initiation. Details of these sequences can be found in the compilation of Stormo *et al.* (1982). In addition to the AUG initiation codon, three more or less conserved structural features have been recognized. First and foremost is the S-D sequence which contains all or part of the polypurine sequence 5'-UAAGGAGGU-3'. Second, in polycistronic messengers at least, there may be a requirement for one or more termination codons in the ribosome binding site. Finally, the ribosome binding site of genes encoding highly expressed proteins, e.g. phage capsid protein and ribosomal proteins, contains all or part of the sequence PuPuUUUPuPu.

Using a gene expression system which includes a portable Shine-Dalgarno region, de Boer *et al.* (1983b) have examined the effect on translation of varying the sequence of the four bases that follow the S-D region. The presence of four A residues or four T residues in this position gave the highest translational efficiency. Translational efficiency was 50% or 25% of maximum when the region contained, respectively, four C residues or four G residues.

The composition of the triplet immediately preceding the AUG start codon also affects the efficiency of translation. For translation of β-galactosidase mRNA the most favourable combinations of bases in this triplet are UAU and CUU. If UUC, UCA or AGG replaced UAU or CUU the level of expression was 20-fold less (Hui *et al.* 1984).

Based on this information the reader can be forgiven for thinking that it would be easy to construct a consensus ribosome binding site using oligonucleotide synthesis. The first problem is that for maximal expression of a cloned gene the ribosome binding site for that gene must be close to the promoter. The second problem is highlighted by experiments where the distance between the gene and the promoter was varied.

Gene–promoter separation: the effect of mRNA secondary structure

Cleavage of λ*plac* 5.1 DNA with a combination of *Eco* RI and *Alu* I restriction endonucleases produces a mixture of fragments one of which is 95 base pairs long and contains the *lac* promoter and β-galactosidase gene S-D sequence. Roberts *et al.* (1979b) constructed a series of recombinant plasmids in which this promoter-bearing fragment was located at varying distances in front of the λ*cro* gene. Nine different recombinant plasmids were selected, transformed into E. *coli*, and the level of *cro* protein in each clone measured. The DNA sequence across each of the *lac–cro* fusion was also determined. The results obtained are summarized in Fig. 9.6. The most striking feature of these results was the enormous difference (>2000-fold) in the levels of *cro* protein produced by the different recombinants. Since the same promoter is being used in each plasmid, it is uniform in each case. Iserentant & Fiers (1980) have constructed secondary structure

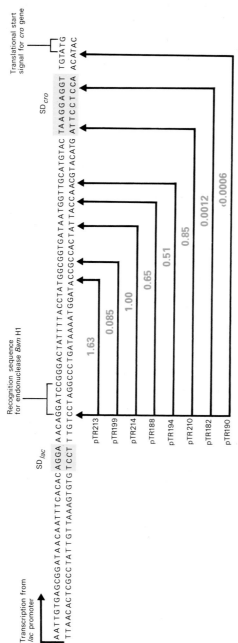

Fig. 9.6 Effect on *cro* protein production of gene promoter separation. Shown is a portion of the sequence of pTR161 extending from the *lac* promoter to the startpoint of translation of the *cro* gene. Also shown is the extent of the deletion in eight derivatives of pTR161. The figures on the brackets indicate the amount of *cro* protein synthesized relative to pTR161. The deletions were created after *Bam* HI digestion by the method outlined in Fig. 9.8.

models for the RNA transcripts produced by the plasmids of Roberts *et al.* (1979b). From an analysis of these structures they concluded that for good expression the initiation and, to a lesser extent, the S-D site must be accessible (Fig. 9.7). Thus the initiation of translation involves interaction between an activated 30s ribosomal subunit and the 5′ terminal region of a mRNA *which is already folded* in a specific secondary structure. Confirmation that the sequence between the promoter and the ATG codon affects the level of gene expression has been provided by Chang *et al.* (1980), Shepard *et al.* (1982) and Gheysen *et al.* (1982). These latter workers were able to confirm the conclusions of Iserentant and Fiers (1980) that the varying levels of expression are a reflection of secondary structure of the mRNA.

From an analysis of 123 naturally occurring prokaryotic mRNAs Ganoza *et al.* (1987) have concluded that sequences 5′ to the initiation codon have little self-pairing and do not pair extensively with the proximal coding region.

From the foregoing it might be thought that mRNA secondary structure alone accounts for the observation of Roberts *et al.* (1979b) that the distance between the gene and its promoter affects the efficiency of translation. However, length *per se* also may be important. Bingham & Busby (1987) measured the effect on translation of varying the distance between the 5′ end of the message and the S-D sequence. When this distance fell below 15 bases there was a marked decrease in translational efficiency.

Based on the above work a strategy for maximizing gene expression has been proposed: namely, a bank of clones is established in which the gene under study is placed at varying distances from the promoter. These recombinant clones are then screened and the one giving the highest yield

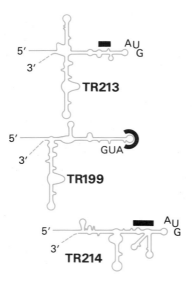

Fig. 9.7 Postulated secondary structure on the *cro* mRNA synthesized by three of the plasmids shown in Fig. 9.6. Note that for clone TR199, in which *cro* is poorly expressed, the initiation codon is base-paired in a stem structure. The solid bars indicate the location of the S-D sequence.

is selected. Roberts *et al.* (1979b) have described a general method whereby this might be achieved (Fig. 9.8). The gene is cloned in a plasmid such that a unique restriction site is located within 100 base pairs of the 5' end of the gene. The plasmid is opened at that site and varying amounts of DNA excised with exonuclease III and S1 nuclease. The promoter fragment is then inserted, in this case the 95 base pair *lac* promoter fragment, and the plasmid closed by T4 DNA ligase. This produces a set of plasmids bearing the promoter separated by varying distances from the cloned gene. However, there is a limitation with this method: its use is restricted to genes whose product can be assayed readily. This problem led Guarente *et al.* (1980) to devise a method to maximise expression of cloned genes in the absence of assays for their gene products.

The method in question makes use of a *lacI-lacZ* fusion in which the carboxyterminal end of the *lacZ* gene is fused in phase to a fragment of the *lacI* gene. The *lacZ* fragment, which comprises most of the gene,

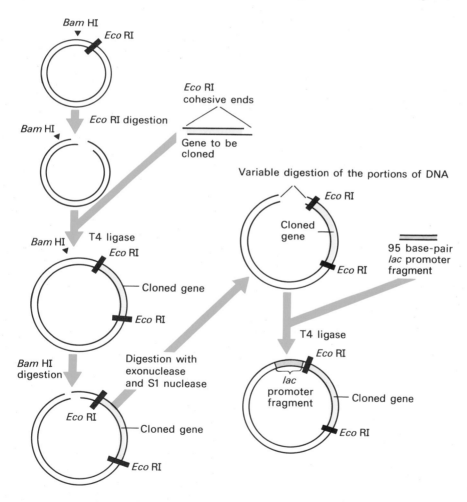

Fig. 9.8 General method for varying the distance between the *lac* promoter and a cloned gene. The example shown uses a hypothetical plasmid carrying unique closely spaced sites for *Eco* RI and *Bam* HI.

encodes a protein fragment that has β-galactosidase activity. This activity is not affected by additional protein sequences at its amino-terminus. The *lacI-Z* fused gene is not expressed because the *lacI* promoter at the end of the I gene has been deleted. Now suppose that the gene being studied, and whose protein product is difficult to study, is gene X. A fragment of DNA bearing the amino-terminal end of gene X is then joined to the *lacI-Z* fusion (Fig. 9.9) so that the two are in the same translational reading frame. The *lac* promoter and S-D sequence are then placed at varying distances in front of the ATG initiation codon, as before. Those promoter placements which favour efficient expression of gene X are identified by their β-galactosidase activity. Gene X is then reconstituted intact, with the portable promoter in place, by recombination *in vitro* or *in vivo*.

Stability of mRNA

Provided the various elements of the translational machinery are present in excess, the rate of synthesis of a particular protein will depend on the steady-state level of mRNA in the cell. This level is a reflection of the balance between synthesis and degradation. By using strong promoters the rate of synthesis is maximized, but a similar effect could be achieved by reducing the degradation of messenger. Degradation of mRNA usually proceeds by a combination of endonuclease and 3'-exonuclease attack. The stability of a given message *in vivo* reflects its susceptibility to digestion by these enzymes which, in turn, depends upon mRNA sequence and structure, as well as the association of macromolecules (e.g. ribosomes) with the message (for review see Belasco & Higgins 1988).

In *E. coli* the average half-life of an mRNA molecule is only 1–2 minutes. By contrast, the mRNA from gene 32 of bacteriophage T4 has a half-life of 20

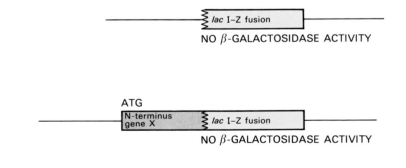

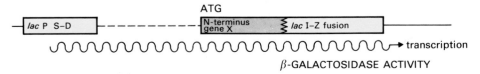

Fig. 9.9 Method for maximizing expression of cloned genes in the absence of assays for their gene products. (See text for details.)

minutes or longer. The stability of this transcript involves specific sequences upstream from the initiation codon of gene 32 and a *trans*-acting phage function (Gorski *et al.* 1985). This 5′ leader sequence can stabilize other mRNAs in phage-infected and -uninfected cells. These properties have been exploited to construct a plasmid with a gene 32 expression cassette that produces foreign proteins at high levels in both infected and uninfected cells (Duvoisin *et al.* 1986). The hybrid transcripts that are produced are relatively long-lived, the half-life varying from 4 to 10 minutes depending on the exact sequence inserted.

The effect of codon choice

The genetic code is degenerate, and hence for most of the amino acids there is more than one codon. However, in all genes, whatever their origin, the selection of synonymous codons is distinctly non-random (for reviews see Ernst 1988, Kurland 1987, and McPherson 1988). The bias in codon usage has two components: correlation with tRNA availability in the cell, and non-random choices between pyrimidine ending codons. Ikemura (1981a) measured the relative abundance of the 26 known tRNAs of *E. coli* and found a strong positive correlation between tRNA abundance and codon choice. Later, Ikemura (1981b) noted that the most highly expressed genes in *E. coli* contain mostly those codons corresponding to major tRNAs but few codons of minor tRNAs. By contrast, genes that are expressed less well use more suboptimal codons. Kurland (1987) has detailed how the preferential use of a codon subset matching the major tRNA isoacceptor species is a sensible cellular strategy for optimizing translational kinetics at high growth rates. This explains why there is a major codon preference strategy and requires no explanation for which codons are preferred. McPherson (1988) believes that the codons are utilized according to mistranslational constraints. This is done in a fashion that favours replacement of one amino acid with a functionally conservative substitute in the event of codon−anticodon mispairing. It should be noted that the bias in codon usage even extends to the stop codons (Sharp & Bulmer 1988). UAA is favoured in genes expressed at high levels whereas UAG and UGA are used more frequently in genes expressed at a lower level.

Where maximal expression of a chemically synthesized gene is desired it would be an easy matter to optimize codon choice in accordance with the observations of Ikemura (1981a,b). However, the correlation of cellular tRNA level with codon frequency may be more complex. A shortage of a charged tRNA during protein synthesis can lead to misincorporation of amino acids and frame-shifting. Thus the arrangement of the successive codons in the coding region of a natural mRNA as well as its corresponding overlapping triplets may have evolved in order to minimize such frame-shifts *in vivo*. Not surprisingly, a study by Grosjean & Fiers (1982) of the codon usage of several highly expressed mRNAs showed that most codons that are absent in the normal reading frame frequently are present in the two non-reading frames.

Analysis of the genetic code shows that triplets X_1X_2C and X_1X_2U

always code for the same amino acid. Generally speaking, these two triplets are decoded by the same tRNA. Gouy & Gautier (1982) and Grosjean & Fiers (1982) have pointed out that there can be bias in the choice of pyrimidine at the third position of the codon and that the bias is correlated with gene expressivity. Thus, in highly expressed genes, if the first two bases of the codon are both A or U, then C is the favoured third base, whereas if the first two bases are G or C, the preferred third base is U. The explanation for this preference for a particular pyrimidine is equalization of the codon—anticodon interaction energy. Selection of codons which have a very strong or very weak energy of interaction with their corresponding anticodon leads to a decrease in the efficiency of translation.

Given the constraints on codon usage and distribution noted above it might be anticipated that the rate of translation, and hence the accumulation of product, might be limited when attempts are made to get high level expression from foreign or synthetic genes. Robinson *et al.* (1984) found that replacing rare codons with more common ones could improve expression. By contrast, Hoekema *et al.* (1987) observed only a threefold decrease in the amount of protein made from each mRNA molecule when they substituted rare codons for up to 39% of the optimal codons in a highly expressed gene.

The accuracy of translation, as opposed to its rate, also may be affected by codon choice. Significant substitution of lysine for arginine occurred at the rare AGA codon during high-level expression of heterologous genes in *E. coli* (McPherson 1988).

Gene synthesis revisited

Improvements in the technology for the chemical synthesis of oligonucleotides means that today it is a very simple and quick matter to construct a gene with any desired sequence. However, from the foregoing discussion it should be apparent that generating the optimal sequence is much more difficult. With promoters, it is not just the -35 and -10 regions that are important, so too are the base sequences surrounding these sites. If the promoter is a controllable one then the base sequence can influence the fine-control mechanism known as attenuation (Yanofsky & Colter 1982). The base sequence in the 5' untranslated region can affect the rate of translation by its effect on mRNA stability and ribosome binding. Finally, the rate and fidelity of translation can be influenced by both codon choice and codon context. Clearly, gene synthesis may be easy, optimization is much more difficult, particularly since we don't know all the rules!

The effect of plasmid copy number

A major rate-limiting step in protein synthesis is the binding of ribosomes to mRNA molecules. Since the number of ribosomes in a cell far exceeds any one class of messenger one way of increasing the expression of a cloned gene is to increase the number of the corresponding transcripts. Two factors affect the rate of transcription: first, the strength of the promoter as described above, and second, the number of gene copies. The easiest way of increasing the gene dosage is to clone the gene of interest on a high copy-number plasmid.

Most of the high copy-number vectors in common use, e.g. pBR322, are derivatives of Col E1. Two negatively acting components are known to be involved in the control of Col E1 replication (Fig. 9.10). One is a 108 nucleotide untranslated RNA molecule called RNAI (Tomizawa & Itoh 1981), and the other is a protein repressor called ROP (Cesarini *et al.* 1982). RNA II is a plasmid-encoded RNA molecule which is processed by RNase H to give a 555 nucleotide primer for the initiation of Col E1 replication (Tomizawa & Itoh 1982). RNA I inhibits DNA replication by base pairing with a complementary sequence on RNA II, thereby preventing processing by RNase H. The ROP repressor is a 63 residue polypeptide which controls the initiation of transcription of RNA II. Deletion of the ROP gene, as in pAT153 (Twigg & Sheratt 1980), or mutations in RNA I (Muesing *et al.* 1981) result in increased copy numbers. However, it should be noted that the host cell genetic background can also affect copy number (Nugent *et al.* 1983, Seelke *et al.* 1987).

Plasmids have been constructed in which copy number is controlled by a regulatable promoter (Yarranton *et al.* 1984). These plasmids consist of two origins of replication, one of which is active at 30°C and associated with a partition sequence (see p. 164) to ensure maintenance at low copy number. The other origin has the promoter for RNA II synthesis replaced by the λ PL promoter. In strains carrying the thermolabile λ repressor gene mutation cI857, the second origin of replication is functional only when the repressor is inactivated by temperature induction. At temperatures above 38°C uncontrolled plasmid DNA replication occurs, resulting in the accumulation of a high plasmid copy number per cell.

Mutant strains may also arise which result in a drop in copy number of plasmids that they harbour. These 'low cop' mutations are chromosomal in origin and may be counter-selected using increased antibiotic concentrations. The use of high antibiotic concentrations, whilst ensuring a population of 'high cop' plasmids, will not ensure that a high cop plasmid which no longer produces the recombinant protein does not arise.

Plasmid low cop mutants may arise which have a slower replication rate than normal. If such a mutation arose *in vivo* in one plasmid molecule

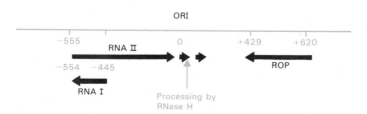

Fig. 9.10 The replication control region of plasmid Col E1. ROP controls the transcription of RNA II and RNA I inhibits the processing of RNA II by RNase H. The figures represent the base pair co-ordinates measured from the origin of replication.

out of a total cellular population of at least 50 (in the case of pBR322) then it is very unlikely that this mutant would ever take over in the population, since the normal plasmids would replicate to compensate for the mutant and maintain the normal cellular copy number of pBR322. Under some circumstances, however, low cop plasmid mutants may be constructed *in vitro*.

Transcription termination

The presence of transcription terminators at the ends of cloned genes is important for a number of reasons. First, the synthesis of unnecessarily long transcripts will increase the energy drain on a cell which is expected to produce large amounts of non-essential protein. Second, undesirable secondary structures may form in the transcript, which could reduce the efficiency of translation. Finally, promoter occlusion may occur, i.e. transcription from the promoter of the cloned gene may interfere with transcription of another essential or regulatory gene. Thus Stueber & Bujard (1982) found that certain strong promoters led to a decrease in plasmid copy number because of read-through into the ROP gene and interference with plasmid replication. Copy number control, and hence gene expression, was restored to normal by the inclusion of transcriptional terminator at an appropriate point. In some instances strong terminators are required for shotgun cloning foreign DNA in *E. coli*. In their absence, strong promoters on the cloned fragments read through into vector sequences and cause plasmid instability (see also below) by interfering with replication (Stassi & Lacks 1982, Chen & Morrisson 1987).

Plasmid stability

Having maximized the expression of a particular gene it is important to consider what effects this will have on the bacterium harbouring the recombinant plasmid. Increases in the levels of expression of recombinant genes lead to reductions in cell growth rates and may result in morphological changes such as filamentation and increased cell fragility. If a mutant arises which has either lost the recombinant plasmid, or has undergone structural rearrangement so that the recombinant gene is no longer expressed, or has a reduced plasmid copy number, then this will have a faster growth rate and may quickly take over and become predominant in the culture (Fig. 9.11).

Segregative instability

The loss of plasmids due to defective partitioning is called segregative instability. Naturally occurring plasmids are stably maintained because they contain a partitioning function, *par* which ensures that they are accurately segregated at each cell division. Such *par* regions are essential for the stability of low copy-number plasmids. The higher copy-number plasmid ColE1 also contains a *par* region, but this region is deleted in pBR322 which is segregated randomly at cell division. Although the copy

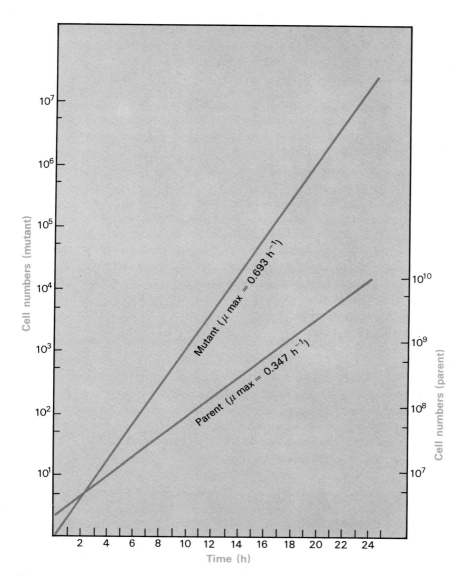

Fig. 9.11 Competition between a slow-growing production strain and a faster-growing mutant generated from it. At time zero there are 2×10^6 cells ml^{-1} of the parent strain and a single mutant cell ml^{-1}. Notice that when the parent cell density is 10^{10} cells ml^{-1} the mutant cell density is 10^7 cells ml^{-1}. On the next subculture the mutant cells will become predominant.

number of pBR322 is high and the probability of plasmid-free cells arising is very low, under certain conditions such as nutrient limitation or during rapid host cell growth, plasmid-free cells may arise (Jones *et al.* 1980, Nugent *et al.* 1983). This problem can be obviated by maintaining antibiotic selection. However, this may not be a desirable solution for large-scale culture because of cost and waste disposal considerations. The *par* region from a plasmid such as pSC101 may be cloned into pBR322-type vectors, thus stabilizing the plasmid (Primrose *et al.* 1983, Skogman *et al.* 1983). The *par* region of pSC101 has been sequenced and does not appear to encode any proteins or to contain transcriptional or translational start signals. Regions of the *par* locus appear capable of forming regions of intra-strand secondary structure and these may play a role in partitioning (Miller *et al.* 1983). The *par* locus is a specific binding site for DNA gyrase (Wahle & Kornberg 1988) but it is not known if the enzyme mediates its effect through its supercoiling activity or by an influence on DNA replication.

An alternative strategy to counter segregative instability has been described by Rosteck & Hershberger (1983). Unlike the use of a *par* region which *prevents* plasmid loss, their method *counter-selects* plasmid-free cells. In their method the plasmid carrying the gene of interest also carries the λcI gene which encodes the λ repressor. Host cells carrying the chimaeric plasmids are then lysogenized with a repressor-defective mutant of bacteriophage λ. Loss of the plasmid from the lysogens causes concomitant loss of the λ repressor and hence cell death because the prophage is induced to enter the lytic growth cycle.

Plasmid instability may also arise due to the formation of multimeric forms of a plasmid. The mechanism that controls the copy number of a plasmid ensures a fixed number of plasmid origins per bacterium. Cells containing multimeric plasmids have the same number of plasmid origins but fewer plasmid molecules, which leads to greater instability if these plasmids lack a partitioning function. These multimeric forms are not seen with Col E1 which has a natural method of resolving the multimers back to monomers. It contains a highly recombinogenic resolution site that resolves multimers in a way analogous to the resolution of transposon-induced co-integrates (Summers & Sherrat 1984). The resolution site of Col E1 has been cloned into pBR322-type plasmids and has eliminated problems due to multimerization.

Structural instability

Structural instability of plasmids may arise by deletion, insertion, or rearrangements of DNA. Some of the earliest reports of deletions were in chimaeric plasmids which can replicate in both *E. coli* and *B. subtilis* (for reviews see Ehrlich *et al.* 1981, Kreft & Hughes 1981). Spontaneous deletions have now been observed in a wide range of plasmid, virus and chromosomal DNAs. A common feature of these deletions is the involvement of homologous recombination between short direct repeats (Jones *et al.* 1982). Artificial plasmids with multiple tandem promoters are particularly prone to deletion formation (Nugent *et al.* 1983) as are palindromes (Hagan & Warren 1983).

As well as homologous recombination between sites on a plasmid, structural instability may be mediated by insertion (IS) elements or transposons resident in the host chromosome or on plasmids. Both of these elements can cause spontaneous mutations by their insertion, adjacent deletion or inversion of DNA. There are many reports of plasmid instability due to insertion of IS elements from the chromosome; for example, transformation of *tyr*R strains of *E. coli* with multicopy plasmids carrying the tyrosine operon gave rise to modified plasmids with either insertions or deletions (Rood *et al.* 1980). These effects were due to IS1 insertion.

To prevent such problems of instability on scale-up it is desirable to minimize expression of the cloned gene until the organism is introduced into the final fermentation vessel. One strategy is to use controllable promoters, another is to use R1-derived runaway replication mutants (see Chapter 3). Better still, both techniques can be combined.

Host cell physiology can affect the level of expression

All of the factors affecting gene expression which we have discussed so far can be explained in terms of the sequence of bases in a nucleic acid and the way these sequences interact with a particular protein or protein complex. However, gene expression also is controlled by a less tangible factor — the physiology of the host cell. Factors which will be important include the choice of nutrients and the way in which they are supplied to the culture and environmental parameters such as temperature and dissolved oxygen. So far there has been no systematic study of the effect of different growth conditions on the synthesis of foreign proteins in *E. coli*. The few studies which have been done are outlined below and emphasize the importance of physiology.

One of the problems associated with the overproduction of proteins in *E. coli* is the sequestration of the product into insoluble aggregates or 'inclusion bodies' (see p. 350). However, this effect is dependent on the growth temperature. Thus, inclusion bodies of interferon D occur when the producing *E. coli* is grown at 37°C but not when grown at 23–30°C (Schein & Noteborn 1988). Similar results were obtained by Takagi *et al.* (1988). They found that denatured pro-subtilisin accumulated as aggregates in the periplasmic space when the pro-subtilisin gene was fused to a signal sequence and expressed at a high level. By reducing the expression level and the growth temperature a 16-fold increase in active subtilisin was obtained.

A different effect of physiology on gene expression was noted by Klotsky & Schwartz (1987). They found that plasmid copy number varied by a factor of 3–4 when cells were grown on a mineral salts medium supplemented with different carbon sources. The level of β-lactamase activity synthesized from a plasmid-borne gene mimicked the plasmid copy number.

Section 3
Cloning in Organisms Other than *E. coli*

10 Cloning in Bacteria Other than *E. coli*

For many experiments it is convenient to use *E. coli* as a recipient for genes cloned from eukaryotes or other prokaryotes. Transformation is easy and there is available a wide range of easy-to-use vectors with specialist properties, e.g. regulatable high-level gene expression. However, use of *E. coli* is not always practicable because it lacks some auxiliary biochemical pathways that are essential for the phenotypic expression of certain functions, e.g. degradation of aromatic compounds or plant pathogenicity. In such circumstances the genes have to be cloned back into species similar to those whence they were derived. Yet another reason for cloning in other organisms is that *E. coli* is not used in any of the traditional industrial fermentations. Rather the bacteria used tend to be Gram-positive species such as *Bacillus* (amylases, proteases), actinomycetes (antibiotics) or coryneforms (amino acids, steroid transformations). However, cloning in these organisms presents a different set of problems compared to cloning in *E. coli*.

An essential prerequisite for cloning is a method, preferably transformation, for introducing recombinant plasmids into the organism of choice. Failure to detect transformants can be due to failure of the plasmids to enter the cell and in such instances electroporation provides a suitable alternative. However, failure can be due to other causes. For example, the plasmid DNA might be taken up by the cell but fail to replicate. Alternatively, the plasmid-borne markers might not be expressed in their new environment. The solution to the latter problem is to use a plasmid carrying an easily selectable marker and which is indigenous to the chosen organism. However, in many instances the indigenous plasmids are cryptic, as with most *Bacillus* spp., and/or too large to handle easily. Furthermore, considerable time and effort is required to develop a useful vector from a newly isolated plasmid. In the case of Gram-negative bacteria it is fortunate that plasmids belonging to three different incompatibility groups have a broad host range and can be subjugated as vectors. This eliminates the need to construct a vector system specific to every species of interest. In the case of Gram-positive *Bacillus* spp. rapid progress in the development of vectors was facilitated by the discovery that antibiotic-resistance plasmids from *S. aureus* can replicate in them too and express their antibiotic resistance.

BROAD HOST RANGE VECTORS FOR CLONING IN GRAM-NEGATIVE BACTERIA

Plasmids belonging to any one of incompatibility groups P, Q or W have a broad host range among Gram-negative bacteria. Properties of represen-

172

*Section 3
Cloning in
Organisms
Other than
E. coli*

tative plasmids from each group are shown in Table 10.1. Members of groups P and W are self-transmissible, i.e. they carry *tra* genes encoding the conjugative apparatus. Considerable DNA coding capacity is required to specify conjugal transfer, which to a great extent explains the much larger size of RP4 and Sa relative to RSF1010. Group Q plasmids are not self-transmissible and usually are introduced into the recipient by transformation. However, if the donor cell also carries a narrow host range conjugative plasmid it can be transferred to the recipient by conjugation. Thus group Q plasmids have a *mob* (mobilization) function. Vectors derived from different incompatibility groups can be stably maintained in the same cell. This permits a variety of genetic analyses, e.g. complementation studies between cloned sequences in a wide range of Gram-negative hosts.

Although efficient methods for the transformation of *E. coli* with plasmid DNA have been developed there are few documented examples of transformation of non-enteric Gram-negative bacteria. For plasmid transformation a major factor in the efficient transformation of *E. coli* is the availability of restriction-defective recipients. Such mutants of non-enteric bacteria generally are not available and it is extremely difficult, if not impossible, to transform them with heterologous DNA. For example, the frequency of transformation of *P. putida* is 10^5 transformants/µg of plasmid RSF 1010 DNA, a frequency 10-fold lower than in *E. coli*, but only if the plasmid is prepared from another *P. putida* strain. Otherwise no transformants are obtained (Bagdasarian *et al.* 1979).

The problem of poor transformation efficiency makes it unrealistic to attempt direct transformation of non-enteric bacteria with DNA from a ligation. The most satisfactory approach is to make use of *E. coli* as an intermediate host for transformation of the ligation mix and screening for recombinant plasmids. Once identified, the recombinant DNA of interest can be purified and transformed or conjugated into the desired host. The isolation of the recombinant DNA in this way acts as an amplification step, although it suffers from the disadvantage that screening based on gene function may not be possible.

Vectors derived from Q group plasmid RSF 1010

Plasmid RSF1010 is a multicopy replicon which specifies resistance to two antimicrobial agents, sulphonamide and streptomycin. The plasmid DNA which is 8684 base pairs long has been completely sequenced (Scholz *et al.* 1989). A detailed physical and functional map has been constructed

Table 10.1 Properties of representative broad host range plasmids.

Plasmid	Incompatibility group	Size (kb)	Copy no.	Self-transmissible	Markers
RP4 (RK2, RP1)	P	60	1–3	yes	$Ap^R Km^R Tc^R$
RSF 1010	Q	8.68	15–40	no	$Sm^R Su^R$
Sa	W	29.6	3–5	yes	$Km^R Sm^R Cm^R Sp^R$

(Bagdasarian *et al.* 1981, Scherzinger *et al.* 1984). The features mapped are the restriction endonuclease recognition sites, RNA polymerase binding sites, resistance determinants, genes for plasmid mobilization (*mob*), three replication proteins (*Rep* A, B and C) and the origins of vegetative (*ori*) and transfer (*nic*) replication.

Plasmid RSF 1010 has unique cleavage sites for *Eco* RI, *Bst* EII, *Hpa* I, *Dra* II, *Nsi* I and *Sac* I and from the nucleotide sequence data is predicted to have unique sites for *Afl* III, *Ban* II, *Not* I, *Sac* II, *Sfi* I and *Spl* I. There are two *Pst* I sites, about 750 base pairs apart, which flank the sulphonamide-resistance determinant (Fig. 10.1). None of the unique cleavage sites is located within the antibiotic-resistance determinants and none is particularly useful for cloning. Before the *Sst*, *Eco* and *Pst* sites can be used another selective marker must be introduced into the RSF1010 genome. This need arises because the Sm^r and Su^R genes are transcribed from the same promoter (Bagdasarian *et al.* 1981). Insertion of a DNA fragment between the *Pst* sites inactivates both resistance determinants. Although the *Eco* and *Sst* sites lie outside the coding regions of the Sm^R gene, streptomycin resistance is lost if a DNA fragment is inserted at these sites unless the fragment provides a new promoter. Furthermore, the Su^R determinant which remains is a poor selective marker.

A whole series of vectors has been derived from plasmid RSF1010. The earliest vectors contained additional unique cleavage sites and more useful antibiotic resistance determinants. For example, plasmids KT230 and KT231 encode Km^R and Sm^R and have unique sites for *Hind* III, *Xma* I, *Xho* RI and *Sst* I which can be used for insertional inactivation. These two vectors have been used to clone in *P. putida* genes involved in the catabolism of aromatic compounds (Franklin *et al.* 1981). More recently vectors for the regulated expression of cloned genes have been described. Some of these make use of the *tac* promoter (Bagdasarian *et al.* 1983, Deretic *et al.* 1987) which will function in *P. putida* as well as *E. coli*. Another makes use of positively activated twin promoters from a plasmid specifying catabolism of toluene and xylenes (Mermod *et al.* 1986). Expression of cloned genes can be obtained in a wide range of Gram-negative bacteria following induction with micromolar quantities of benzoate, and the product of the cloned gene can account for 5% of total cell protein.

Fig. 10.1 The structure of plasmid RSF1010. The pale red tinted areas show the approximate positions of the Sm^R and Su^R genes. The region marked *ori* indicates the location of the origin of replication. The *mob* function is required for conjugal mobilization of a compatible self-transmissible plasmid. A, B and C are the regions encoding the three replication proteins.

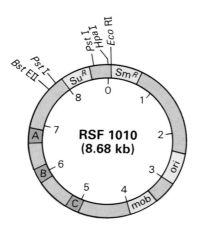

174

*Section 3
Cloning in
Organisms
Other than
E. coli*

Plasmid vectors that contain the *cos* site (cosmids, see p. 70) and which can be packaged *in vitro* into bacteriophage λ particles are of considerable utility for cloning large segments of DNA and for constructing gene banks. Bagdasarian *et al.* (1981) have produced a cosmid derivative of RSF1010 called pKT247. Whereas normal RSF1010-derived vectors can be used to transform the recipient directly, *E. coli* has to serve as an intermediate host if cosmid cloning is employed. Cosmid pKT249 encodes Ap^R, Su^R and Sm^R and has unique sites for endonucleases *Eco* RI and *Sst* I. Frey *et al.* (1983) have described two improved cosmids. These new cosmids permit the selective cloning into their unique *Bam* HI site of 36 kb DNA fragments by a strategy that avoids the formation of polycosmids but does not require the cleaved vector to be dephosphorylated.

As indicated earlier, RSF1010-derived cloning vectors can be mobilized to other bacteria by certain conjugative plasmids. This characteristic is advantageous when the ultimate recipient is not transformable. However, for certain experiments regulatory authorities prefer vectors which have a low or non-existent frequency of conjugal transfer. For this reason Bagdasarian *et al.* (1981) constructed mobilization-defective derivatives of pKT230, pKT231 and pKT247, called pKT262, pKT263 and pKT264 respectively, by deleting the *mob* function (see Fig. 10.1).

Plasmid R300b is similar to, if not identical with, plasmid RSF1010. By cloning R300b in pBR322 a vector called pSS515 was constructed which encodes Ap^R and Tc^R while retaining the broad host range properties of R300b. This new vector has been used successfully to clone and express in the obligate methylotroph *Methylophilus methylotrophus* eukaryotic genes encoding dihydrofolate reductase (Hennan *et al.* 1982) and interferon (IFN)-α_1 (de Maeyer *et al.* 1982).

Broad-host-range vectors derived from P group plasmids

The best-studied P group plasmid is RP4 (also known as RP1 and RK2). It specifies resistance to Ap, Km and Tc and has single cleavage sites for *Eco* RI, *Bam* HI (in the Ap^R gene), *Bgl* II, *Hpa* I and *Hind* III (in the Km^R gene). Relatively little use has been made of RP4 as a cloning vehicle but Jacob *et al.* (1976) have used it for cloning DNA from *Rhizobium leguminosarum* and *Proteus mirabilis*, and Windass *et al.* (1980) for cloning the *E. coli* glutamate dehydrogenase gene in *M. methylotrophus*.

The basic problem with P group plasmids such as RP4 (RP1, RK2) is their size. Attempts to produce smaller derivatives have been hindered by the fact that non-contiguous regions are required for replication (Fig. 10.2). Originally it was thought that three dispersed functions *ori*V, *trf*A and *trf*B were required. More recently it has been shown that a 0.7 kb segment containing the replication origin and a 1.8 kb fragment designated *trf*A* are the minimal requirements for replication (Schmidhauser *et al.* 1983). Smaller derivatives of RK2 have been obtained by partial digestion with *Hae* II. One derivative, pRK248, is only 6.2 Mdal. in size and specifies Tc^R (Thomas *et al.* 1979). A more useful derivative, pRK2501, has been constructed by inserting a *Hae* II fragment specifying Km^R into pRK248. Plasmid pRK2501 has unique cleavage sites for *Sal* I, *Hind* III, *Xho* I, *Bgl* II

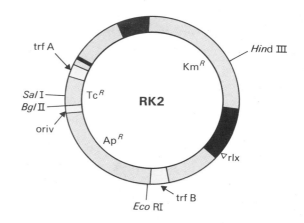

Fig. 10.2 The structure of P group plasmid RK2. The red tinted areas indicate those regions of RK2 originally believed to be essential for replication. The black areas indicate the location of genes encoding self-transmission of RK2. The site designated *rlx* must be present if conjugation-deficient derivatives of RK2 are to be mobilized by a compatible self-transmissible plasmid.

and *Eco* RI, the first three being in either the Tc^R or Km^R genes. There are two problems associated with pRK2501. First, it is not stably maintained (Thomas *et al.* 1981). Second, it can only be transformed into recipients as the conjugative functions of RK2 have been deleted. It cannot even be mobilized from *E. coli* to other recipients by other conjugative plasmids because a region of RK2 called *rlx*, a *cis*-acting function necessary for conjugal transfer, also has been removed.

An improved RK2-derived vector system has been constructed by Ditta *et al.* (1980). In this system the transfer and replication functions are separated on different plasmids. Plasmid pRK290 contains a functional RK2 replicon and can be mobilized at high frequency by a helper plasmid because it retains the *rlx* locus. It encodes Tc^R and has single *Eco* RI and *Bgl* II sites for insertion of foreign DNA. The Km^R plasmid pRK203 consists of the RK2 transfer genes cloned onto a Col E1 replicon and its function is to mobilize RK290 into other hosts. Where the intended recipient of cloned genes is transformable, RK290 alone is used. If the recipient is not trans-formable, the vector RK290 containing the foreign DNA insert first is transformed into an *E. coli* strain carrying pRK2013 and then conjugated into the desired recipient. Many derivatives of RK290 have been constructed which have additional selectable markers and additional sites for gene cloning (for review see Schmidhauser *et al.* 1988). Cosmid vectors also have been constructed.

Olsen *et al.* (1982) discovered a small (2 × 10^6 mol. wt) multicopy, broad-host-range plasmid which had arisen spontaneously from the P group plasmid RP1. Presumably this plasmid retains the *oriV* and *trfA** functions which appear to be the minimum requirements for replication of RK2. From this plasmid two derivatives were constructed. The first has two *Pst* I sites and can be used for cloning DNA where there is direct selection for the acquired trait, e.g. acquisition of antibiotic resistance or reversal of auxotrophy. The second plasmid contains an entire pBR322 molecule and if genes are inserted at the unique *Hind* III or *Bam* HI sites they can be detected by insertional inactivation of the Tc^R marker. Like pRK2501, these two vectors can be used only if the intended host is transformable.

176
Section 3
Cloning in
Organisms
Other than
E. coli

Vectors derived from the broad host range group W plasmid Sa

Although a group W plasmid such as Sa (Fig. 10.3) can infect a wide range of Gram-negative bacteria, it has been developed mainly as a vector for use with the oncogenic bacterium *Agrobacterium tumefaciens* (see p. 239). Two regions of the plasmid have been identified as involved in conjugal transfer of the plasmid and one of them has the unexpected property of suppressing oncogenicity by *A. tumefaciens* (Tait *et al.* 1982). Sufficient information for the replication of the plasmid in *E. coli* and *A. tumefaciens* is contained within a 4 kb DNA fragment. Leemans *et al.* (1982) have described four small (5.6−7.2 Mdal.), multiply marked derivatives of plasmid Sa. The derivatives contain single target sites for a number of the common restriction endonucleases and at least one marker in each is subject to insertional inactivation. Although these Sa derivatives are non-conjugative they can be mobilized by other conjugative plasmids. Tait *et al.* (1983) also have constructed a set of broad host range vectors from plasmid Sa. The properties of their derivatives are similar to those of Leemans *et al.* (1982b) but one of them also contains the bacteriophage λ *cos* sequence and hence is a cosmid. Specialist vectors for use in *Agrobacterium* and which are derived from a natural *Agrobacterium* plasmid have been described by Gallie *et al.* (1988).

Factors affecting vector utilization

Quantitative data on the stability of maintenance and general utility of broad-host-range vectors are limited given the number of Gram-negative bacteria actually studied and the frequent use of these plasmids under selective conditions. These analyses also have been made difficult by the differential expression of plasmid-borne markers in different bacteria (Mermod *et al.* 1986). Where regulatable promoters are used, different levels of inducer are needed in different bacteria (Mermod *et al.* 1986).

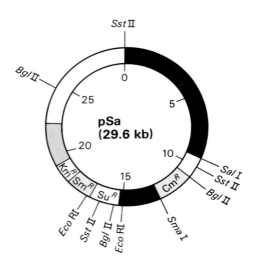

Fig. 10.3 The structure of plasmid pSa. The grey area encodes the functions essential for plasmid replication. The black areas represent the regions containing functions essential for self-transmission of pSa, the one between the *Sst* and *Sal* sites being responsible for suppression of tumour induction by *Agrobacterium tumefaciens*.

Finally, there are indications that plasmid copy number can vary significantly from host to host but few copy number determinations have been
done in bacteria other than E. coli.

177
Chapter 10
Cloning in
Bacteria Other
than E. coli

In certain important Gram-negative bacteria, e.g. the pathogenic *Bacter-*
oides and the nitrogen-fixing cyanobacteria, broad-host-range vectors are
not stably maintained. The solution is to isolate cryptic plasmids from
such species and clone them into a suitable broad-host-range vector.

Transposons as broad-host-range vectors

Transposons are mobile genetic elements which can insert at random into
plasmids or the bacterial chromosome independently of the host cell
recombination system. In addition to genes involved in transposition,
transposons carry genes conferring new phenotypes on the host cell, e.g.
antibiotic resistance. Grinter (1983) devised a broad-host-range cloning
vector based on transposon Tn7 which encodes Tp^R and Sm^R. Unlike the
vectors just described, the transposon vector permits the cloned genes to
be stably inserted into the chromosome. Such a vector could have useful
industrial applications for it does not put an extragenomic genetic load on
the recipient cell.

Two compatible broad-host-range plasmids form the basis of the original
vector system. One of the plasmids is derived from RP1 but only encodes
Tc^R. This plasmid is unstable and in the absence of selection for Tc^R is lost
rapidly from host cells. A Tn7 derivative was inserted into this plasmid to
generate pNJ5073. The Tn7 retains the original Tp^R and Sm^R markers but
the DNA sequence which encodes transposition functions was replaced
by a *Hin*d III fragment containing the *E. coli trp*E gene. The second part of
the vector system is plasmid pNJ9279 which encodes Km^R and is derived
from plasmid R300b ($\equiv$ RSF 1010, see p. 172). It is used to provide the
transposition functions missing from the Tn7 derivative.

If pNJ5073 alone is introduced into a bacterium then, as noted above,
in the absence of antibiotic selection it will not be maintained stably. Not
even the Tp^R and Sm^R markers will be retained, for the variant Tn7 is
unable to transpose. If pNJ9297 also is present then it can provide in *trans*
the necessary transposition functions: selection of Tp^R and Sm^R followed
by screening for Tc^S will identify those cells in which the defective transposon has hopped into the chromosome. In this way Grinter (1983) was
able to select strains of *M. methylotrophus* and *P. aeruginosa* which had
incorporated the *E. coli trp*E gene into their chromosome. In theory any
gene can be inserted into the chromosome by replacing the *trp*E gene with
an appropriate *Hin*d III-generated DNA fragment.

The original vectors of Grinter (1983) have been much improved (Barry
1988). The Tn7 element has been marked with the *lac*YZ genes and transferred to the pUC-based high copy number vectors (see p. 56). Additional
modifications include removal of restriction sites in the vector portion, the
deletion of non-essential Tn7 and other sequences, the addition of new
cloning sites within the Tn7-*lac* element and the incorporation of different
promoters upstream from the *lac* genes to facilitate selection for transposition in diverse bacteria. The range of options to deliver DNA to the

178

*Section 3
Cloning in
Organisms
Other than
E. coli*

chromosome has been expanded with new carrier and delivery plasmids. The biocomponent system has been improved by employing IncQ plasmids as both helper and carrier and using a combination of incompatibility and instability to easily isolate chromosomal transposition derivatives.

CLONING IN *B. SUBTILIS*

There are a number of reasons for cloning in *B. subtilis*. First, *Bacillus* spp. are Gram-positive and generally obligate aerobes compared with *E. coli* which is a Gram-negative facultative anaerobe. Thus the two groups of organisms may have quite different internal environments. Second, *Bacillus* spp. are able to sporulate and consequently are used as models for prokaryotic differentiation. The use of gene manipulation is facilitating these studies. Third, *Bacillus* spp. are widely used in the fermentation industry particularly for the production of exoenzymes. They can be tailored to secrete the products of cloned eukaryotic genes. Finally, from a biohazard point of view, *B. subtilis* is an extremely safe organism for it has no known pathogenic interactions with man or animals. Indeed it is consumed in large quantities in the East. Furthermore, in the literature between 1912 and 1989 there is only one authentic report of a human infection due to *B. subtilis* and that was in a severely compromised host—a drug addict.

Plasmid transformation in *B. subtilis*

An essential feature of any cloning experiment involving plasmids is transformation of a recipient cell with recombinant DNA. Unlike *E. coli*, the genetically developed strains of *B. subtilis* are naturally transformable and the conditions for maximizing competence are well established. The mechanism of DNA uptake and incorporation is also understood in some detail (Venema 1979). Double-stranded DNA molecules are adsorbed to the cell surface of competent cells where they are subjected to exonuclease and endonuclease processing. One strand of the DNA molecules, about 6000–13 000 bases long, is internalized. Subsequent establishment of the DNA within the cell can occur by recombination with a homologous replicon in recombination-proficient (Rec$^+$) strains. Alternatively a Rec-independent pathway can be utilized which probably involves annealing of overlapping complementary strands derived from the same parental molecule, followed by repair synthesis (Michel *et al.* 1982).

Although it is very easy to transform *B. subtilis* with fragments of chromosomal DNA there are problems associated with transformation by plasmid molecules. Ehrlich (1977) first reported that competent cultures of *B. subtilis* can be transformed with CCC plasmid DNA from *S. aureus* and that this plasmid DNA is capable of autonomous replication and expression in its new host. The development of competence for transformation by plasmid and chromosomal DNA follows a similar time course and in both cases transformation is first order with respect to DNA concentration suggesting that a single DNA molecule is sufficient for successful transformation (Contente & Dubnau 1979). However transformation of *B. subtilis* with plasmid DNA is very inefficient by comparison with chromosomal transformation for only one transformant is obtained per 10^3-10^4 plasmid molecules.

As will be seen later, much cloning in *B. subtilis* is done with bifunctional vectors that replicate in both *E. coli* and *B. subtilis*. Van Randen & Venema (1984) have shown that such bifunctional vectors can be transferred by replica-plating *E. coli* colonies containing them onto a lawn of competent *B. subtilis* cells. However, plasmid transformation by replica plating differed in one respect from plasmid transformation in liquid. Whereas chromosomal integration of plasmid-borne chromosomal alleles with concomitant loss of plasmids occurred frequently during regular plasmid transformation of Rec$^+$ *B. subtilis*, this was a rare event during plasmid transfer by replica plating.

An explanation for the poor transformability of plasmid DNA molecules was provided by Canosi *et al.* (1978). They found that the specific activity of plasmid DNA in transformation of *B. subtilis* was dependent on the degree of oligomerization of the plasmid genome. Purified monomeric CCC forms of plasmids transform *B. subtilis* several orders of magnitude less efficiently than do unfractionated plasmid preparations or multimers. Furthermore, the low residual transforming activity of monomeric CCC DNA molecules can be attributed to low level contamination with multimers (Mottes *et al.* 1979). Using a recombinant plasmid capable of replication in both *E. coli* and *B. subtilis* (pHV14, see p. 183). Mottes *et al.* (1979) were able to show that plasmid transformation of *E. coli* occurs regardless of the degree of oligomerization, in contrast to the situation with *B. subtilis*. Oligomerization of linearized plasmid DNA by DNA ligase resulted in a substantial increase of specific transforming activity when assayed with *B. subtilis* and caused a decrease when used to transform *E. coli*. An explanation of the molecular events in transformation which generate the requirement for oligomers has been presented by de Vos *et al.* (1981). Basically, the plasmids are cleaved into linear molecules upon contact with competent cells just as chromosomal DNA is cleaved during transformation of *Bacillus*. Once the linear plasmid enters the cell it is not reproduced unless it can circularize, hence the need for multimers to provide regions of homology which can recombine. Michel *et al.* (1982) have shown that multimers, or even dimers, are not required provided part of the plasmid genome is duplicated. They constructed plasmids carrying direct internal repeats 260−2000 base pairs long and found that circular or linear monomers of such plasmids were active in transformation.

Canosi *et al.* (1981) have shown that plasmid monomers will transform recombination proficient *B. subtilis* if they contain an insert of *B. subtilis* DNA. However the transformation efficiency of such monomers is still considerably less than that of oligomers. One consequence of the requirement for plasmid oligomers for efficient transformation of *B. subtilis* is that there have been very few successes in obtaining large numbers of clones in *B. subtilis* recipients (Keggins *et al.* 1978, Michel *et al.* 1980). The potential for generating multimers during ligation of vector and foreign DNA is limited. If the ratio of foreign to vector DNA is elevated in order to increase the proportion of recombinant molecules generated, the yield of

180

*Section 3
Cloning in
Organisms
Other than
E. coli*

transformants will decrease rapidly due to competition between vector—vector and vector—foreign DNA ligations. However, if transformants are obtained, the chances are that the plasmids will replicate as high mol. wt multimers. Apparently, insertion of foreign DNA into the commonly-used vectors can impair normal replication, leading to multimer formation (Gruss & Ehrlich 1988). Although this feature is of no value in the initial selection of recombinants it should facilitate any subsequent transformation steps.

Transformation by plasmid rescue

An alternative strategy for transforming *B. subtilis* has been suggested by Gryczan *et al.* (1980a). If plasmid DNA is linearized by restriction endonuclease cleavage, no transformation of *B. subtilis* results. However, if the recipient carries a homologous plasmid and if the restriction cut occurs within the homologous moiety, then this same marker transforms efficiently. Since this rescue of donor plasmid markers by a homologous resident plasmid requires the *B. subtilis* recE gene product, it must be due to recombination between the linear donor DNA and the resident plasmid. Since DNA linearized by restriction endonuclease cleavage at a unique site is monomeric this rescue system (*plasmid rescue*) bypasses the requirement for a multimeric vector. The model presented by de Vos *et al.* (1981) to explain the requirement for oligomers (see previous section) can be adapted to account for transformation by monomers by means of plasmid rescue. In practice, foreign DNA is ligated to monomeric vector DNA and the *in-vitro* recombinants used to transform *B. subtilis* cells carrying a homologous plasmid. Using such a 'plasmid rescue' system, Gryczan *et al.* (1980a) were able to clone various genes from *B. licheniformis* in *B. subtilis*. One disadvantage of this system is that transformants contain both the recombinant molecule and the resident plasmid and they have to be streaked several times to allow plasmid segregation. Alternatively, the recombinant plasmids must be retransformed into plasmid-free cells.

Transformation of protoplasts

A third method for plasmid DNA transformation in *B. subtilis* involves polyethylene glycol (PEG) induction of DNA uptake in protoplasts and subsequent regeneration of the bacterial cell wall (Chang & Cohen 1979). The procedure is highly efficient and yields up to 80% transformants, making the method suitable for the introduction even of cryptic plasmids. In addition to its much higher yield of plasmid-containing transformants the protoplast transformation system differs in two respects from the 'traditional' system using physiologically competent cells. First, linear plasmid DNA and non-supercoiled circular plasmid DNA molecules constructed by ligation *in vitro* can be introduced at high efficiency into *B. subtilis* by the protoplast transformation system, albeit at a frequency $10^1–10^3$ lower than the frequency observed for CCC plasmid DNA. Second, while competent cells can be transformed easily for genetic determinants located on the *B. subtilis* chromosome, no detectable transformation with chromosomal DNA is seen using the protoplast assay. Until recently a disadvantage of the protoplast system was that the regeneration medium was nutritionally complex, necessitating a two-step selection procedure for auxotrophic markers. Recently details have been presented of a defined regeneration medium (Puyet *et al.* 1987).

The advantages and disadvantages of the three transformation systems are summarized in Table 10.2. It should be noted that although shotgun cloning of chromosomal DNA is difficult, it is relatively easy to isolate recombinants containing fragments of bacteriophage DNA inserted into plasmid DNA. This is simply because of the small size of these genomes. Another point of interest is that in competent cells of *B. subtilis*, transforming DNA is taken up in a single-stranded form. A consequence of this is that it is not possible to clone recombinant molecules freshly constructed *in vitro* using homopolymer tailing. The reason is that recombinant molecules prepared in this way contain single-stranded regions because the homopolymer tails on the two components are generally of different lengths. For similar reasons it is not possible to use phosphatase-treated vectors when cloning directly into competent cells. These problems do not arise if protoplasts are transformed, because the DNA is taken up in a double-stranded form.

Plasmid vectors for cloning in *B. subtilis*

The development of *B. subtilis* cloning systems was hindered by the absence of suitable vector replicons; *E. coli* plasmids do not replicate efficiently in *B. subtilis* nor are their markers expressed. A large number of cryptic plasmids had been identified in *B. subtilis* (Le Hegarat & Anagnostopoulos 1977) but in the absence of a detectable phenotype they could not be used for cloning. Ehrlich (1977) showed that plasmids isolated from *S. aureus* that code for resistance to tetracycline (pT127) or chloramphenicol (pC194) can be transformed into *B. subtilis* where they replicate and express antibiotic resistance normally. Almost all the vectors used for cloning in *B. subtilis* are derived from plasmids like pT127 and pC194, which originated in *S. aureus*. Two antibiotic-resistance plasmids have been identified in *Bacillus* sp., pBC16 in *B. cereus* (Bernhard *et al.* 1978) and pAB124 in a thermophilic bacillus. Both confer tetracycline resistance in *B. subtilis*. Characterization of pAB124 has indicated that this plasmid could be used as a cloning vector (Bingham *et al.* 1980).

How do *S. aureus* plasmids rate as cloning vehicles? Well, they are small and are maintained in multiple copies like many *E. coli* plasmids, but are not chloramphenicol enrichable. The properties of these plasmids are shown in Table 10.3. In general, these plasmids are stable in *B. subtilis* but stability is greatly reduced following the insertion of exogenous DNA. However, pE194 and pSAO501 are temperature sensitive and do not replicate above 35°C. There have been reports that recombinants derived from plasmid pC194 can suffer deletions (Gryczan & Dubnau 1978) but this could be a consequence of the DNA fragments cloned. In *E. coli* the incidence of deletion formation in recombinant molecules depends on the origin and structure of the insert and the vector.

Table 10.2 Comparison of the different methods of transforming *B. subtilis*.

System	Efficiency (transformants/μg DNA)		Advantages	Disadvantages
competent cells	unfractionated plasmid	2×10^4	Competent cells readily prepared. Transformants can be selected readily on any medium. Recipient can be Rec$^-$	Requires plasmid oligomers or internally duplicated plasmids which makes shotgun experiments difficult unless high DNA concentrations and high vector/donor DNA ratios are used. Not possible to use phosphatase-treated vector
	linear	0		
	CCC monomer	4×10^1		
	CCC dimer	8×10^3		
	CCC multimer	2.6×10^5		
plasmid rescue	unfractionated plasmid	2×10^6	Oligomers not required can transform with linear DNA. Transformants can be selected on any medium	Transformants contain resident plasmid and incoming plasmid and these have to be separated by segregation or retransformation. Recipient must be Rec$^+$
protoplasts	unfractionated plasmid	3.8×10^6	Most efficient system gives up to 80% transformants. Does not require competent cells. Can transform with linear DNA and can use phosphatase-treated vector.	Efficiency lower with molecules which have been cut and religated. Efficiency also very size-dependent, and declines steeply as size increases.
	linear	2×10^4		
	CCC monomer	3×10^6		
	CCC dimer	2×10^6		
	CCC multimer	2×10^6		

Table 10.3 *S. aureus* plasmids which have been transformed into *B. subtilis.*

Plasmid	Phenotype conferred on host cell	Molecular weight
pC194	chloramphenicol resistance	1.8×10^6
pE194	erythromycin resistance	2.3×10^6
pSA0501	streptomycin resistance	2.7×10^6
pUB110	kanamycin resistance	3.0×10^6
pT127	tetracycline resistance	2.9×10^6

Improved vectors for cloning in *B. subtilis*

As indicated in Chapter 3, the ideal vector confers more than one selectable phenotype on host cells and in addition has unique restriction endonuclease cleavage sites in the corresponding genes. Since none of the *S. aureus* plasmids isolated so far carries more than one selectable marker, improved vectors have been constructed by gene manipulation *in vitro*. An example is the formation of pHV11 (Ehrlich 1978).

Plasmid pC194 has a single *Hind* III site whereas pT127 has three sites. *Hind* III cleavage destroyed the transforming activity of both plasmids, and transforming activity could be restored by DNA ligase treatment of cleaved pC194 but not pT127. However, if the two cleaved DNAs were ligated together some Tc^R colonies could be recovered and these were Cm^R as well. The plasmid extracted from one of these Tc^{R}Cm^R clones, designated pHV11, had a mol. wt of 3.3×10^6, equivalent to the sum of pC194 and the largest *Hind* III fragment of pT127. In addition, *Hind* III treatment of pHV11 DNA gave rise to two DNA segments that matched the *Hind* III cleavage of pC194 and the slowest of the pT127 segments in electrophoretic mobility.

The properties of some improved vectors for cloning in *B. subtilis* are shown in Table 10.4. In each case the unique sites that do not permit insertional inactivation of an antibiotic-resistance determinant have been shown not to occur in essential plasmid genes since the sites can receive foreign DNA without interfering with replication of the chimaera. However, some of these sites may be in essential genes in the parental plasmids since the parental plasmids are not composite replicons.

Construction of *E. coli*–*B. subtilis* hybrid replicons

The work of Ehrlich (1978) described above indicated that pC194 can be used as a *Hind* III cloning vector in *B. subtilis*. Earlier (p. 49) we described how pBR322 can be used as a *Hind* III vector in *E. coli*. Thus by ligating pBR322 with pC194 it should be possible to construct a hybrid plasmid which could be used as a vector in both *E. coli* and *B. subtilis*. The beauty of such a vector is that with it it is possible to test the expression of a cloned gene in either host. Two such hybrid replicons, pHV14 and pHV15, have been constructed and these differ only in the orientation of pC194 with respect to pBR322.

184

*Section 3
Cloning in
Organisms
Other than
E. coli*

Table 10.4 Some improved vectors for cloning in *B. subtilis*.

Plasmid	Origin	Size (Mdal.)	Markers	Single sites	Markers inactivated	Reference
pBD6	pSA0501 + pUB110	5.8	$Km^R Sm^R$	*Bam* HI *Tac* I *Hind* III *Bgl* II	none none Sm^R Km^R	Gryczan *et al.* (1980b)
pBD9	pE 194 + pUB110	5.4	$Em^R Km^R$	*Pst* I *Eco* RI *Bam* HI *Tac* I *Bgl* II *Hpa* I	none none none none Km^R Em^R	Gryczan *et al.* (1980b)
pBD12	pUB110 + pC194	4.5	$Cm^R Km^R$	*Hind* III *Eco* RI *Xba* I *Bam* HI *Tac* I *Bgl* II	none none none none none Km^R	Gryczan *et al.* (1980b)
pSL103	pUB110 + *trp* fragment	5.0	$Km^R Trp^+$	*Hind* III	Trp	Keggins *et al.* (1979)

To construct pHV14 and pHV15, pC194 and pBR322 were digested with *Hind* III, mixed and ligated, and the mixture used to transform *E. coli* to chloramphenicol resistance (Fig. 10.4). Cm^R transformants were obtained and as expected were Tc^S, due to insertion of pC194, but still Ap^R. The plasmids isolated from these transformants were of a size equivalent to the sum of the parental plasmids and were cleaved by *Hind* III into segments having mobilities that matched those of the cleaved parental DNAs.

In all cases the hybrid DNAs could transform *E. coli* to both ampicillin resistance and chloramphenicol resistance, confirming the observed phenotype of the colonies selected originally. The hybrid replicons also transformed *B. subtilis* to chloramphenicol resistance, but the Ap^R gene of pBR322 was not expressed. The failure to get expression of ampicillin resistance was not due to fragmentation of the hybrid DNA, for the *B. subtilis* transformants had acquired plasmids matching those from *E. coli* in both size and *Hind* III restriction pattern. Nor was the lack of expression of the Ap^R gene related to its orientation relative to the pC194 part of the hybrid plasmid, because neither pHV14 or pHV15 conferred ampicillin resistance in *B. subtilis*. The question concerning why the Ap^R gene from *E. coli* is not expressed in *B. subtilis* when the Cm^R gene from *B. subtilis* (originally *S. aureus*) is expressed in *E. coli* is particularly interesting and is discussed in more detail later in this chapter (see p. 191).

Plasmids pHV14 and pHV15 fail to express tetracycline resistance because the continuity of the promoter for the tetracycline-resistance gene is destroyed upon insertion of pC194 at the *Hind* III site of pBR322 (see Fig. 3.9). When *E. coli* cells carrying pHV14 are plated on tetracycline-containing media, rare Tc^R-derivatives can be isolated (Primrose & Ehrlich 1981). The plasmids isolated from these Tc^R cells confer resistance to Ap, Cm and Tc on *E. coli* recipients, i.e. tetracycline resistance is plasmid borne. These same plasmids still confer only Cm resistance on *B. subtilis* recipients. Two types of pHV14-derivative have been isolated in this way (Fig. 10.5): a derivative (pHV33) that appears to carry a point mutation which restores tetracycline resistance, and a derivative (pHV32) that carries a small deletion.

The deletion derivative pHV32 is very convenient to use for genetic labelling of replication regions active in *B. subtilis* since:

1 the deletion destroys the replication functions of pC194 which are essential for replication of pHV14 in *B. subtilis*,

2 it can be prepared in large amounts free from any contaminating. *B. subtilis* replicons in an appropriate *E. coli* strain;

3 the chloramphenicol-resistance gene it contains can transform *B. subtilis* at a high efficiency.

Using this plasmid Niaudet & Ehrlich (1979) were able to label genetically *in vitro* the cryptic *B. subtilis* plasmid, pHV400.

A direct selection vector for *B. subtilis*

Earlier we described the development of direct selection vectors for use in *E. coli* (see p. 57). Gryczan & Dubnau (1982) have developed a similar

186

*Section 3
Cloning in
Organisms
Other than
E. coli*

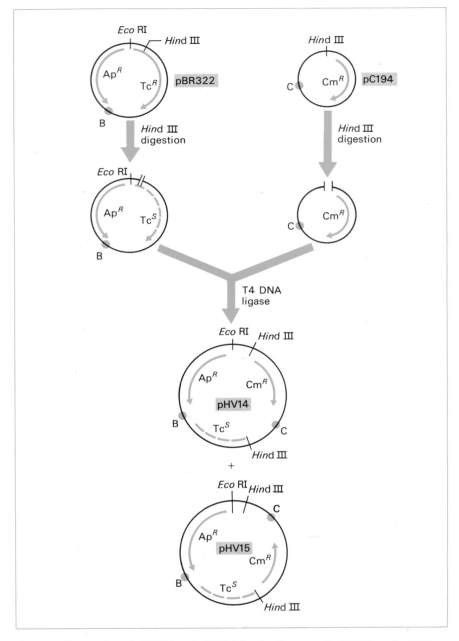

Fig. 10.4 Formation of pHV14 and pHV15 by the ligation of pBR322 and pC194. Note particularly the inactivation of the Tc^R gene and the orientations of the Cm^R gene in pHV14 and pHV15. B_B and C_C represent the origins of replication of pBR322 and pC194 respectively.

system for *B. subtilis*. It is based on plasmid pBD214 which encodes Cm^R and carries a *thy*⁺ gene, and on *B. subtilis* BD393, a highly competent thymine auxotroph. Thy⁻ strains are resistant to trimethoprim (Tp) and Thy⁺ strains are sensitive. Inactivation of the pBD214 *thy*⁺ gene by insertion

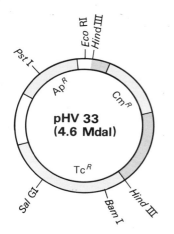

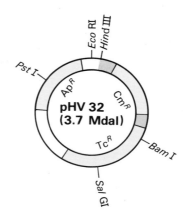

Fig. 10.5 The structure of pHV32 and pHV33. The grey areas represent DNA originally from pC194.

of a foreign DNA fragment at unique *Eco* RI, *Bcl* I, *Pvu* II or *Eco* RV sites permits selection of Cm^R Tp^R clones, all of which carry recombinant plasmids.

Low copy number vectors have been developed for use in *B. subtilis* (Bron *et al.* 1987, Imanaka *et al.* 1986). As with similar vectors in *E. coli* (see p. 57) they are used to facilitate the cloning of genes whose gene product is toxic to the host cell.

Problems Associated with Direct Cloning in *B. subtilis*

The use of plasmids for direct cloning in *B. subtilis* has been relatively unsuccessful. Structural and segregational instability have been a major problem. Segregational instability appears to be associated with a decrease in the copy number of plasmids containing inserts. Bron & Luxen (1985) found that the segregational instability of a series of plasmids was proportional to their copy number. Copy number decreased as the insert size increased. Such a tendency toward segregational instability under conditions of selective pressure may also contribute to structural instability. This structural instability is independent of the host recombination systems for it still occurs in Rec^- strains (Peijnenburg *et al.* 1987).

Haima *et al.* (1987) have shown that if the endogenous restriction system of *B. subtilis* is functional it can have a profound effect on the size, stability and number of recombinant plasmids obtained. In a restriction-deficient but modification-proficient mutant of *B. subtilis*, clones were obtained at a high frequency. Large inserts were relatively abundant, over 26% of the recombinants having insert sizes between 6 and 15 kb. In an isogenic restriction-proficient host these large inserts were seldom found. Transformation of *B. subtilis* with *E. coli*-derived individual recombinant plasmids was affected by restriction in two ways. First, by comparison with the vector, the transforming activities of recombinant plasmids

188

Section 3
Cloning in
Organisms
Other than
E. coli

carrying inserts larger than 4 kb were reduced to varying degrees in the restricting host. The levels of the reduction increased with insert length resulting in a 7800-fold reduction for the largest plasmid used. Second, over 80% of the cells transformed with the largest plasmid were found to contain a deleted plasmid when examined subsequently. In the non-restricting strain the transforming activities of the plasmids were fairly constant as a function of insert length and no structural instability was observed.

Structural instability is much more common in *B. subtilis* than in *E. coli*. Much of this instability is due to uncontrolled transcription of the cloned DNA rather than to structural features of the DNA. The latter merely permits the cell's various recombination systems to generate deletions. Now that transcription terminators have been identified which function efficiently in *B. subtilis* (Peschke *et al.* 1985) it is possible to minimize deletions resulting from the presence of strong promoters in the insert DNA. Vectors are used in which transcription terminators are positioned and oriented to prevent read-through into the replication region(s) of the plasmid.

The fate of cloned DNA in *B. subtilis*

When DNA from an organism which has little or no homology with *B. subtilis*, e.g. *B. licheniformis*, is cloned in *B. subtilis* the chimaeric plasmid appears to be stable. However, if the cloned insert has homology with *B. subtilis* the outcome is quite different. The homologous region may be excised and incorporated into the chromosome by the normal recombination process, i.e. substitution occurs via a double cross-over event. However, if only a single cross-over occurs the entire recombinant plasmid is integrated into the chromosome. It is possible to favour integration by using vectors which cannot replicate in *B. subtilis*.

B. subtilis phage φ3T contains a gene specifying the enzyme thymidylate synthetase and this gene has homology with the corresponding bacterial gene. The φ3T gene is carried on a particular *Eco* RI fragment which has been cloned in the *E. coli* vector pMB9 to generate the recombinant pCD1 (Fig. 10.6). When pCD1 is transformed into *E. coli* it complements *thy* auxotrophs and is maintained as a plasmid. When transformed into *B. subtilis* pCD1 cannot replicate and the *thy* gene is excised from the plasmid and integrated into the chromosome (Duncan *et al.* 1978). In fact CCC pCD1 DNA transforms *B. subtilis* as well as linear pCD1 DNA. One spin-off from this is that it is possible to study the integration of only one gene under carefully controlled conditions.

The extent of the homology of pCD1 with the *B. subtilis* chromosome can be greatly reduced by excising a 1.5 Mdal. *Bgl* II fragment carrying the *thy* gene. This generates pCD2 (Fig. 10.6) in which the region homologous with the *B. subtilis* chromosome is only 0.5 Mdal. long. Like φ3T, phage β22 DNA also encodes thymidylate synthetase but the corresponding gene has no detectable homology with *B. subtilis*. The *thy* gene from β22 has been cloned in pCD2, using *Bgl* II, to generate pCD4 (Fig. 10.6). In *E. coli* pCD4 again complements *thy* auxotrophs and is maintained as a plasmid.

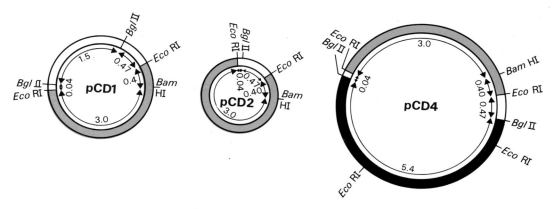

Fig. 10.6 The structures of pCD1, pCD2 and pCD4. The red areas represent material from pMB9, the unshaded areas material from φ3T and the black areas material from β22. In pCD1, the *thy* gene is carried on the 1.5 Mdal. fragment. In pCD4 the *thy* gene is carried on the 5.4 Mdal. fragment between the two *Eco* RI sites.

The plasmid also transforms Thy$^-$ auxotrophs of *B. subtilis* to Thy$^+$ but in this case the entire plasmid is integrated into the genome, not just the *thy* gene. When the same *Bgl* II fragment from β22 was inserted into pMB9 rather than pCD2, Thy$^+$ transformants were obtained in *E. coli* but not *B. subtilis*, i.e. integration of the plasmid is dependent on the 0.5 Mdal. region of homology.

The *E. coli* chromosomal *thy* gene has been cloned in pMB9 and the resulting chimaera, pER2, has no detectable homology to the *B. subtilis* chromosome by Southern blotting. There must be some very low level of homology, however, for pER2 can transform thymine auxotrophs of *B. subtilis* to thymine independence. In all these Thy$^+$ transformants pER2 is integrated into the *B. subtilis* chromosome (Rubin *et al.* 1980).

Clearly, it is possible to integrate foreign genes into the chromosome of *B. subtilis* through the development of chimaeric plasmids with limited regions of homology. The addition of a foreign segment of DNA should provide an opportunity for the introduction of additional segments of foreign DNA through recombination in the new region of homology. This principle of 'scaffolding' is shown in Fig. 10.7. For example, the first segment of foreign DNA introduced could be pER2, as described above. Subsequently, another gene cloned in pBR322 could be used to transform the first recipient. In this way any gene previously cloned in pBR322 could be readily integrated into the chromosome of *B. subtilis*.

If a foreign gene is introduced into the *B. subtilis* chromosome by scaffolding it should be flanked by identical sequences. Such constructs can be amplified (Janniere *et al.* 1985). For example, suppose a kanamycin-resistance determinant has been introduced into the chromosome. If strains bearing this insert are grown in the presence of increasing levels of kanamycin, derivatives can be isolated in which multiple repeats of the gene have occurred. Amplification factors as high as 30 have been noted. Surprisingly, such amplified sequences are sometimes stable in the absence of selection.

190

*Section 3
Cloning in
Organisms
Other than
E. coli*

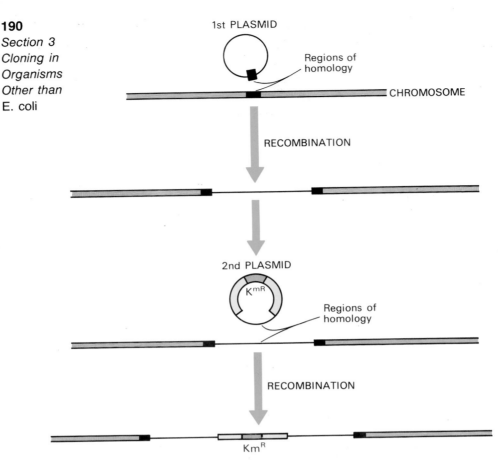

Fig. 10.7 The principle of 'scaffolding'.

Cloning sequences adjacent to the site of insertion

Once a recombinant plasmid has integrated into the chromosome it is relatively easy to clone adjacent sequences. Suppose, for example, that a pHV32 derivative carrying *B. subtilis* DNA in the *Bam* HI site (Fig. 10.8) has recombined into the chromosome. If the recombinant plasmid has no *Bgl* II sites it can be recovered by digesting the chromosomal DNA with *Bgl* II, ligating the resulting fragments and transforming *E. coli* to Ap^R. However, the plasmid which is isolated will be larger than the original one because DNA flanking the site of insertion will also have been cloned. In this way Niaudet *et al.* (1982) used a plasmid carrying a portion of the *B. subtilis ilv*A gene to clone the adjacent *thy*A gene.

Deleting regions of the chromosome

In the experiments described above the integration of non-replicating vectors into the chromosome of *B. subtilis* occurred via a single cross-over event (see Figs 10.7 and 10.8). However, bearing in mind that linear molecules may be formed during transformation of *B. subtilis* with plasmids,

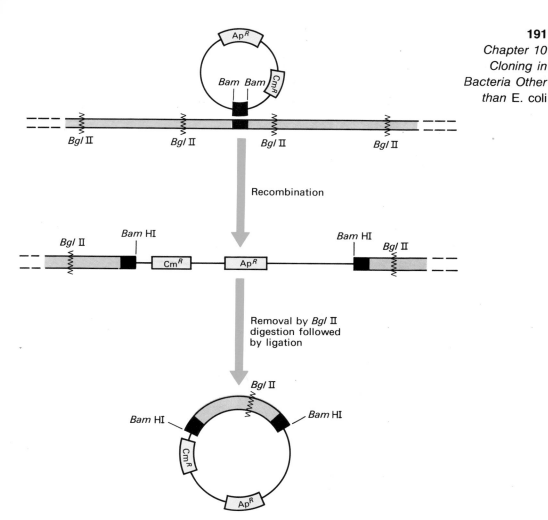

Fig. 10.8 Cloning DNA sequences flanking the site of insertion. The solid black bar on the plasmid represents a *Bam* HI fragment of *B. subtilis* chromosomal DNA. Note that the plasmid has no *Bgl* II sites. (See text for details.)

insertion of plasmid DNA into the chromosome can occur by a double cross-over. If more than one fragment of *B. subtilis* DNA is cloned, and these fragments come from non-adjacent regions of the chromosome, then the intervening DNA may be deleted during integration of the plasmid (Niaudet *et al.* 1982). The basic principles of this technique, which are shown in Fig. 10.9, could be applied to many organisms.

Expression of cloned genes in *B. subtilis*

E. coli is promiscuous in its ability to recognize transcription and translation signals from a wide variety of organisms, e.g. the expression of drug-resistance genes from Gram-positive bacteria (Cohen *et al.* 1972, Ehrlich 1978) and amino acid biosynthetic genes from yeast (Struhl *et al.* 1976). By

192

Section 3
Cloning in
Organisms
Other than
E. coli

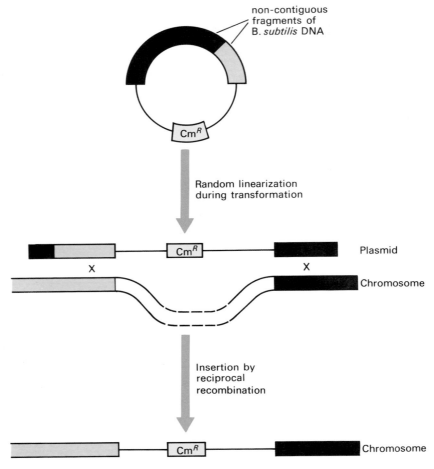

Fig. 10.9 Formation of deletions during insertion of a recombinant plasmid.

contrast, *B. subtilis* is restricted in its ability to express genes from other genera (Ehrlich 1978a, Ehrlich *et al.* 1978b). With one exception (Rubin *et al.* 1980), the only foreign genes expressed in *B. subtilis* are from other Gram-positive bacteria, e.g. drug resistance genes from *S. aureus* (Ehrlich 1977) and *Streptococcus* (Yagi *et al.* 1978). The lack of expression of *E. coli* genes in *B. subtilis* is not due to structural alterations, for these genes can be isolated from *B. subtilis* and are still functional when reintroduced into *E. coli*. What, then, are the barriers to the expression of *E. coli* genes in *B. subtilis*?

The first step in gene expression is the transcription of DNA by RNA polymerase. A comparison of the RNA polymerases from the two organisms has not revealed gross differences but the template specificity of the two polymerases differs in a very interesting way. The *E. coli* enzyme can transcribe equally well DNAs of *E. coli* bacteriophage T4 and of *B. subtilis* bacteriophage φe. By contrast, the *B. subtilis* polymerase is much less active with the *E. coli* than with the *B. subtilis* phage DNAs (Shorenstein & Losick 1973). This behaviour is consistent with the observation that *B.*

subtilis genes can function in *E. coli* but the *E. coli* genes cannot function in *B. subtilis*. Moran *et al.* (1982) determined the nucleotide sequence of two *B. subtilis* promoters and found no obvious differences with those of *E. coli*. Since then over 50 *B. subtilis* promoters, recognized by σ factors associated with vegetative growth, have been sequenced and the consensus −35 and −10 sequences are identical to those from *E. coli*. From detailed studies (for review see Mountain 1989) it would appear that recognition of a promoter in *B. subtilis* involves other structural features, but what these features are is not clear.

The translational apparatus of *E. coli* and various *Bacillus* spp. are considered to be essentially similar. Nevertheless, intriguing differences have been reported. *E. coli* ribosomes were found to support protein synthesis when directed by four out of five mRNAs from Gram-negative cells and six out of six mRNAs from Gram-positive cells. By contrast, *B. subtilis* ribosomes were not active with any of the mRNAs from Gram-negative bacteria nor three out of four mRNAs from Gram-positive bacteria (Stallcup *et al.* 1974) McLaughlin *et al.* (1981) and Moran *et al.* (1982) have analysed the mRNA sequences located downstream from a number of *B. subtilis* chromosomal and phage promoters. They have found that the sequence complementarity between the mRNA and the 3′ terminal region of the 16s ribosomal RNA is far greater than with *E. coli* Shine−Dalgarno sequences. Confirmation of this has been provided by Band and Henner (1984). In addition, they constructed a series of plasmids differing only in the sequence of the Shine−Dalgarno region preceding an IFN-α gene, and used this series of plasmids to test the efficiency of interferon expression in both *B. subtilis* and *E. coli*. In *B. subtilis* interferon expression was much more sensitive to changes in the sequence of the Shine−Dalgarno region than in *E. coli*.

Plasmid expression vectors for *B. subtilis*

Because of the problems associated with heterologous gene expression in *B. subtilis*, expression vectors developed for use in *E. coli* (see p. 144) cannot be used in this organism. Consequently specific *B. subtilis* expression vectors have been constructed. Williams *et al.* (1981a) generated a plasmid which can be used to clone promoters active in *B. subtilis*. This plasmid, pPL603, specifies Km^R and carries a *Bacillus* structural gene for chloramphenicol acetyl transferase (CAT) which lacks a promoter. There is a unique *Eco* RI site upstream of the CAT coding sequence (Fig. 10.10). Promoter-containing DNA is identified by cloning *Eco* RI-generated fragments of DNA from *B. subtilis* spp. into pPL603 with subsequent selection of Cm^R transformants. In such transformants the CAT is inducible by chloramphenicol regardless of the promoter fragment cloned because regulation seems to occur at the level of translation. A similar construction has been utilized by Goldfarb *et al.* (1981) but because their Cm^R gene was derived from *E. coli* the CAT is not inducible by Cm.

Plasmid pPL608 is a derivative of pPL603 containing a 0.2 Mdal. *Eco* RI* promoter fragment from *Bacillus* phage SPO2. This promoter fragment permits expression of the CAT gene enabling pPL608 to confer Cm^R on *B.*

194

Section 3
Cloning in
Organisms
Other than
E. coli

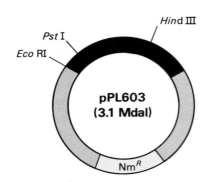

Fig. 10.10 The structure of plasmid pPL603. The black bar represents a DNA fragment from *B. pumilus* which includes a promoterless gene specifying chloramphenicol acetyl transferase (CAT). The initiation codon for the CAT gene lies between the unique *Pst* I and *Hind* III sites.

subtilis. Williams *et al.* (1981b) cloned the *E. coli trpC* gene in the *Hind* III site of pPL608 and obtained complementation of a *B. subtilis* tryptophan auxotroph. Expression of the *trp* gene was Cm inducible but the Cm^R phenotype was destroyed. Proof that the *trp* gene was under the control of the SPO2 promoter was provided by the observation that removal of the promoter fragment or reversing the orientation of the *E. coli* fragment did not yield Trp^+ transformants.

Williams *et al.* (1981b) also cloned a mouse dihydrofolate reductase (DHFR) gene in pPL608 but used the *Pst* I site instead of the *Hind* III site. Mammalian DHFR is resistant to trimethoprim (Tp) whereas the bacterial enzyme is Tp sensitive. Expression of the mouse DHFR in *B. subtilis* was detected by the Tp^R phenotype, which is conferred. However, the transformants were Cm^R and Tp^R and the Tp^R was not Cm-inducible, indicating that the *Pst* I site lies between the promoter fragment and the start of the CAT gene. Schoner *et al.* (1983) have identified a plasmid mutation lying between the *Eco* RI and *Pst* I site which increases expression of the mouse DHFR tenfold. Nucleotide sequence analysis suggests that the SPO2 fragment contains sequences that correspond to a ribosome binding site.

Similar constructions to those of Williams *et al.* (1981b) have been made by Hardy *et al.* (1981). They fused a DNA fragment encoding all but the eight N-terminal amino acids of the foot-and-mouth disease virus (FMDV) VP1 antigen into the Em^R gene of a *B. subtilis* vector. *B. subtilis* cells transformed with the chimaeric plasmid synthesized a fusion protein consisting of the N-terminus of the Em^R gene product and VP1 protein minus the first eight amino acids. Just as expression of Em resistance is inducible by Em, so too was synthesis of the fusion protein. In addition, they placed the gene encoding the hepatitis B core antigen just downstream from the termination codon of the Em^R gene and in phase with it. Again, expression of the cloned gene was inducible by Em.

One of the main problems which has delayed the development of stable, efficient *B. subtilis* expression systems has been the lack of well-characterized, strong controllable promoters analogous to λpL, *trp*, *lac* and *tac* promoters which have proven so useful in *E. coli*. One solution is to use promoters which naturally show growth phase-dependent expression such as those for the *Bacillus* α-amylase and protease genes (Lehtovaara *et al.* 1984, Ferrari *et al.* 1987, Nicholson *et al.* 1987).

A different kind of repression system has been devised by Yansura & Henner (1984). They placed the *E. coli lac* operator on the 3′ side of the promoter of a *Bacillus* penicillinase gene thus creating a hybrid promoter controllable by the *E. coli lac* repressor. The *E. coli lac* repressor gene was placed under the control of a promoter and binding site functional in *B. subtilis*. When the penicillinase gene that contained the *lac* operator was expressed in *B. subtilis* on a plasmid that also produces the *lac* repressor, the expression of the penicillinase gene was modulated by IPTG, an inducer of the *lac* operon in the *E. coli*.

Plasmid secretion vectors for use in *B. subtilis*

As discussed in Chapter 9, most secretory proteins are synthesized initially as preproteins which have additional amino acids, the signal sequence, at their N-termini. Proteins secreted by *Bacillus* spp. are no different in that they have a signal sequence with a hydrophilic region at the N-terminus followed by a hydrophobic region. However, *Bacillus* signal sequences are longer than those from Gram-negative bacteria and eukaryotes. This is reflected in an increase in size of the hydrophilic region and the presence of a peptide extension beyond the hydrophobic core (Kroyer & Chang 1981, Neugebauer *et al.* 1981, Palva *et al.* 1981). Despite these differences between Gram-positive and Gram-negative signal sequences they can be used in an identical way (see p. 149) to promote secretion of cloned-gene products. Thus Palva *et al.* (1982) fused an *E. coli*-derived β-lactamase gene to the promoter and signal sequence region of the α-amylase gene from *B. amyloliquefaciens*. The β-lactamase gene was expressed in *B. subtilis* and over 95% of the enzyme activity was secreted to the growth medium. Later Palva *et al.* (1983) showed that the amylase signal would also permit secretion of interferon from *B. subtilis*.

Smith *et al.* (1987) have constructed two plasmid vectors that can be used to probe for export signal-coding regions in *B. subtilis*. The factors contain genes coding for extracellular proteins, the α-amylase gene of *B. licheniformis* and the β-lactamase gene from *E. coli*, which lack a functional signal sequence. By shotgun cloning of restriction fragments from *B. subtilis* chromosomal DNA a variety of different export-coding regions were selected. These regions were functional in both *E. coli* and *B. subtilis*.

Bacteriophage vectors for use in *B. subtilis*

Although plasmid vectors have been used extensively in *B. subtilis* they do have some disadvantages. Not the least of these is the instability of plasmids bearing inserts, although the factors contributing to instability are being identified. Many of the plasmids are maintained at a greater copy number than the host chromosome, and with some cloned genes the resultant gene-dosage effects can be deleterious for the cell (Banner *et al.* 1983). Finally, homologous inserts of *B. subtilis* DNA cannot be maintained

196

Section 3
Cloning in
Organisms
Other than
E. coli

on plasmids except in recombination-deficient hosts. Such Rec⁻ hosts are difficult to construct and maintain and, for those studying sporulation, have the additional disadvantage that they do not sporulate well. Vectors based on bacteriophage φ105 appear to be free from these defects.

Bacteriophage φ105 is temperate and thus has a chromosomal attachment site analogous to that of λ in *E. coli*. Thermo-inducible mutants are available. The wild-type genome is 39.2 kb long but, like λ, it can package larger genomes. The upper size limit for efficient packaging is 40.2 kb which is only 1 kb larger than the wild-type genome (Errington & Pughe 1987). However, because some phage DNA can be deleted it is possible to isolate vectors which permit the cloning of up to 5 kb of foreign DNA (Jones & Errington 1987).

Bacteriophage vectors are used in a manner analogous to plasmid vectors (Errington 1984) as shown in Fig. 10.11. DNA is isolated from a suitable vector phage such as φ105J9, which has unique sites for *Bam* HI and *Xba* I. As with λ, the phage DNA has cohesive ends and multimers are obtained by incubating in the presence of DNA ligase. Treatment of the multimers generates linear molecules containing an entire phage genome flanked by *Bam* HI ends. After ligation with chromosomal DNA fragments, circularized recombinants are selected by transfection of protoplasts. The resultant 'transducing' phages can infect and stably lysogenize *B. subtilis*.

CLONING IN GRAM-POSITIVE BACTERIA OTHER THAN *B. SUBTILIS*

The only Gram-positive organisms other than *B. subtilis* in which there has been extensive *in-vitro* gene manipulation are the streptomycetes and lactic streptococci. Much of the interest in cloning in these organisms is because of their industrial applications. Lactic streptococci are used in the production of fermented milks, e.g. yoghurt and cheeses, and the requisite metabolic functions are plasmid borne. The techniques of gene manipulation for dairy organisms such as *Streptococcus lactis* are little different from those used with *B. subtilis* (for reviews see Kondo & McKay 1985 and Gasson & Davies 1985). The only technique not described so far is plasmid transfer by the use of protoplast fusion between related species (Okamoto *et al.* 1985) or unrelated species such as *S. lactis* and *B. subtilis* (Baigori *et al.* 1988).

Over 60% of known naturally occurring antibiotics come from *Streptomyces*, making the genus the major taxonomic class sought by pharmaceutical companies in screening programmes for new naturally occurring isolates. Recombinant DNA technology is being used to generate novel antibiotics and to improve the titres of naturally occurring antibiotics.

Fig. 10.11 The use of the phage vector φ105J9 for the construction of genomic libraries in *B. subtilis*. (See text for details.) The cohesive ends of the phage DNA are indicated by *cos*.

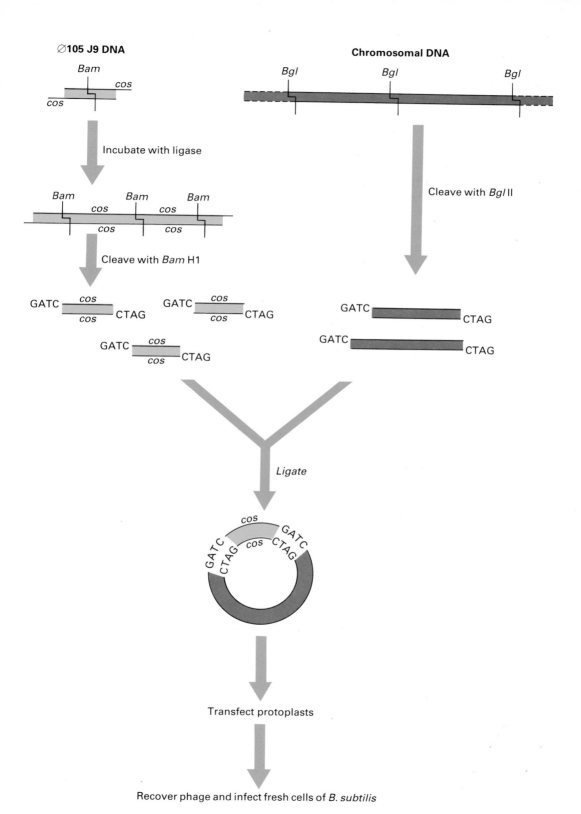

198

Section 3
Cloning in
Organisms
Other than
E. coli

These topics are covered in a later chapter (see p. 356). As with the lactic streptococci, the methods for handling recombinant DNA in *Streptomyces* are similar to those already described and will not be detailed here (for review see Hopwood *et al.* 1987).

One feature of *Streptomyces* does merit further discussion. Many species spontaneously amplify specific chromosomal DNA sequences, giving rise to several hundred tandem copies which can account for 10% of total DNA. Altenbuchner & Cullum (1987) linked a thiostrepton resistance gene and a tyrosinase gene to an amplifiable sequence from *S. lividans*. This construct was transformed into *S. lividans* where it integrated into the chromosome by homologous recombination. When DNA amplification was induced the cloned genes were co-amplified to give stable high copy number clones.

11 Cloning in Yeast and Other Microbial Eukaryotes

The analysis of eukaryotic DNA sequences has been facilitated by the ease with which DNA from eukaryotes can be cloned in prokaryotes using the vectors described in previous chapters. Such cloned sequences can be obtained easily in large amounts and can be altered *in vivo* by bacterial genetic techniques and *in vitro* by specific enzyme modifications. To determine the effects of these experimentally induced changes on the function and expression of eukaryotic genes, the rearranged sequences must be taken out of the bacteria in which they were cloned and reintroduced into a eukaryotic organism, preferably the one from which they were obtained. Despite the overall unity of biochemistry there are many functions common to eukaryotic cells which are absent from prokaryotes, e.g. localization of ATP-generating systems to mitochondria, association of DNA with histones, mitosis and meiosis, and obligate differentiation of cells. The genetic control of such functions must be assessed in a eukaryotic environment. In this chapter we will discuss the potential for cloning in the yeast *Saccharomyces cerevisiae* and in later chapters we will discuss the possibilities for cloning in animal and plant cells. It should be borne in mind that as well as using yeast as a host for cloned genes from other eukaryotes it also can be used as a host for analysing cloned yeast genes, i.e. surrogate yeast genetics.

Transformation in yeast

Transformation of yeast was first achieved by Hinnen *et al.* (1978) who fused yeast spheroplasts (i.e. wall-less yeast cells) with polyethylene glycol in the presence of DNA and $CaCl_2$ and then allowed the spheroplasts to regenerate walls in a stabilizing medium containing 3% agar. The transforming DNA used was plasmid pYeLeu 10 which is a hybrid composed of the *E. coli* plasmid Col E1 and a segment of yeast DNA containing the Leu 2^+ gene. Spheroplasts from a stable Leu 2^- auxotroph were transformed to prototrophy by this DNA at a frequency of 1×10^{-7}. Untreated spheroplasts reverted with a frequency of $< 1 \times 10^{-10}$. When 42 Leu$^+$ transformants were checked by hybridization, 35 of them contained Col E1 DNA sequences. Genetic analysis of the remaining seven transformants indicated that there had been reciprocal recombination between the incoming Leu 2^+ and the recipient Leu 2^- alleles.

Of the 35 transformants containing Col E1 DNA sequences, genetic analysis showed that in 30 of them the Leu 2^+ allele was closely linked to the original Leu 2^- allele whereas in the remaining 5, the Leu 2^+ allele was located on another chromosome. These results can be confirmed by restriction endonuclease analysis since pYeLeu 10 contains no cleavage

200

*Section 3
Cloning in
Organisms
Other than
E. coli*

sites for *Hin*d III. When DNA from the Leu 2⁻ parent was digested with endonuclease *Hin*d III and electrophoresed in agarose, multiple DNA fragments were observed but only one of these hybridized with DNA from pYeLeu 10. With the 30 transformants in which the Leu 2⁻ and Leu2⁺ alleles were linked, only a single fragment of DNA hybridized to pYeLeu 10 but this had an increased size consistent with the insertion of a complete pYeLeu 10 molecule into the original fragment. This data is consistent with there being a tandem duplication of the Leu 2 region of the chromosome (Fig. 11.1). With the remaining five transformants, two DNA fragments which hybridized to pYeLeu 10 could be found on electrophoresis. One fragment corresponded to the fragment seen with DNA from the recipient cells, the other to the plasmid genome which had been inserted in another chromosome (see Fig. 11.1). These results represented the first unambiguous demonstration that foreign DNA, in this case cloned Col E1 DNA, can integrate into the genome of a eukaryote. A plasmid such as pYeLeu 10 which can do this is known as a YIp-*yeast integrating plasmid*.

The development of yeast vectors

Common principles. Since the first demonstration of transformation in yeast a number of different kinds of yeast vector have been constructed. All of them have features in common. First, all of them can replicate in *E. coli*, often at high copy number. This is important because for many experiments it is necessary to amplify the vector DNA in *E. coli* before transformation of the ultimate yeast recipient. Second, all employ markers, e.g. Leu 2⁺, His⁺, Ura 3⁺, Trp 1⁺, that can be selected readily in yeast and

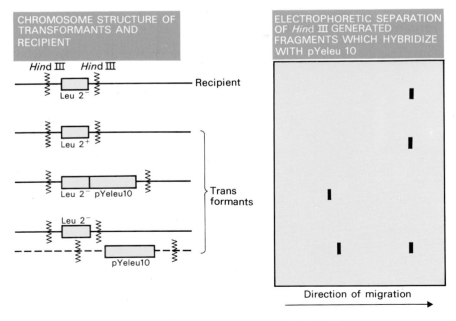

Fig. 11.1 Analysis of yeast transformants. (See text for details.)

which often will complement the corresponding mutations in *E. coli* as
well. In addition to these selectable markers most of the vectors also carry
antibiotic resistance markers for use in *E. coli*. Finally, all of them contain
unique target sites for a number of restriction endonucleases.

201
*Chapter 11
Cloning in
Yeast and
Other Microbial
Eukaryotes*

Yeast episomal plasmids (YEp). The first YEps were constructed by Beggs
(1978) using a naturally occurring yeast plasmid. The properties of this
plasmid have been reviewed by Murray (1987) and can be summarized as
follows. The plasmid which is 2 μm long (6.3 kb) is found in many strains
of *S. cerevisiae* and has no known function. There are 50–100 copies of the
plasmid per haploid cell which represents 2–4% of the total yeast genome.
The plasmid is divided by perfect inverted repeats of 599 base pairs into
two unique regions, each with a pair of genes transcribed from a divergent
promoter (Fig. 11.2). One of these genes, called FLP for its 'flipping'
activity, encodes a site-specific recombinase which catalyses genetic cross-
over between sequences in the inverted repeats. When this occurs the
orientation of the non-repeated sequences is inverted and hence the 2 μm
plasmid exists as two isomeric forms. Beggs (1978) constructed chimaeric
plasmids containing this 2 μm yeast plasmid, fragments of yeast nuclear
DNA and the *E. coli* vector pMB9. These chimaeras were able to replicate
in both *E. coli* and yeast, transformed yeast with high frequency, and
some were able to complement auxotrophic mutations in yeast. The chimaeric plasmids were constructed in two stages. First, plasmid pMB9 (Fig.
11.2) and the 2 μm plasmid were joined by ligation of the DNA fragments
produced by *Eco* RI endonuclease digestion. Tc^R clones were selected after
transforming *E. coli* with the ligated DNAs and these were screened for a
hybrid plasmid large enough to carry the complete yeast 2 μm plasmid
sequence. Yeast–pMB9 hybrid plasmids theoretically have eight possible
configurations. These are determined by the orientation of the insertion
into pMB9, which of the *Eco* RI sites on the yeast plasmid is used for
insertion and whether the yeast sequence is in the A or B configuration.
Five of the eight possible configurations, which can be distinguished by
restriction mapping, were found among seven complete hybrid plasmids
examined. The exact configuration of the yeast 2 μm plasmid–pMB9
hybrids is probably not important.

For the second stage Beggs (1978) sheared nuclear DNA isolated from
S. cerevisiae and linked it by means of poly (dA–dT) tails to a mixture of
pMB9-yeast 2 μm plasmid hybrid plasmids which had been linearized by
digestion with *Pst* I. Tc^R transformants were selected in *E. coli* and of the
21 000 obtained, two were found which complemented a Leu⁻ mutant.
The plasmids from these clones, pJDB219 and pJDB248, transformed a *leuB*
mutant of *E. coli* to Leu⁺ or tetracycline resistance at the same frequency.

Both pJDB219 and pJDB248 transformed Leu 2 yeast mutants to Leu⁺
with a frequency of 5×10^{-4} to 3×10^{-3} transformants per viable cell.
This is several orders of magnitude greater than the frequency of transformation obtained by Hinnen *et al.* (1978). However, chromosomal integration
of the transforming fragment was essential with the system of Hinnen *et
al.* (1978). By contrast pJDB219 and pJDB248 could be recovered as plasmids
from yeast cells and their inheritance in yeast was non-Mendelian.

202

*Section 3
Cloning in
Organisms
Other than
E. coli*

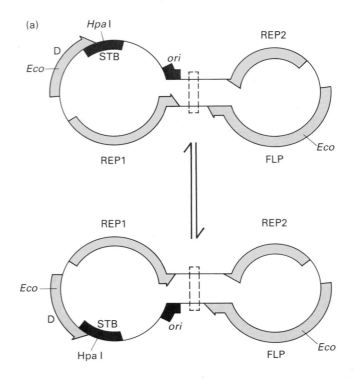

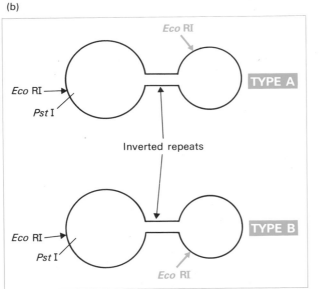

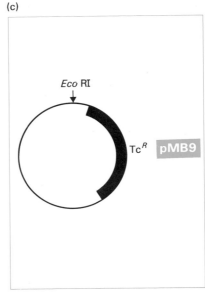

Fig. 11.2 Molecular (a) and schematic (b) structures of the yeast 2 μm plasmid. Only a limited number of restriction sites are shown. REP1 and REP2 are proteins involved in replication of the plasmid and STB is a partition locus. Ori is the origin of replication, FLP the gene for a site-specific recombinase, and D a locus which promotes plasmid amplification. The inverted repeats which contain the FLP recognition sequences are shown boxed. The structure of pMB9 is shown in (c).

203

Chapter 11
Cloning in
Yeast and
Other Microbial
Eukaryotes

Many other YEp vectors based on the 2 μm plasmid have been developed, e.g. Gerbaud *et al.* 1979, Storms *et al.* 1979, Struhl *et al.* 1979. They possess the same advantages as those of Beggs (1978), viz., they have a high copy number (25−100 copies cell^{-1}) in yeast and they transform yeast very well. Indeed, the transformation frequency can be increased 10- to 20-fold if single-stranded plasmids are used (Singh *et al.* 1982).

Recently Heinemann & Sprague (1989) have shown that broad-host-range bacterial plasmids can mediate conjugative transmission of plasmids such as YEps from *E. coli* to yeast. Use of conjugation should facilitate cloning in yeast since it avoids the need for spheroplasts.

Yeast replicating plasmids (YRp). Struhl *et al.* (1979) constructed a useful vector which consists of a 1.4 kb yeast DNA fragment containing the *trp*-1 gene inserted into the *Eco* RI site of pBR322. This vector transformed *trp*-1 yeast protoplasts to Trp$^+$ at high frequency and transforming sequences were always detected as CCC DNA molecules in yeast. No transformants were found in which the vector had integrated into the chromosomal DNA. Since pBR322 alone cannot replicate in yeast cells (Beggs 1978) a yeast chromosomal sequence must permit the vector to replicate autonomously and to express yeast structural genes in the absence of recombination with host chromosomal sequences. A similar vector based on pBR313 was developed by Kingsman *et al.* (1979). In both cases the yeast gene was linked to a centromere and initially it was thought that this was important. Since then it has been shown that a centromere is not essential; rather, the vector carries an *autonomously replicating sequence (ars)* derived from the chromosome.

Although plasmids containing an *ars* transform yeast very efficiently the resulting transformants are exceedingly unstable. For unknown reasons, YRp plasmids tend to remain associated with the mother cell and are not efficiently distributed to the daughter cell (N.B. *S. cerevisiae* does not undergo binary fission but buds off daughter cells instead). Occasional stable transformants are found and these appear to be cases in which the entire YRp has integrated into a homologous region on a chromosome in a manner identical to that of YIps (Stinchcomb *et al.* 1979, Nasmyth & Reed 1980).

Yeast centromere plasmids (YCp). Using a YRp vector Clarke & Carbon (1980) isolated a number of hybrid plasmids containing DNA segments from around the centromere-linked *leu*2, *cdc*10 and *pgk* loci on chromosome III of yeast. As expected for plasmids carrying an *ars* most of the recombinants were unstable in yeast. However, one of them was maintained stably through mitosis and meiosis. The stability segment was confined to a 1.6 kb region lying between the *leu*2 and *cdc*10 loci and its presence on plasmids carrying either of 2 *ars* tested resulted in those plasmids behaving like minichromosomes (Clarke & Carbon 1980, Hsiao & Carbon 1981). Genetic markers on the minichromosomes acted as linked markers segregating in the first meiotic division as centromere-linked genes and were unlinked to genes on other chromosomes. Stinchcomb *et al.* (1982) and Fitzgerald-Hayes *et al.* (1982) have isolated the centromeres from chromosomes IV

204

Section 3
Cloning in
Organisms
Other than
E. coli

and XI of yeast and found that they confer on plasmids similar properties to the centromere of chromosome III.

Since then centromeres from another nine chromosomes have been cloned and found to have a similar structure and function. Structurally, plasmid-borne centromere sequences have the same distinctive chromatin structure that occurs in the centromere region of yeast chromosomes (Bloom & Carbon 1982). Functionally YCps exhibit three characteristics of chromosomes in yeast cells. First, they are mitotically stable in the absence of selective pressure. Second, they segregate during meiosis in a Mendelian manner. Finally, they are found at low copy number in the host cell.

The low copy number of YCp vectors is not altered if they include the yeast 2 μm plasmid amplification system that normally drives plasmids to high copy number (Tschumper & Carbon 1983). This suggests that the centromere is dominant over the 2 μm plasmid amplification system. However, if the FLP gene product is synthesized, high copy number plasmids are generated *in vivo* in which the centromere sequence has been deleted or inactivated.

By placing an inducible promoter adjacent to a cloned centromere it is possible to modulate centromere function by controlling promoter activity. Such conditional constructs have been used to study centromere structure and function relationships (Hill & Bloom 1987), loss and breakage of dicentric chromosomes (Futcher & Carbon 1986), and to produce regulated copy number plasmids (Chlebowicz-Sledziewska & Sledziewski 1985). When transcription is induced plasmid stability is decreased, unless a second centromere is present, and copy number is increased (Apostol & Greer 1988).

When wild-type yeast cells are forced to maintain multiple YCps bearing independently selectable markers, the cells grow slowly and cell viability is decreased, indicating a toxic effect from the presence of excess centromeres (Futcher & Carbon 1986). A plasmid system has been constructed which permits the selection of cells which tolerate high copy number YCps. The plasmids carry a gene conferring resistance to antibiotic G418 but which has a weak promoter ensuring low-level resistance. Selection is made for cells which can grow in the presence of high levels of antibiotic. Three different events can give rise to high level resistance (Tschumper & Carbon 1987). In some cells the centromere sequence has been inactivated and plasmid replication is under the control of the 2 μm plasmid. In other cells mutants have been selected in which the strength of the $G418^R$ gene promoter has been increased. The most interesting class of high-level resistance mutants is that in which chromosomal mutants have arisen which enable yeast cells to tolerate high copy number YCps.

Yeast artifical chromosomes (YAC). All three autonomous plasmid vectors are maintained in yeast as circular DNA molecules—even the YCp vectors which possess yeast centromeres. Thus none of these vectors resembles the normal yeast chromosomes, which have a linear structure. The ends of all yeast chromosomes, like those of all other linear eukaryotic chromosomes, have unique structures that are called *telomeres*. Telomeres have a unique structure (see later) that has evolved as a device to preserve the integrity of the ends of

DNA molecules which often cannot be finished by the conventional mechanisms of DNA replication (see Watson 1972 for detailed discussion). Szostak & Blackburn (1982) were able to clone yeast telomeres by developing a linear yeast vector.

205
Chapter 11
Cloning in
Yeast and
Other Microbial
Eukaryotes

The linear yeast vector was constructed from two components. The first was a YRp, pSZ213 (Fig. 11.3) which has no *Bam* HI sites and a single *Bgl* II site. The second component was fragments of an unusual, linear, rRNA-encoding plasmid found in the protozoan *Tetrahymena*. This plasmid DNA was cleaved with *Bam* HI and the end fragments were ligated to *Bgl* II-cut pSZ213. Since *Bam* HI-cut and *Bgl* II-cut ends of DNA are complementary they can be joined by ligase but the product of ligation is not a substrate for either enzyme. Consequently both *Bam* HI and *Bgl* II endonucleases were present in the ligation mixture to cut circularized or dimerized vector molecules and dimerized end fragments. In this way the desired linear vector carrying *Tetrahymena* telomeres accumulated in the reaction mixture and it was purified by agarose gel electrophoresis.

The linear plasmid containing the *Tetrahymena* telomeres retains the Leu 2^+ marker and this was used for selection of transformants in yeast. Since YRp vectors are capable of integration, those transformants in which the linear plasmid was replicating autonomously were detected by their mitotic instability. Restriction mapping showed that the plasmid in the unstable transformants is a linear molecule identical in structure to the linear molecule constructed *in vitro* and used in the transformation. Furthermore, the ends of *Tetrahymena* rDNA have three unusual structural

(a)

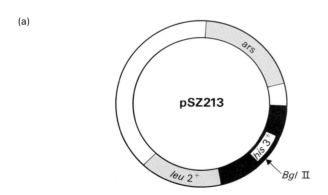

(b)

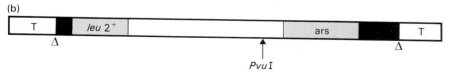

Fig. 11.3 (a) Simplified map of plasmid pSZ213 used for cloning yeast telomeres. The plasmid has no *Bam* HI sites and insertion of telomeric DNA at the unique *Bgl* II site inactivates the *his3*$^+$ gene. (b) Structure of a linear plasmid constructed from pSZ213 showing the asymmetric location of the unique *Pvu* I site. The open triangles indicate the location of *Bam/Bgl* joints created when the telomeric DNA (indicated with a T) was added to *Bgl* II-digested pSZ213.

206

*Section 3
Cloning in
Organisms
Other than
E. coli*

features: a variable number of short, 5'-CCCCAA-3' repeat units, specific single-strand interruptions within the repeated sequences and a cross-linked terminus. All three structural features were maintained when the *Tetrahymena* sequences were cloned in yeast.

The linear plasmid carrying the *Tetrahymena* telomeres has a single, asymmetrically placed target site for endonuclease *Pvu* I. This enzyme cuts yeast DNA into approximately 2000 fragments and since yeast has 17 chromosomes 34 of these fragments should contain telomeres. After ligating *Pvu* I-digested chromosomal and plasmid DNA unstable *Leu*[+] transformants were selected. Many of the plasmids generated in this way would have multiple yeast *Pvu* I fragments ligated onto the vector followed by either a yeast end or an end derived from the vector. These plasmids would be as large as or larger than the original vector. In contrast, the ligation of a single *Pvu* I telomere fragment from yeast, smaller than the *Tetrahymena*-containing fragment being replaced, would yield a linear plasmid smaller than at the start. After size screening of the plasmids from the transformants three plasmids carrying a yeast telomere at one end were identified. Analysis of these plasmids shows that yeast telomeres have at least some of the structural features of *Tetrahymena* telomeres. Each yeast telomere ends with approximately 100 base pairs of irregularly repeated sequences of the form

$$5'\text{-CCCA}\ldots\ldots\ldots$$
$$3'\text{-GGGT}\ldots\ldots\ldots$$

attached to specific X and Y sequences. Within these specific terminal X and Y sequences are additional internal sections of the repetitive elements.

Stability of yeast cloning vectors. As noted earlier, YRp vectors are not stably maintained by yeast cells and, in the absence of selection, are quickly lost from the population. By contrast, the segregational stability of YCp vectors which carry a yeast centromere is considerably greater than that of YRp vectors. Murray & Szostak (1983a) have found that YRp vectors have a strong bias to segregate to the mother cell at mitosis. This segregation bias explains how the fraction of plasmid-bearing cells can be small despite the high average copy number of YRp vectors. The presence of a centromere eliminates segregation bias, thus accounting for the increased stability of YCp vectors relative to YRp vectors. YEp vectors are stably maintained in yeast cells provided the strain contains endogenous intact 2 μm circles. In the absence of endogenous 2 μm circles YEp vectors show maternal segregation bias.

Despite the fact that YCp vectors are relatively stable, they are still 1000 times less stable than *bona fide* yeast chromosomes. However, YCp vectors and YRp vectors are circular molecules whereas chromosomes are linear molecules. A linear plasmid vector carrying a centromere would be much more representative of a yeast chromosome. Dani & Zakian (1983) constructed linear yeast plasmids but found that stability was reduced relative to the circular plasmid. However, Murray & Szostak (1983b) found that stability of linear yeast plasmids was related to size. Thus artificial yeast chromosomes which were 55 kb long and contained cloned genes, *ars,*

207
Chapter 11
Cloning in
Yeast and
Other Microbial
Eukaryotes

centromeres and telomeres had many of the properties of natural yeast chromosomes. When the artificial chromosomes were less than 20 kb in size, centromere function was impaired.

Whereas the stability of linear YCp vectors is dependent on size, this is not true of YRp vectors. Murray & Szostak (1983a) constructed a linear YRp vector and found that it did not exhibit maternal segregational bias. The model used by them to explain these results is as follows. During DNA replication there is an association between *ars* elements and fixed nuclear sites. Replicated molecules remain attached to this site which is destined to segregate to the mother cell. The replicated plasmids will exist initially as catenated dimers which subsequently are resolved. If the dimers are attached to the putative segregation site by only one of their constituent monomers, their resolution by topoisomerase activity would release one monomer from each dimer and this monomer would be free to segregate at random to either the mother or the daughter cell. With linear plasmids, replication will produce two linear molecules which are not interconnected. Thus, prior to mitosis there will be at least one freely segregating molecule and this explains the increased stability of linear YRp vectors.

Retrovirus-like vectors. The genome of *S. cerevisiae* contains 30–40 copies of a 5.9 kb mobile genetic element called Ty (for review see Fulton *et al.* 1987). This transposable element shares many structural and functional features with retroviruses (see p. 291) and the copia element of *Drosophila* (see p. 309). Ty consists of a central region containing two long, open reading frames (ORF) flanked by two identical terminal 334 base pair repeats called *delta* (Fig. 11.4). Each delta element contains a promoter as well as sequences recognized by the transposing enzyme. New copies of the transposon arise by a replicative process in which the Ty transcript is converted to a progeny DNA molecule by a Ty-encoded reverse transcriptase. The complementary DNA can transpose to many sites in the host DNA.

The Ty element has been modified *in vitro* by replacing its delta promoter sequence with promoters derived from the phosphoglycerate kinase or galactose-utilization genes (Mellor *et al.* 1985, Garfinkel *et al.* 1985). When such constraints are introduced into yeast on high copy number vectors the Ty element is over-expressed. This results in the formation of large numbers of virus-like particles (VLPs) which accumulate in the cytoplasm (Fig. 11.5). The particles, which have a diameter of 60–80 nm, have

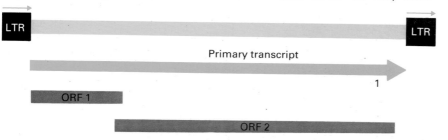

Fig. 11.4 Structure of a typical Ty element. ORF 1 and ORF 2 represent the two open reading frames. The delta sequences are indicated by LTR (long-terminal repeats).

208
Section 3
Cloning in
Organisms
Other than
E. coli

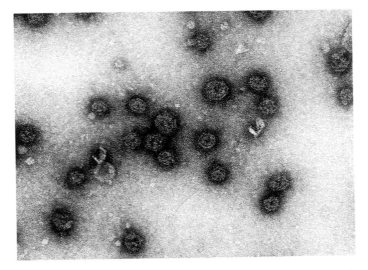

Fig. 11.5 Ty virus-like particles (magnification 80 000) carrying the entire HIV1 TAT coding region. (Photograph courtesy of Dr S. Kingsman.)

reverse transcriptase activity. The major structural components of VLPs are proteins produced by proteolysis of the primary translation product of ORF 1. Adams *et al.* (1987) have shown that fusion proteins can be produced in cells by inserting part of a gene from human immunodeficiency virus (HIV) into ORF 1. Such fusion proteins formed hybrid HIV:Ty-VLPs.

The Ty element also can be subjugated as a vector for transposing genes to new sites in the genome. The gene to be transposed is placed between the 3′ end of ORF 2 and the 3′ delta sequence (Fig. 11.6). Providing the inserted gene lacks transcription termination signals, transcription of the 3′ delta sequence will occur, which is a prerequisite for transposition. Such constructs act as amplification cassettes for once introduced into yeast transposition of the new gene occurs to multiple sites in the genome (Boeke *et al.* 1988, Jacobs *et al.* 1988).

Choice of vector for cloning

There are three reasons for cloning genes in yeast. The first of these relates to the potential use of yeast as a cloning host for the overproduction of proteins of commercial value. Yeast offers a number of advantages such as the ability to glycosylate proteins during secretion (see p. 219) and the absence of pyrogenic toxins. Commercial production demands overproduction and the factors affecting expression of genes in yeast are discussed in a later section (see p. 217).

A second reason for cloning genes in yeast is the ability to clone large pieces of DNA. Although there is no theoretical limit to the size of DNA which can be cloned in a bacterial plasmid, large recombinant plasmids exhibit structural and segregative instability. In the case of bacteriophage λ vectors the size of the insert is governed by packaging constraints. Many DNA sequences of interest are much larger than this. For example the

209
*Chapter 11
Cloning in
Yeast and
Other Microbial
Eukaryotes*

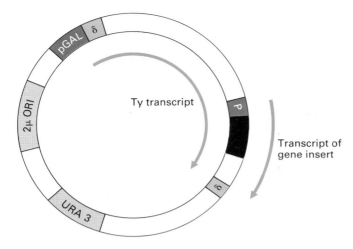

Fig. 11.6 Structure of the multicopy plasmid used for inserting a modified Ty element, carrying a cloned gene, into the yeast chromosome. pGAL and P are yeast promoters, δ represents the long-terminal repeats (delta sequences) and the black region represents the cloned gene. (See text for details.)

gene for blood factor VIII covers about 190 kbp or about 0.1% of the human X chromosome, and the Duchenne muscular dystrophy gene spans more than a megabase. Long sequences of cloned DNA will also facilitate efforts to sequence the human genome. Yeast artificial chromosomes offer a convenient way to clone large DNA fragments, since there is no practical size limitation to YACs (see p. 204). The method developed by Burke *et al.* (1987) is shown in Fig. 11.7.

For many biologists the primary purpose of cloning is to understand what particular genes do *in vivo*. Thus most of the applications of yeast vectors have been in the surrogate genetics of yeast. One advantage of cloned genes is that they can be analysed easily, particularly with the advent of DNA sequencing methods. Thus nucleotide sequencing analysis can reveal many of the elements which control expression of a gene as well as identifying the sequence of the gene product. In the case of the yeast actin gene (Gallwitz & Sures 1980, Ng & Abelson 1980) and some yeast tRNA genes (Abelson 1979, Olson 1981) this kind of analysis revealed the presence within these genes of non-coding sequences which are copied into primary transcripts. These *introns* subsequently are eliminated by a process known as *splicing* and this is discussed in more detail on p. 320. Nucleotide sequence analysis also can reveal the direction of transcription of a gene although this can be determined *in vivo* by other methods; for example, if the yeast gene is expressed in *E. coli* using bacterial transcription signals, the direction of reading can be deduced by observing the orientation of a cloned fragment required to permit expression. Finally, if a single transcribed yeast gene is present on a vector the chimaera can be used as a probe for quantitative solution hybridization analysis of transcription of the gene.

The availability of different kinds of vectors with different properties (see Table 11.1) enables yeast geneticists to perform manipulations in

Table 11.1 Properties of the different yeast vectors.

Vector	Transformation frequency	Loss in non-selective medium	Disadvantages	Advantages
YIp	1–10 transformants per µg DNA	much less than 1% per generation	(1) low transformation frequency (2) can only be recovered from yeast by cutting chromosomal DNA with restriction endonuclease which does not cleave original vector containing cloned gene	(1) of all vectors, this kind give most stable maintenance of cloned genes (2) an integrated YIp plasmid behaves as an ordinary genetic marker, e.g. a diploid heterozygous for an integrated plasmid segregates the plasmid in a Mendelian fashion. (3) most useful for surrogate genetics of yeast, e.g can be used to introduce deletions, inversions and transpositions (see Botstein & Davis 1982)
YEp	10^3–10^5 transformants per µg DNA	1% per generation	(1) novel recombinants generated *in vivo* by recombination with endogenous 2 µm plasmid	(1) readily recovered from yeast (2) high copy number (3) high transformation frequency (4) very useful for complementation studies
YRp	10^2–10^3 transformants per µg DNA	much greater than 1% per generation but can get chromosomal integration	(1) instability of transformants	(1) readily recovered from yeast (2) high copy number. Note that the copy number is usually less than that of YEp vectors but this may be useful if cloning gene whose produc deleterious to the cell if p d in excess.

211

Chapter 11
Cloning in
Yeast and
Other Microbial
Eukaryotes

Vector	Transformation frequency	Stability	Advantages	Disadvantages
			(3) high transformation frequency (4) very useful for complementation studies (5) can integrate into the chromosome	
YCp	10^2–10^3 transformants per µg DNA	less than 1% per generation	(1) low copy number is useful if product of cloned gene is deleterious to cell (2) high transformation frequency (3) very useful for complementation studies (4) at meiosis generally shows Mendelian segregation	(1) low copy number makes recovery from yeast more difficult than that of YEp or YRp vectors
YAC		depends on length: the longer the YAC the more stable it is	(1) high-capacity cloning system permitting DNA molecules greater than 40 kb to be cloned (2) can amplify large DNA molecules in a simple genetic background	(1) difficult to map by standard techniques
Ty	Depends on vector used to introduce Ty into cell	Stable, since integrated into chromosome	(1) get amplification following chromosomal integration	(1) needs to be introduced into cell in another vector

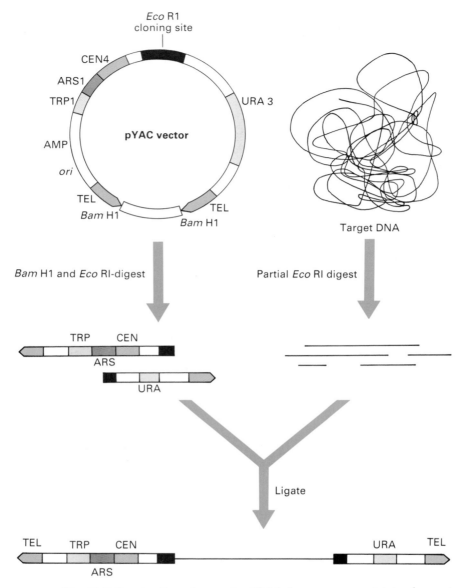

Fig. 11.7 Construction of a yeast artificial chromosome containing large pieces of cloned DNA. Key regions of the pYAC vector are as follows: TEL, yeast telomeres; ARS 1, autonomously replicating sequence; CEN 4, centromere from chromosome 4; URA 3 and TRP 1, yeast marker genes; Amp, ampicillin-resistance determinant of pBR322; ori, origin of replication of pBR322.

yeast like those long available to *E. coli* geneticists with their sex factors and transducing phages.

Thus cloned genes can be used in conventional genetic analysis by means of recombination using YIp vectors or linearized YRp vectors (Orr-Weaver *et al*. 1981). Complementation can be carried out using YEp, YRp, YCp or YAC vectors but there are a number of factors which make YCps the vectors of choice (Rose *et al*. 1987). For example, YEps and YRps exist at high copy number in yeast and this can prevent the isolation of genes whose products are toxic when overexpressed, e.g. the genes for actin and tubulin. In other cases the over-expression of genes other than the gene of interest can suppress the mutation used for selection (Kuo & Campbell 1983). All the yeast vectors can be used to create partial diploids or partial polyploids and the extra gene sequences can be integrated or extra-chromosomal. Deletions, point mutations and frame shift mutations can be introduced *in vitro* into cloned genes and the altered genes returned to yeast and used to replace the wild-type allele. Excellent reviews of these techniques have been presented by Botstein & Davis (1982), Hicks *et al.* (1982) and Struhl (1983).

213

Chapter 11
Cloning in
Yeast and
Other Microbial
Eukaryotes

Plasmid construction by homologous recombination in yeast

During the process of analysing a particular cloned gene it often is necessary to change the plasmid's selective marker. Alternatively it may be desired to move the cloned gene to a different plasmid, e.g. from a YCp to a YEp. Again, genetic analysis may require many different alleles of a cloned gene to be introduced to a particular plasmid for subsequent functional studies. All these objectives can be achieved by standard *in vitro* techniques, but Ma *et al.* (1987) have shown that methods based on recombination *in vivo* are much quicker. The underlying principle is that linearized plasmids are efficiently repaired during yeast transformation by recombination with a homologous DNA restriction fragment.

Suppose we wish to move the HIS 3 gene from pBR328, which cannot replicate in yeast, to YEp420 (see Fig. 11.8). Plasmid pRB328 is cut with *Pvu* I and *Pvu* II and the HIS 3 fragment selected. The HIS 3 fragment is mixed with YEp420 which has been linearized with *Eco*RI and the mixture transformed into yeast. Two cross-over events occurring between homologous regions flanking the *Eco*RI site of YEp420 will result in integration of the HIS3 gene. This gene can be selected directly. If this were not possible, selection could be made for the URA3 gene, for a very high proportion of the clones will also carry the HIS3 gene.

Many other variations of the above method have been described by Ma *et al.* (1987), to whom the interested reader is referred for details.

Expression of cloned genes in yeast

As might be expected, most cloned yeast genes are expressed when re-introduced into yeast. More surprising, some bacterial genes are also expressed in yeast (Cohen *et al.* 1980, Jimenez & Davies 1980) and in one instance expression was dependent on a bacterial promoter (Breunig *et al.* 1982). Since Struhl & Davis (1980) showed that a yeast promoter is functional in *E. coli* it might be thought that transcription signals such as promoters can be active in prokaryotes and eukaryotes. However, this clearly is not the case, for a number of workers failed to get expression of foreign genes in yeast. Thus Rose *et al.* (1981) obtained expression of β-galactosidase in *E. coli* when it was under the control of either the *E. coli* Tc^R promoter or the yeast *ura3* promoter but achieved expression in yeast

214

Section 3
Cloning in
Organisms
Other than
E. coli

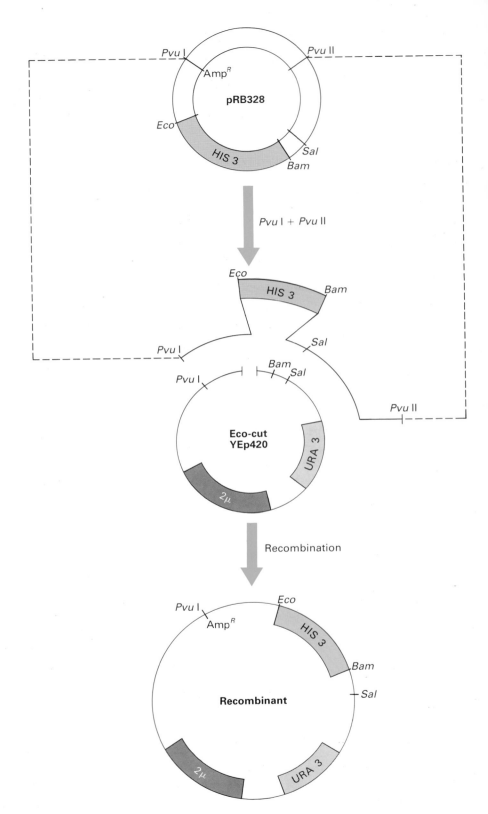

215

Chapter 11
Cloning in
Yeast and
Other Microbial
Eukaryotes

only with the latter promoter. When Beggs *et al.* (1980) introduced the rabbit β-globin gene into yeast, β-globin-specific transcripts were obtained but transcription started at a position downstream from the usual initiation site. Finally, even though a *Drosophila* gene corresponding to the yeast *ade* 8 locus has been identified by complementation, *Drosophila* genes complementing mutants at other yeast loci have not been obtained (Henikoff *et al.* 1981).

Use of yeast promoters. Because of difficulties in obtaining heterologous gene expression in yeast a number of groups have turned to the use of yeast promoters and translation initiation signals. Thus expression of the *E. coli* β-galactosidase gene was obtained by fusing it to the N-terminus of the *ura3, cyc1* and *arg3* genes (Guarente & Ptashne 1981, Rose *et al.* 1981, Crabeel *et al.* 1983). Expression in yeast of an interferon-alpha gene was obtained by fusing it to either the *pgk* or *trp1* genes (Tuite *et al.* 1982, Dobson *et al.* 1983). In many instances, expression of a mature protein rather than a fusion protein is required. To achieve this Hitzeman *et al.* (1981) started with a plasmid carrying the promoter and part of the coding sequence of the *adh*1 gene (Fig. 11.9). This plasmid was cut with endonuclease *Xho* I for which there is a unique cleavage site downstream from the initiating ATG codon. The linearized vector was digested with nuclease *Bal* 31 to remove 30–70 base pairs of DNA from each end of the molecule. The DNA then was incubated with the Klenow fragment of DNA polymerase and deoxynucleoside triphosphates to fill in the ends and synthetic *Eco* RI linkers added. After cleavage with endonucleases *Eco* RI and *Bam* HI the assorted fragments containing variously deleted *adh*1 promoter sequences were isolated by preparative electrophoresis. In this way six different promoter fragments were isolated, joined to an interferon-α gene and used to direct the synthesis of mature interferon in yeast. In a similar fashion synthesis of mature hepatitis B virus surface antigen was achieved from the *adh*1 promoter (Valenzuela *et al.* 1982) and mature interferon γ from the *pgk* promoter (Derynck *et al.* 1983).

The structure of yeast promoters

When the experiments described above were carried out little was known about the structure of yeast promoters. The general failure of bacterial and higher eukaryotic genes to be expressed in yeast when using their own promoters would suggest that yeast promoters have a unique structure. This is indeed the case (Guarante 1987). Four structural elements can be recognized in the average yeast promoter (Fig. 11.10). First, several consensus sequences are found at the transcription initiation site. Two of

Fig. 11.8 Plasmid construction by homologous recombination in yeast. pRB328 is digested with *Pvu* I and *Pvu* II and the HIS 3-containing fragment transformed into yeast along with the *Eco* RI-cut YEp420. Homologous recombination occurs between pBR322 sequences, shown as thin lines, to generate a new plasmid carrying both HIS 3 and URA 3.

216

*Section 3
Cloning in
Organisms
Other than
E. coli*

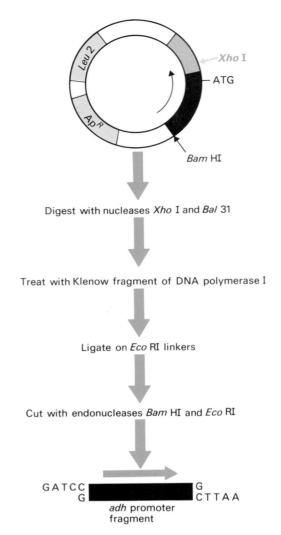

Fig. 11.9 The procedure used to generate *adh* promoter fragments. The *adh* promoter region in the starting plasmid is indicated by the black area. The grey area represents another gene fused to the N-terminus of the *adh* gene. The distance between the unique *Xho* I site and the initiating ATG codon is 17 bases. The arrow inside the plasmid circle (top) and above the *adh* promoter fragment (bottom) indicates the direction of transcription.

these sequences, TC(G/A)A and PuPuPyPuPu, account for more than half of the known yeast initiation sites (Hahn *et al.* 1985, Rudolph & Hinnen 1987). These sequences are not found at transcription initiation sites in higher eukaryotes which implies a mechanistic difference in their transcription machinery compared with yeast.

The second motif in the yeast promoter is the TATA box (Dobson *et al.* 1982). This is an AT-rich region with the canonical sequence (TATAT/AAT/A located 60–120 nucleotides before the initiation site. Functionally it can be considered equivalent to the Pribnow box of E. coli promoters (see p. 52).

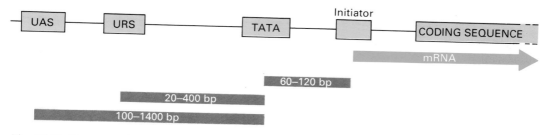

Fig. 11.10 Structure of typical yeast promoters. (See text for details.)

The third and fourth structural elements are upstream activating sequences (UASs) and upstream repressing sequences (URSs). These are found in genes whose transcription is regulated. Binding of positive control proteins to UASs turns up the rate of transcription and deletion of the UASs abolishes transcription. An important structural feature of UASs is the presence of one or more regions of dyad symmetry (Rudolph & Hinnen 1987). Binding of negative control proteins to URSs turns down the transcription rate of those genes that need to be negatively regulated.

Controlling gene expression in yeast

By analogy with *E. coli*, a number of factors can control the level of expression of genes in yeast. These are: (a) the copy number of the gene of interest and the stability of that copy number, (b) the nature of the promoter used, and (c) the sequence of the mRNA and the impact this has on mRNA structure and stability. One factor which does not appear to affect gene expression is codon bias. Bennetzen & Hall (1982) showed that there was an extreme codon bias in highly expressed yeast genes but high levels of expression have been obtained from foreign genes containing rare yeast codons (Cousens *et al.* 1987, Chen & Hitzeman 1987).

Copy number. The range of yeast vectors now available means that it is possible to vary gene copy number almost at will. Using YACs, YCps and YIps it is possible to stably maintain a single copy of a gene in yeast, or two copies if an existing functional yeast gene is cloned. Multiple copies can be obtained using YEps, YRps and Ty vectors, the latter providing stability of copy number as a result of their chromosomal dispersion.

Promoter choice. The structure of promoters and the mechanism of gene regulation in yeast differ considerably from *E. coli*. As a consequence many yeast promoters have been studied in order to elucidate essential structural features. Thus many different promoters have been cloned and sequenced. Some of these promoters are regulatable (Table 11.2) and this makes them more useful. If maximum expression of a protein is required, e.g. for commercial production, then a strong promoter is required. As yet there have been no comparative studies of the strength of yeast promoters. For such studies a reporter gene will be necessary and chloramphenicol acetyl transferase would appear to be particularly suitable (Hadfield *et al.*

218

*Section 3
Cloning in
Organisms
Other than
E. coli*

Table 11.2 Regulation of some yeast promoters.

Gene	Changes in growth medium for activation
acid phosphatase (PHO 5)	phosphate depletion
alcohol dehydrogenase (ADH 1)	glucose depletion
metallothionein (CUP 1)	addition of heavy metal
galactose utilization (GAL 1)	addition of galactose

1987, Mannhaupt *et al.* 1988). Each copy of the gene, whether present in multiple copies or as a single copy, gives rise to approximately 0.1% of the total soluble protein as chloramphenicol acetyl transferase.

The impact of mRNA sequence. From a survey of the translational initiation regions of 131 yeast genes Cigan & Donahue (1987) were able to conclude that yeast leader regions are rich in adenine nucleotides, have an average length of 52 nucleotides, and are void of secondary structure. There also is a bias in the sequences flanking the AUG start codons with the preferred sequence being 5'-(A/Py)A(A/U)A *AUG* UCU-3'-. Clearly, these gene features should be preserved when making new constructs in which genes are fused to yeast promoters.

There is considerable evidence that the sequence of the 3' untranslated region of yeast mRNA greatly affects the level of translation. When the 3' untranslated region of the yeast pyruvate kinase messenger RNA was truncated, or modified by the inclusion of sequences permitting the formation of hairpin-loop structures, the level of translation was greatly reduced (Purvis *et al.* 1987). These modifications had no effect on the stability of the mRNA indicating that it is the efficiency of translation that is affected.

Secretion of proteins by yeast

In yeast, proteins destined for the cell surface or for export from the cell are synthesized on, and translocated into, the endoplasmic reticulum. From there they are transported to the Golgi body for processing and packaging into secretory vesicles. Fusion of the secretory vesicles with the plasma membrane then occurs constitutively or in response to an external signal. Of the proteins naturally synthesized and secreted by yeast, only a few end up in the growth medium, e.g. the mating pheromone α factor and the killer toxin. The remainder, such as invertase and acid phosphatase, cross the plasma membrane but remain within the periplasmic space or become associated with the cell wall.

Polypeptides destined for secretion have a hydrophobic amino-terminal extension which is responsible for translocation to the endoplasmic reticulum (Blobel & Dobberstein 1975). The extension is usually composed of about 20 amino acids and is cleaved from the mature protein within the endoplasmic reticulum. Such signal sequences precede the mature yeast invertase and acid phosphatase sequences. Rather longer leader sequences precede the mature forms of the α mating factor and the killer toxin

(Kurjan & Herskowitz 1982, Bostian *et al.* 1984). The initial 20 or so amino acids are similar to the conventional hydrophobic signal sequences but cleavage does not occur in the endoplasmic reticulum. In the case of α factor, which has an 89 amino acid leader sequence, the first cleavage occurs after a Lys−Arg sequence at positions 84 and 85 and happens in the Golgi body (Julius *et al.* 1983, 1984).

219
*Chapter 11
Cloning in
Yeast and
Other Microbial
Eukaryotes*

To date, a large number of non-yeast polypeptides have been secreted from yeast cells containing the appropriate recombinant plasmid but the rules governing secretion still are not clear. For example, Hitzeman *et al.* (1983) obtained secretion of interferon when the construct carried the endogenous mammalian signal sequence. Sequencing of interferon purified from the growth medium showed it to have the same amino-terminus as natural mature interferon, indicating that the yeast cells had recognized and correctly processed the mammalian signal sequence. Similar results have been obtained with many other mammalian and plant proteins. By contrast, no secretion into the medium was observed with calf prochymosin (Mellor *et al.* 1983) or human α_1-antitrypsin (Cabezon *et al.* 1984) with their natural signal sequences. In the latter case, when the normal signal sequence was replaced with that from the yeast invertase gene, secretion of α_1-antitrypsin was observed (Moir & Durmais 1987).

Interpreting the above results is difficult. First, a whole series of different gene constructs has been used and in many instances little attention has been paid to the level of expression. If a cloned gene is expressed poorly, secretion might not be detectable, particularly if the medium contains proteases. Secondly, most workers have failed to accurately quantitate the levels of protein inside the cell, in the periplasmic space and in the growth medium. Nor have they measured the leakage of intracellular or periplasmic proteins into the medium. Without this information it is difficult to measure the efficiency of secretion.

Targeting of proteins to the nucleus. The yeast nucleus contains a discrete set of proteins which are synthesized in the cytoplasm. In order to elucidate the mechanism governing protein localization Hall *et al.* (1984) constructed a set of hybrid genes by fusing the yeast MAT α gene, encoding a presumptive nuclear protein, and the *E. coli lacZ* gene. A segment of the MAT gene product which was 13 amino acids long was sufficient to localize β-galactosidase activity in the nucleus. The nuclear location of the β-galactosidase was confirmed by immunofluorescence.

Glycosylation of proteins

Many eukaryotic proteins are glycosylated. The mechanism for glycosylation resides in the endoplasmic reticulum and Golgi apparatus so glycosylation can occur only if the proteins are directed to these sites via the secretory pathway. In both yeast and mammalian cells an inner core of sugars is added to the protein in the endoplasmic reticulum. This involves the addition of a dolichol-linked oligosaccharide to asparagine residues which are part of the sequence Asn−x−Ser/Thr. Subsequently the protein is transferred to the Golgi apparatus where an outer core of sugars is as-

220

*Section 3
Cloning in
Organisms
Other than
E. coli*

sembled. In yeast this occurs by stepwise addition of mannose residues. In mammalian cells the outer core can be complex and branched, with sugars other than mannose included in the chain after trimming of the outer core (Dunphy & Rothman 1985).

From the foregoing discussion it should come as no surprise that many of the foreign proteins secreted by yeast are glycosylated. Where a detailed analysis of the secreted proteins has been made glycosylation has occurred at the expected sites (Moir & Dumais 1987, Penttila *et al.* 1988, Yoshizumi & Ashikari 1987) but the composition and sequence of the attached oligosaccharide differs from that occurring naturally. In most instances the unusual glycosylation has no influence on the properties of the protein *in vitro*. However, for proteins which are to be used therapeutically, the difference in glycosylation could affect stability, immunogenicity, and tissue distribution *in vitro*.

Excision of introns by yeast

Many eukaryotic genes contain non-coding regions called introns. Introns have two common structure features: their sequences begin with the dinucleotide 5'-GT-3' and end with the dinucleotide 5'-AG-3'. Besides these invariant nucleotides there is limited structural similarity at and around the intron−exon junction and consensus sequences have been derived from comparison of more than 100 junctions. Because of the similarity of these junction sequences in widely different species, e.g. yeast and man, it was thought that the mechanism of RNA processing to remove introns might be universal. Support for this idea came from the observation that monkey cells can splice out introns from mouse and rat genes, and mouse cells correctly splice transcripts of rabbit and chicken genes. This raises the question whether yeast cells can remove introns from genes of higher eukaryotes. If so this would be of great practical value, for the presence of introns prevents shotgun cloning of functional eukaryotic genes in *E. coli*.

To test the ability of yeast cells to excise introns from foreign genes Beggs *et al.* (1980) transformed *S. cerevisiae* with a hybrid plasmid containing a cloned rabbit chromosomal DNA segment including a complete β-globin gene with two intervening sequences and extended flanking regions. Yeast cells transformed with this chimaera produced β-globin-specific mRNA. However, these globin transcripts were about 20−40 nucleotides shorter at the 5' end than normal globin mRNA, contained one intron and extended only as far as the first half of the second intron. This result could be taken to indicate that the splicing mechanisms in yeast and rabbit differ, but it could be argued that a complete transcript is a prerequisite for RNA splicing and that the prematurely terminated globin RNA was not a substrate for the yeast splicing enzyme(s). Consequently, Langford *et al.* (1983) inserted into the intron-containing yeast actin gene an intron-containing fragment from either *Acanthamoeba* or duck. In both instances yeast cells removed the natural yeast intron but not the foreign intron from the chimaeric transcript.

Why are introns in foreign genes not removed by yeast cells? The sequence of the coding regions surrounding the splice site cannot be important, for Teem & Rosbash (1983) have inserted a yeast intron into the *E. coli* β-galactosidase gene and found that it is removed correctly in yeast cells. The most likely explanation is that sequences within the intron are required for correct splicing in yeast. In this context Langford & Gallwitz (1983) have found an octanucleotide (5'-TACT AACA-3') which occurs 20–55 nucleotides upstream from the 3' splice site in all split protein-coding genes of *S. cerevisiae* analysed and which is absent from most introns of higher eukaryotes. Single A→C transversions in the fourth or eighth position of this sequence prevent splicing from occurring (Langford *et al.* 1984).

221

Chapter 11
Cloning in
Yeast and
Other Microbial
Eukaryotes

Cloning in filamentous fungi and other yeasts

There is an interest in cloning in filamentous fungi and other yeasts for a variety of reasons. For example, to understand the biochemical mechanisms involved in conidiation in fungi (Timberlake & Marshall 1988), to understand the basis of plant pathogenicity or because the organisms are used industrially (Upshall 1986). An essential step in applying recombinant DNA technology to a species is to develop a DNA mediated transformation system and Table 11.3 lists those fungi where this has been achieved.

Efficient transformation occurs with protoplasts and the transforming DNA becomes integrated into the chromosome by homologous recombination. For *Aspergillus* there is no evidence for replicating extrachromosomal plasmids either in the nucleus or mitochondria, but freely replicating plasmids have been identified in *Neurospora*. Genes of interest can be cloned by complementation of mutations or by making use of positive selection. For example, a plasmid containing the *amdS* gene of *Aspergillus nidulans* may be used to transform *amdS*+ strains by selecting for increased utilization of acetamide as sole nitrogen source. Analysis of transformants (Kelly & Hynes 1987) has shown that multiple tandem copies of the plasmid can be integrated into the chromosome, commonly at sites other than the *amdS* locus. Such transformants are relatively stable through meiosis and mitosis. Finally, reporter genes can be fused to regulatory sequences and used to assay the effects of promoter mutations made *in vitro* (Hamer & Timberlake 1987) and production and secretion of human tissue plasminogen activator can be obtained from *A. nidulans* (Upshall *et al.* 1988). These results suggest that the techniques needed for cloning in fungi in general are little different from those for *S. cerevisiae*.

Table 11.3 Some fungal species known to be transformable.

Saccharomyces cerevisiae	*Pichia pastoris*
Schizosaccharomyces pombe	*Penicillium chrysogenum*
Aspergillus nidulans	*Achlya ambisexualis*
Aspergillus niger	*Glomerella cingulata*
Neurospora crassa	*Coprinus cinereus*
Mucor circinelloides	*Hansenula polymorpha*
Podospora anserina	*Kluyveromyces lactis*

12 Gene Transfer to Plants

The past decade has seen a revolution in our ability to manipulate the genomes of plants. Before the early 1980s there had been no well-substantiated report of expression of a foreign gene in a genetically engineered plant. The situation has changed dramatically since then. Now it is possible to transfer any gene into a plant as a routine procedure. This has come about through very intensive research into vector systems based on the bacterium *Agrobacterium tumefaciens* and through an alternative strategy based on transfection of plant protoplasts (i.e. plant cells from which the cell wall has been removed). This change means that plant molecular biology can advance rapidly and has immense biotechnological implications in the creation of plants with useful genetically engineered characteristics such as herbicide resistance (Chapter 16).

The range of species available for routine transgenic plant production has been limited. This is for two reasons. Firstly, *Agrobacterium*-based gene transfer is efficient only with dicotyledonous plants ('dicots'). Secondly, transgenic plant production by either *Agrobacterium*-based systems or protoplast transfection requires the ability to regenerate plants from single (or a small number of) isolated transfected cells. This regeneration technology has been developed to an advanced state in a small number of 'favourite' experimental species; tobacco, tomato and petunia are the most commonly used. Some agriculturally important dicots are not such well-developed experimental systems in this regard—the legumes are an example. With monocots there is the problem that *Agrobacterium* does not interact productively with them under natural circumstances. With the cereal crop species, members of the Graminae, there has been little success in the regeneration of plants from isolated cells. Therefore until recently the cereals have been inaccessible to DNA transfer. This situation is likely to change with the intensive effort it is receiving.

In order to explain these developments this chapter begins with an account of plant callus and protoplast culture. We then describe *Agrobacterium*-based and protoplast transfection systems, and important developments with monocot species. Finally, we briefly discuss the roles of plant viruses in vector systems.

Plant callus culture

Tissue culture is the process whereby small pieces of living tissue (*explants*) are isolated from an organism and grown aseptically for indefinite periods on a nutrient medium. For successful plant tissue culture it is best to start with an explant rich in undetermined cells, e.g. those from the cortex or meristem, because such cells are capable of rapid proliferation. The usual explants are buds, root tips, nodal segments or germinating seeds, and these are placed on suitable culture media where they grow into an undifferentiated mass known as a *callus* (Fig. 12.1). Because the nutrient media used can support the growth of microorganisms the explant is first washed with a disinfectant such as sodium hypochlorite, hydrogen peroxide or mercuric chloride. Once established, the callus can be propagated indefinitely by subdivision (Fig. 12.2).

For plant cells to develop into a callus it is essential that the nutrient medium contain plant hormones, phytohormones, i.e. an auxin, a cytokinin and a gibberellin (Fig. 12.3). The absolute amounts of these which are required vary for different tissue explants from different parts of the same plant and for the same explant from different genera of plants. Thus there is no 'ideal' medium. Most of the media in common use consist of inorganic salts, trace metals, vitamins, organic nitrogen sources (glycine), inositol, sucrose and growth regulators. Organic nutrients such as casein hydrolysate or yeast extract and a gelling agent are optional extras. The composition of a typical plant growth medium is shown in Table 12.1.

Plant cell culture and protoplasts

When a callus is transferred to a liquid medium and agitated the cell mass breaks up to give a suspension of isolated cells, small clusters of cells and much larger aggregates. Such suspensions can be maintained indefinitely by subculture but, by virtue of the presence of aggregates, are extremely heterogeneous. A high degree of genetic instability adds to this heterogeneity. Some plants such as *Nicotiana tabacum* (tobacco) and *Glycine max* (soybean) yield very friable calluses and cell lines obtained from these species are much more homogeneous and can be cultivated both batchwise and continuously.

When placed in a suitable medium isolated single cells from suspension cultures are capable of division. As with animal cells, for proliferation to occur conditioned medium may be necessary. Conditioned medium is

Fig. 12.1 Close-up view of a callus culture.

224

*Section 3
Cloning in
Organisms
Other than
E. coli*

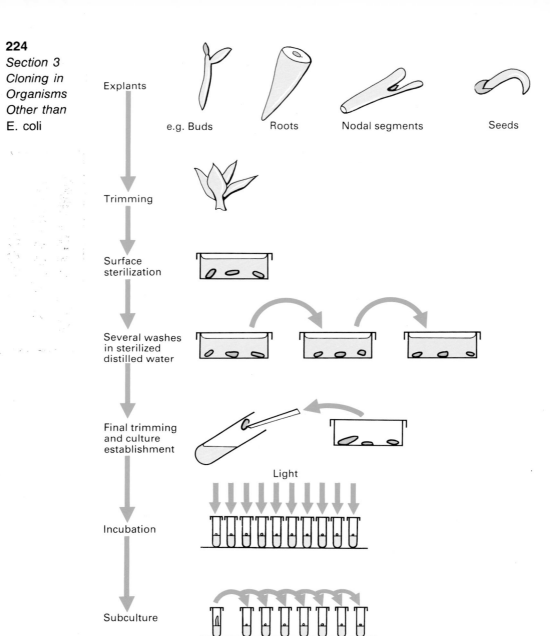

Explants

e.g. Buds   Roots   Nodal segments   Seeds

Trimming

Surface
sterilization

Several washes
in sterilized
distilled water

Final trimming
and culture
establishment

Light

Incubation

Subculture

Fig. 12.2 Basic procedure for establishing and maintaining a culture of plant tissue.

prepared by culturing high densities of cells of the same or different species in fresh medium for a few days and then removing the cells by filter sterilization. Media conditioned in this way contain essential amino acids such as glutamine and serine as well as growth regulators like cytokinins. Provided conditioned medium is used, single cells can be plated out on solid media in exactly the same way as microorganisms; instead of forming a colony as do microbes, plant cells proliferate to give a callus.

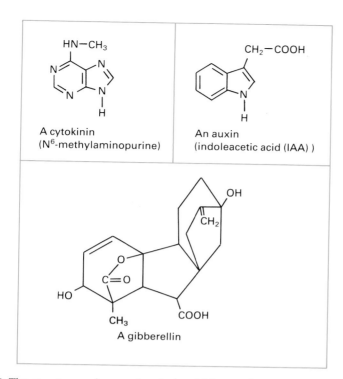

Fig. 12.3 The structures of some chemicals which are plant growth regulators, phytohormones.

Table 12.1 Composition of Murashige & Skoog (MS) culture medium.

Ingredient	Amount (mg/l)
Sucrose	30 000
$(NH_4)NO_3$	1 650
KNO_3	1 900
$CaCl_2.2H_2O$	440
$MgSO_4.7H_2O$	370
KH_2PO_4	170
$FeSO_4.7H_2O$	27.8
Na_2EDTA	37.3
$MnSO_4.4H_2O$	22.3
$ZnSO_4.7H_2O$	8.6
H_3BO_3	6.2
KI	0.83
$Na_2MoO_4.2H_2O$	0.25
$CoCl_2.6H_2O$	0.025
$CuSO_4.5H_2O$	0.025
myo-inositol	100
glycine	2.0
kinetin (a cytokinin)	0.04–10.0
Indoleacetic acid	1.0–30.0

226

*Section 3
Cloning in
Organisms
Other than
E. coli*

Protoplasts. Protoplasts are cells minus their cell walls. They are very useful materials for plant cell manipulations because under certain conditions those from similar and contrasting cell types can be fused to yield somatic hybrids, a process known as *protoplast fusion*. Protoplasts can be produced from suspension cultures, callus tissue or intact tissues, e.g. leaves, by mechanical disruption or, preferably, by treatment with cellulolytic and pectinolytic enzymes. Pectinase is necessary to break up cell aggregates into individual cells and the cellulase to remove the cell wall proper. After enzyme treatment protoplast suspensions are collected by centrifugation, washed in medium without enzyme, and separated from intact cells and cell debris by flotation on a cushion of sucrose (Fig. 12.4). When plated onto nutrient medium protoplasts will in 5–10 days synthesize new cell walls and then initiate cell division.

As indicated earlier, the formation and maintenance of callus cultures requires the presence of a cytokinin and an auxin, whereas only a cytokinin is required for shoot culture and only an auxin for root culture. Therefore it is no surprise that increasing the level of cytokinin to a callus induces shoot formation and increasing the auxin level promotes root formation. Ultimately plantlets arise through development of adventitious roots on the shoot buds formed, or through development of shoot buds from tissues formed by proliferation at the base of rootlets (Fig. 12.5). The formation of roots and shoots on callus tissue is known as *organo-genesis*. The cultural conditions required to achieve organogenesis vary from species to species, and have not been determined for every type of callus.

Under certain cultural conditions calluses can be induced to undergo a different development process known as somatic embryogenesis. In this process the callus cells undergo a pattern of differentiation, similar to that seen in zygotes after fertilization, to produce *embryoids*. Such cells are embryo-like but differ from normal embryos in being produced from *somatic* cells and not from the fusion of two germ cells. These embryoids can develop into fully functional plants without the need to induce root and shoot formation on artificial media. The embryogenic response leading to embryoid formation is stimulated when calluses which were established in medium in which 2,4-dichlorophenoxyacetic acid (2,4-D) was the auxin are transferred to 2,4-D-free medium containing reduced sources of nitrogen, e.g. ammonium salts.

AGROBACTERIUM AND GENETIC ENGINEERING IN PLANTS

Crown gall disease

Crown gall is a plant tumour, a lump of undifferentiated tissue, which can be induced in a wide variety of gymnosperms and dicotyledonous angiosperms by inoculation of wound sites with the Gram-negative soil bacterium *Agrobacterium tumefaciens* (Fig. 12.6). The disease was first described long ago, and the involvement of bacteria was recognized as early as 1907 (Smith & Townsend 1907). It was subsequently shown that the crown gall

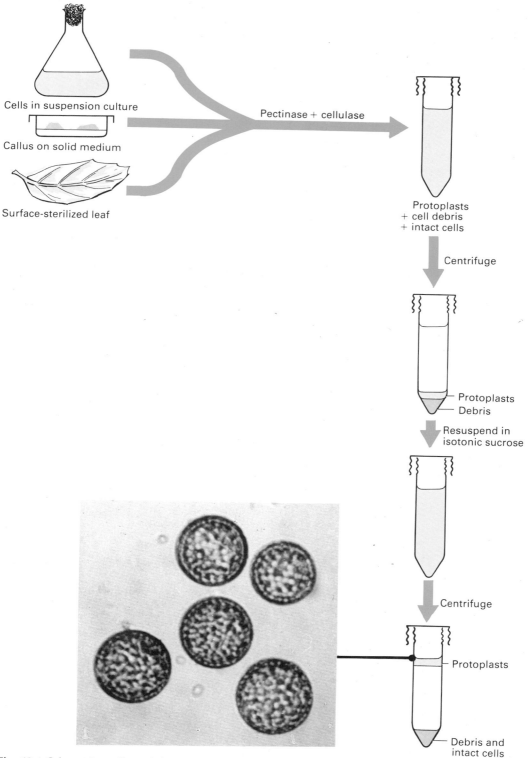

Cells in suspension culture

Callus on solid medium

Surface-sterilized leaf

Pectinase + cellulase

Protoplasts
+ cell debris
+ intact cells

Centrifuge

Protoplasts
Debris

Resuspend in
isotonic sucrose

Centrifuge

Protoplasts

Debris and
intact cells

Fig. 12.4 Schematic outline of the enzymatic procedure used to isolate plant protoplasts. The inset shows a photomicrograph of protoplasts.

228

Section 3
Cloning in
Organisms
Other than
E. coli

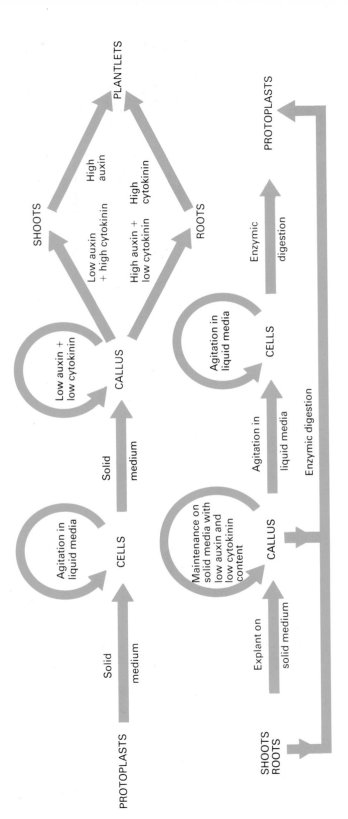

Fig. 12.5 Summary of the different cultural manipulations possible with plant cells, tissues and organs.

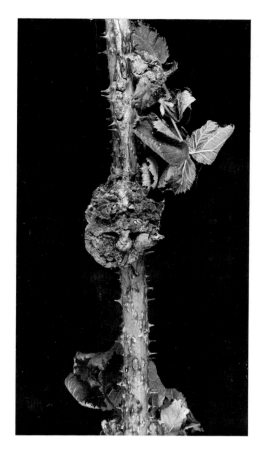

Fig. 12.6 Crown gall on blackberry cane. (Photograph courtesy of Dr C. M. E. Garrett, East Malling Research Station.)

tissue represents true oncogenic transformation; callus tissue can be culti-vated *in vitro* in the absence of the bacterium and yet retain its tumorous properties (Fig. 12.7). These properties include the ability to form an overgrowth when grafted onto a healthy plant, the capacity for unlimited growth as a callus in tissue culture in media devoid of the plant hormones necessary for *in-vitro* growth of normal plant cells, and the synthesis of *opines*, which are unusual amino acid derivatives not found in normal tissue (Fig. 12.8). The most common of these opines are octopine and nopaline. In addition, agropine or the agrocinopine family of opines may be present. Crown gall tumour cells continue to synthesize opines in tissue culture, and shoots or whole plants regenerated from tumour cell lines may also continue to synthesize opine (Braun & Wood 1976, Schell & Van Montagu 1977, Wullems *et al.* 1981a, Wullems *et al.* 1981b).

The metabolism of opines is a central feature of crown gall disease. Opine synthesis is a property conferred upon the plant cell when it is transformed by *A. tumefaciens*. The type of opine produced is determined not by the host plant but by the bacterial strain. In general, the bacterium induces the synthesis of an opine which it can catabolize and use as its sole energy, carbon and/or nitrogen source. Thus, bacteria that utilize

230
Section 3
Cloning in
Organisms
Other than
E. coli

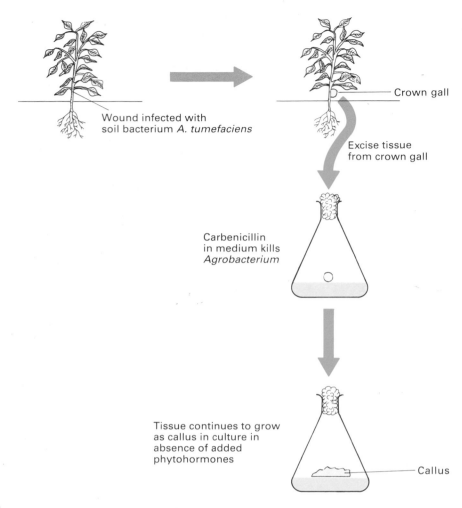

Wound infected with
soil bacterium *A. tumefaciens*

Crown gall

Excise tissue
from crown gall

Carbenicillin
in medium kills
Agrobacterium

Tissue continues to grow
as callus in culture in
absence of added
phytohormones

Callus

Fig. 12.7 *Agrobacterium tumefaciens* induces plant tumours.

octopine induce tumours that synthesize octopine, and those that utilize
nopaline induce tumours that synthesize nopaline (Bomhoff *et al.* 1976,
Montaya *et al.* 1977).

Investigation of the molecular biology of crown gall disease has revealed
that *A. tumefaciens* has evolved a natural system for genetically engineering
plant cells so as to subvert them for its own ends. Recent research, mainly
using tobacco, tomato and petunia as the experimental plants, has enabled
gene manipulators to exploit this natural system.

The tumour-inducing principle and the Ti-plasmid

As it was clear that the continued presence of bacteria is not required for
transformation of the plant cells, attention was focused on the nature of

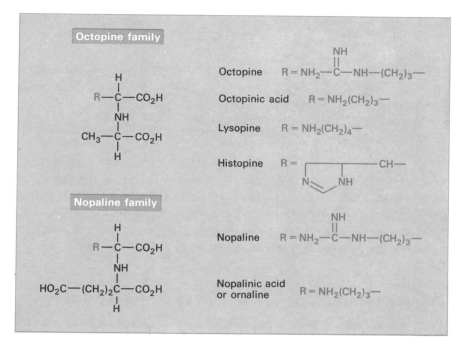

Fig. 12.8 Structures of some opines.

the 'tumour-inducing principle', the name given to the putative genetic element that must be transferred from the bacterium to the plant at the wound site. For a long time it was believed, correctly, that DNA is transferred from the bacterium to the plant cell. Attempts to detect such bacterial DNA in the tumour cells failed, simply because the techniques used were not sufficiently sensitive, until plasmids were detected in *Agrobacterium*. Zaenen *et al.* (1974) first noted that tumour-forming (i.e. virulent) strains of *A. tumefaciens* harbour large plasmids (140–235 kb), and it is now clear that the virulence trait is plasmid borne. Virulence is lost when the bacteria are cured of the plasmid and with at least one strain this can be achieved by growing the cells at 37°C instead of 28°C. Cured strains also lose the capacity to utilize octopine or nopaline (Van Larbeke et al. 1974, Watson *et al.* 1975). Virulence is acquired by avirulent strains when a virulence plasmid is reintroduced by conjugation (Bomhoff *et al.* 1976, Gordon *et al.* 1979). If the plasmid from a nopaline strain is transferred to an avirulent derivative of a previously octopine strain, the avirulent strain then acquires the ability to induce nopaline tumours and catabolize nopaline. The virulence plasmid can also be transferred to the legume symbiont *Rhizobium trifolii* which becomes oncogenic and acquires the ability to utilize either octopine or nopaline, depending on the donor.

From the above information it is clear that plasmids are essential for virulence and for this reason they are referred to as Ti (tumour-inducing) plasmids. Furthermore, the genetic information specifying bacterial utilization of opines, and their synthesis by plants, is also plasmid-borne. It should be remembered, however, that the presence of a plasmid in *A. tumefaciens* does not mean that the strain is tumorigenic. Many strains

232

*Section 3
Cloning in
Organisms
Other than
E. coli*

contain very large cryptic plasmids that do not confer virulence, and in some natural isolates a cryptic plasmid is present together with a Ti-plasmid.

Plasmids in the octopine group have been shown to be closely related to each other while those in the nopaline group are considerably more diverse (Currier & Nester 1976, Sciaky *et al.* 1978). Between these two groups there is little DNA homology, except for four limited regions, one of which includes the genes directly responsible for crown gall formation; T-DNA (Drummond & Chilton 1978, Engler *et al.* 1981).

A related organism, *A. rhizogenes*, incites a disease, hairy root disease, in dicotyledonous plants in a manner very similar to *A. tumefaciens*. A large plasmid, the Ri (root inducing) plasmid, is responsible for pathogenicity and the induction of opine synthesis (White & Nester 1980, Chilton *et al.* 1982). The Ri plasmids share little homology with Ti-plasmids. They are of interest because tissue transformed by *A. rhizogenes* readily regenerates into plantlets which continue to synthesize opine (see p. 242).

Incorporation of T-DNA into the nuclear DNA of plant cells

Complete Ti-plasmid DNA is not found in plant tumour cells but a small, specific segment of the plasmid, about 23 kb in size, is found integrated in the plant nuclear DNA (Chilton *et al.* 1977). This DNA segment is called T-DNA (transferred DNA; Fig. 12.9). The structure and organization of the integrated T-DNA in tumour cells has been studied in detail. The main conclusions (Thomashow *et al.* 1980, Zambryski *et al.* 1980) of these studies are listed as follows.

1 Integration of the T-DNA can occur at many different, apparently random, sites in the plant nuclear DNA.

2 The organization of the integrated nopaline T-DNA is simple. It occurs as a single integrated segment.

3 Integrated octopine T-DNA usually exists as two segments. The left segment, TL, includes genes necessary for tumour formation. The right segment, TR, is not necessary for tumour maintenance but codes for enzymes for agropine biosynthesis (Saloman *et al.* 1984). Usually only one copy of TL-DNA is present per cell but as many as ten copies have been found. TR

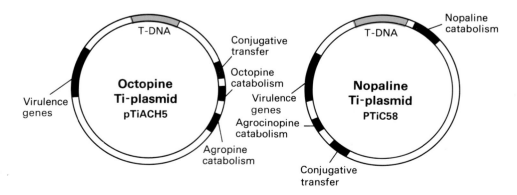

Fig. 12.9 Ti-plasmid gene maps.

may be present in high copy numbers. The significance of these complications in integrated octopine T-DNA structure is not clear. It is likely that rearrangement, amplifications and deletions of T-DNA may follow the initial integration.

As we have noted, the integrated nopaline DNA occurs as a single, 23 kb, segment. Regions including the junctions of nopaline T-DNA with plant DNA have been cloned in phage λ vectors and the DNA sequences at the junctions determined. The righthand junction with plant DNA appears to be rather precise, whereas the left-hand junction can vary by about 100 nucleotides (Yadav *et al.* 1982, Zambryski *et al.* 1982). On the Ti-plasmid itself the T-region is flanked by imperfect, direct 25 base-pair repeats. The repeats are not transferred intact to the plant genome, as the T-DNA endpoints lie within, or immediately internal to, them. These repeats are conserved between nopaline and octopine Ti-plasmids while the sequences surrounding are not. This strongly suggested that the repeats are important in the integration mechanism (Simpson *et al.* 1982, Yadav *et al.* 1982). Deletions that remove the T-region right repeat abolish tumour formation. But, perhaps surprisingly, the left repeat can be removed without affecting tumour formation. Thus only the right repeat is essential, and in addition it has been shown that sequences further to the right and adjacent to the actual repeat sequence are necessary for full activity in transfer and integration into the plant genome (Shaw *et al.* 1984; Peralta *et al.* 1986). The left repeat has little or no transfer activity alone (Jen & Chilton 1986).

The mechanisms by which the T-DNA region of the Ti-plasmid becomes integrated into the plant genome is only partially understood. It appears to closely resemble bacterial conjugation. The genes mediating DNA transfer are not contained in the T-DNA itself but are found in the virulence (*vir*) region of Ti-plasmids. This region contains at least seven transcomplementing genes and large stretches of this region show sequence similarity in octopine and nopaline plasmids. Certain genes of the *vir* region are transcriptionally activated by specific signal molecules that are produced by wounded, metabolically active, but not intact, plant cells. For tobacco these signal molecules are known. They are the phenolic compounds acetosyringone and α-hydroxyacetosyringone (Fig. 12.10) (Stachel *et al.* 1985). It appears that two of the *vir* genes, *vir*A and *vir*G, control the plant-induced activation. *Vir*A is constitutively transcribed and non-inducible. It is thought to encode a transport protein for the signal molecules. *Vir*G is regulated in a complex fashion and encodes a positive regulatory protein that, together with the signal molecule, activates the expression of other *vir* genes, probably by stimulating their transcription. The *vir*G protein product is closely related to the *E. coli omp*^{R} gene product. This protein is required for transcriptional activation of several *E. coli* promoters (Stachel & Zambryski 1986). In addition to these plasmid-borne genes, genes located on the bacterial chromosome have also been identified as essential for T-DNA transfer.

In response to the activating signal molecule, an endonuclease encoded by *vir*D makes two nicks in the T-DNA in the same strand at its border sequences in the Ti-plasmid. The T-DNA single strand is transferred with

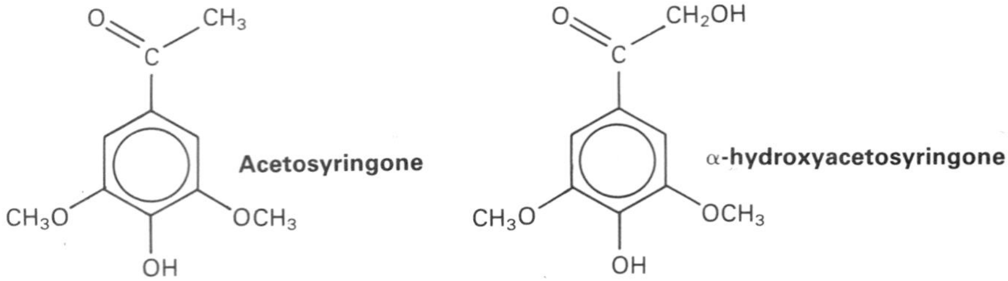

234

Section 3
Cloning in
Organisms
Other than
E. coli

Fig. 12.10 Structures of signal molecules, produced by wounded plant tissue, which activate T-DNA transfer by *Agrobacterium tumefaciens*.

the right border leading (Yanofsky *et al.* 1986, Stachel *et al.* 1986). This probably accounts for the observation that only the right repeat sequence is essential for transfer. The left repeat probably defines the other end of the transferred DNA by virtue of the nick. If the left repeat is absent, random breakage may define the amount of DNA transferred. At some point in the transfer process the single-stranded DNA must be converted into double-stranded form. In bacterial conjugation this conversion occurs in the recipient cell so if the resemblance with T-DNA transfer holds good, we would predict that the T-DNA enters the plant cell in single-stranded form and that complementary strand is synthesized by plant cell enzymes. The integration of the T-DNA is not precise to the nucleotide but, as we have noted, it is more precise at the right border than at the left. The significance of this is unclear, as is the observation that separate T-region circles are formed in *Agrobacterium* induced for gene transfer. These circles contain one copy of the border repeat sequences (Kuokolikova-Nicola *et al.* 1985). Clearly the ability of *Agrobacterium* to transfer DNA right into the nucleus of the plant cell requires a very intimate association between the plant cell and the bacterium. The details of the fascinating relationship have yet to be discovered.

Gene maps and expression of T-DNA

Genetic maps of Ti-plasmids and T-DNA have been obtained by the study of spontaneous deletions and by transposon mutagenesis (Kookman *et al.* 1979, Garfinkel *et al.* 1980, Holsters *et al.* 1980, Ooms *et al.* 1980, De Greve *et al.* 1981, Ooms *et al.* 1981). Such studies first revealed the large *vir* region mapping outside the T-DNA. As expected, regions of the T-DNA itself were found to affect tumour morphology (Fig. 12.11). The loci affecting tumour morphology are designated 'large' (*tml*), 'shooty' (*tms*) and 'rooty' (*tmr*) to indicate the phenotype of the callus obtained when the loci are inactivated. Ooms *et al.* (1981) proposed that certain mutations in T-DNA appear to affect plant hormone concentrations in the resulting tumours so as to produce rooty and shooty types. The basis for this proposal was the observation that although the tumour grows in media lacking added auxins and cytokinins, the callus actually contains high concentrations of these hormones. Indeed, uncloned primary tumours contain some normal, untransformed cells growing as undifferentiated callus owing to the presence of these endogenous hormones (Gordon 1980). The observations

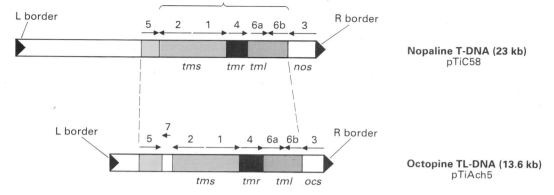

Fig. 12.11 Structure and transcription of T-DNA. The T-regions of nopaline and octopine Ti-plasmids have been aligned to indicate the DNA sequences that are common to both T-regions. The size and orientation of each transcript (numbered) is indicated by arrows. Genetic loci, as defined by deletion and transposon mutagenesis, as follows: *nos* : nopaline synthase

           ocs : octopine synthase

          tms : shooty tumour

          tmr : rooty tumour.

suggested the idea that genes within the T-DNA actually encoded enzymes involved in phytohormone biosynthesis. As we shall see, this idea has been proved correct.

Figure 12.11 shows a transcript map of nopaline and octopine T-DNAs (Willmitzer *et al.* 1982, Willmitzer *et al.* 1983, Winter *et al.* 1984). The transcript encoding octopine synthase (also called lysopine dehydrogenase) has been identified and its gene located at the right end of TL-DNA. Nucleotide sequencing has shown that this gene, *OCS*, has promoter elements which are eukaryotic in character, that it lacks introns, and that it contains a eukaryotic polyadenylation signal near its 3' end (De Greve *et al.* 1982b). Essentially similar results have been obtained for the nopaline synthase gene, *nos* (Depicker *et al.* 1982, Bevan *et al.* 1983a). It is possible that the genes of T-DNA are eukaryotic in origin and have been 'captured' by the Ti plasmid during its evolution. The *ocs* and *nos* gene promote function in a wide variety of plant cells and have been widely used in constructing plant vectors (see below).

Two of the transcripts, numbered 1 and 2, map to genes (*tms*) involved in auxin production. Transcript 4 maps to a gene (*tmr*) involved in cytokinin production. Transcripts 5 and 6 encode products which appear to suppress differentiation only in cells in which they are expressed, i.e. in a non-hormonal manner.

It has been demonstrated directly (Schroder *et al.* 1984, Akiyoshi *et al.* 1984, Klee *et al.* 1984, Weiler & Schroder 1987) that in fact transcripts 1 and 2 from the auxin genes encode two enzymes, tryptophan 2-mono-oxygenase and indoleacetamide hydrolase, which convert tryptophan to indole acetic acid (Fig. 12.12). Transcript 4, from the cytokinin gene, encodes isopentenyl transferase, an enzyme that catalyses synthesis of zeatin-type cytokinins (Fig. 12.13).

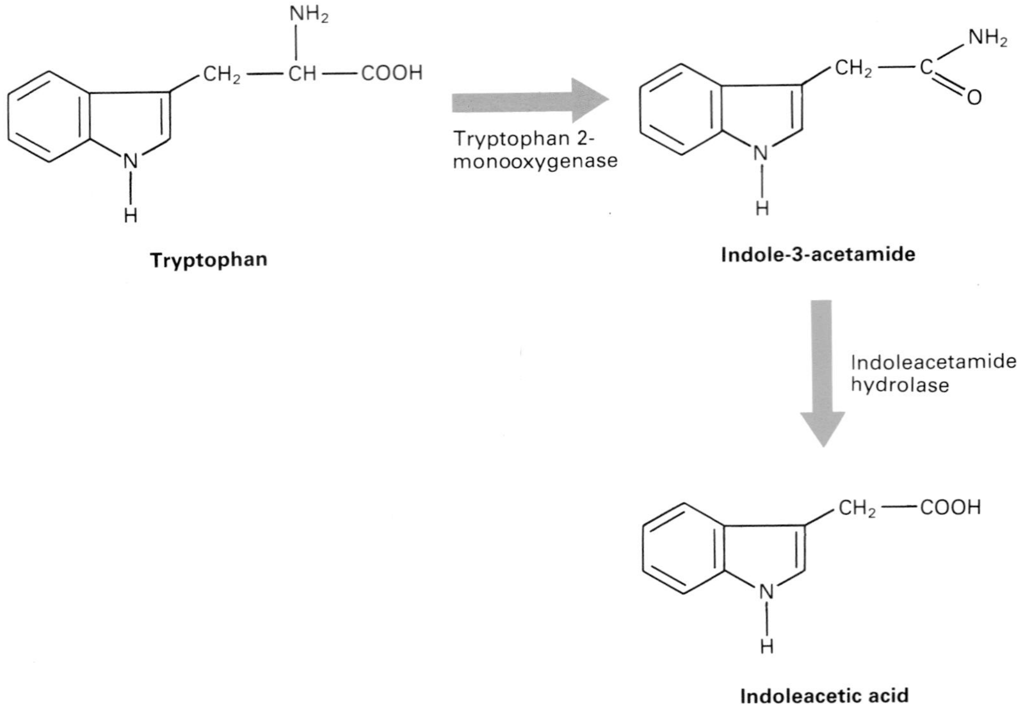

Fig. 12.12 Biosynthesis of the auxin, indoleacetic acid.

It is notable that many other microorganisms produce auxins and cytokinins (reviewed by Morris 1986). For example, the bacterium *Pseudomonas savastanoi* induces galls on olive and oleander by the secretion of these hormones, which are produced by bacterial enzyme reactions that are very similar to those of crown gall tumour cells. But with *P. savastanoi* there is no evidence of gene transfer. Crown gall induction may be a highly evolved refinement of this type of system (Weiler & Schroder 1987).

Disarmed Ti-plasmid derivatives as plant vectors

We have seen that the Ti-plasmid is a natural vector for genetically engineering plant cells because it can transfer its T-DNA from the bacterium to the plant genome. However, wild-type Ti-plasmids are not suitable as general gene vectors because they cause disorganized growth of the recipient plant cells owing to the effects of the oncogenes in the T-DNA.

The tumour cells which result from integration of normal T-DNA have proven recalcitrant to attempts to induce regeneration, either into normal plantlets, or into normal tissue which can be grafted onto healthy plants. However, tobacco callus transformed with wild-type Ti-plasmid does rarely spontaneously regenerate shoots. These shoots have been grafted onto healthy plants for further analysis. Some grafted shoots were fertile and produced seed that developed into apparently normal plants; however,

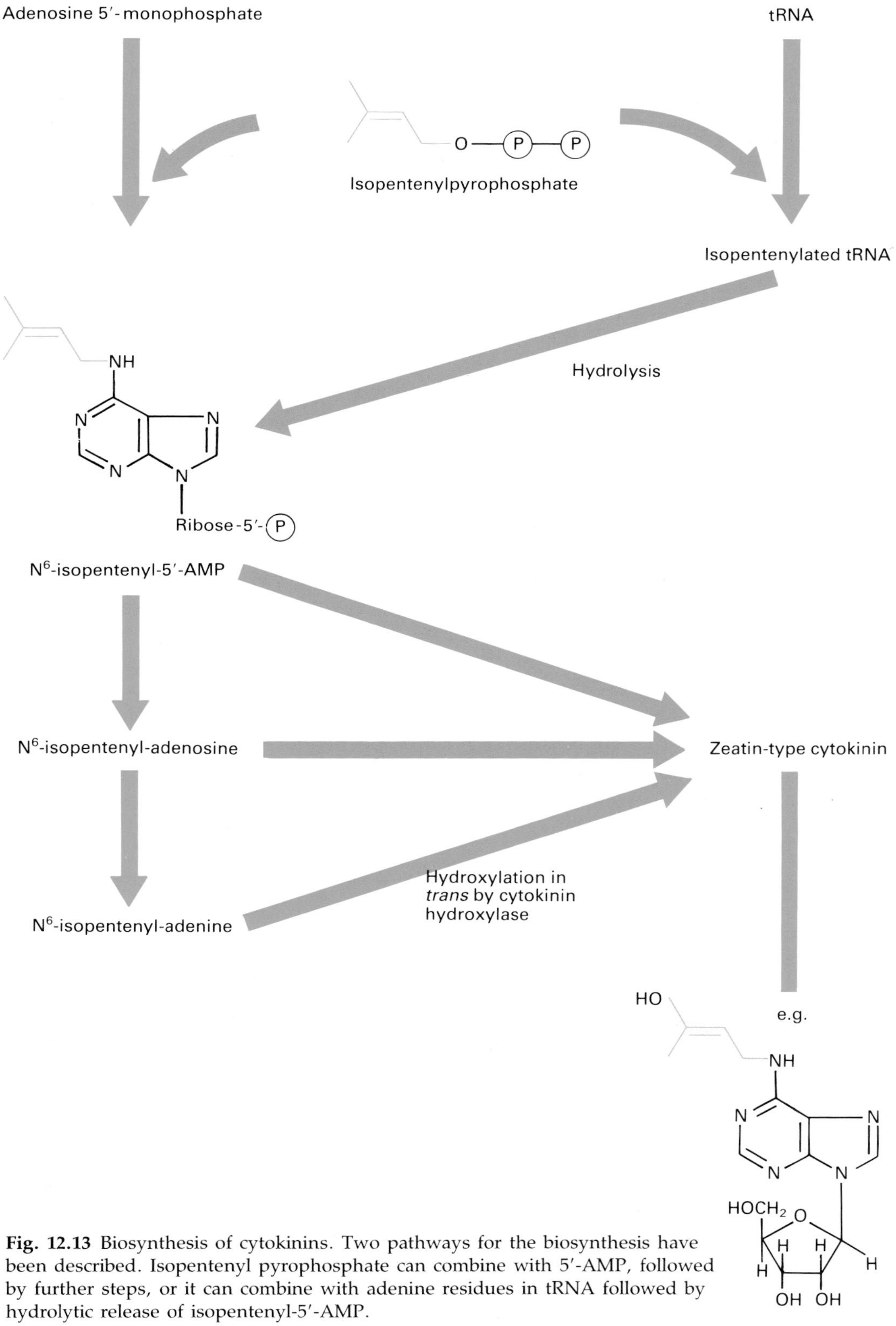

Fig. 12.13 Biosynthesis of cytokinins. Two pathways for the biosynthesis have been described. Isopentenyl pyrophosphate can combine with 5′-AMP, followed by further steps, or it can combine with adenine residues in tRNA followed by hydrolytic release of isopentenyl-5′-AMP.

238

*Section 3
Cloning in
Organisms
Other than
E. coli*

these plants lacked opine and all or most of the T-DNA had been deleted from them (Braun & Wood 1976, Turgeon *et al.* 1976, Lemmers *et al.* 1980, Buins *et al.* 1981, Wullems *et al.* 1981a,b). A single case has been reported of opine-positive complete plants regenerated from a tumour. This tumour had originally been induced by a shooty mutant of octopine T-DNA (Leemans *et al.* 1982a). In breeding experiments these plants transmitted the octopine trait in a simple Mendelian fashion, giving apparently healthy progeny. Analysis of the T-DNA revealed that little T-DNA was present except for the *ocs* gene (De Greve *et al.* 1982a).

These observations, and our knowledge of the oncogenic functions in T-DNA, indicate that in order to be able to regenerate plants efficiently we must use vectors in which the T-DNA has been *disarmed* by making it non-oncogenic. This is most effectively achieved simply by deleting all of its oncogenes. For example, Zambryski *et al.* (1983) substituted pBR322 sequences for almost all of the T-DNA of pTiC58 leaving only the left and right border regions and the *nos* gene. The resulting construct was called pGV3850 (Fig. 12.14). *Agrobacterium* carrying this plasmid transferred the modified T-DNA to plant cells. As expected, no tumour cells were produced, but the fact that transfer had taken place was evident when the cells were screened for nopaline production and found to be positive. Callus tissue could be cultured from these nopaline-positive cells if suitable phytohormones were provided, and fertile adult plants were regenerated by hormone induction of plantlets. The creation of disarmed T-DNA was an important step forward. But the fact that oncogenic transformation does not now occur means that phytohormone-independent callus production cannot be used to self-select recipient plant cells. As we have seen, opine production can be used as a marker but a dominant-acting selectable marker would be much more convenient.

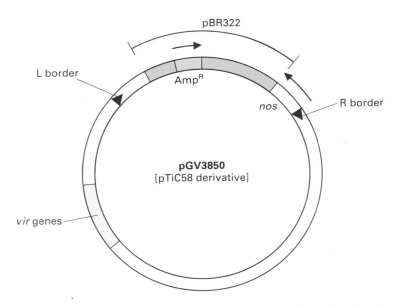

Fig. 12.14 Structure of the Ti-plasmid pGV3850, in which the T-DNA has been disarmed.

Selectable markers for inclusion in T-DNA

In order to provide suitable markers, chimaeric genes were constructed in which *nos* and *ocs* promoters and polyadenylation sites were exploited to provide expression signals for bacterial neomycin phosphotransferase, which confers resistance to the aminoglycoside antibiotics kanamycin and G418 when expressed in plant cells (compare p. 261) or bacterial dihydrofolate reductase, which confers resistance to methotrexate or trimethoprim (compare p. 259) (Bevan *et al.* 1983b, Fraley *et al.* 1983, Herrera-Estrella *et al.* 1983a,b, Horsch *et al.* 1984). Inclusion of such selectable markers in T-DNA opened the way for *Agrobacterium*-mediated transfer of T-DNA becoming an immensely powerful gene delivery system.

Insertion of foreign DNA into T-DNA

Wild-type Ti-plasmids are not suitable as experimental gene vectors because their large size means that it is not possible to find adequate unique restriction sites in the T-region. Their size also makes other procedures cumbersome. Intermediate vectors (abbreviated IV) have therefore been developed in which T-DNA has been subcloned into conventional small, pBR322-based, plasmid vectors of *E. coli* (Matzke & Chilton 1981). Standard procedures can then be used to insert any desired DNA into the T-region of such an IV.

The IV, containing foreign DNA in the T-region, can be transferred to *A. tumefaciens* by conjugation. Since the IVs are conjugation-deficient, conjugation must be mediated by the presence in the donor *E. coli* of a helper, conjugation-proficient plasmid which can mobilize the IV. Suitable helper plasmids are pRK2013 (Ditta *et al.* 1980) which consists of the transfer genes from the naturally occurring plasmid pRK2 (see p. 174) cloned onto a Col E1 replicon, or pRN3 (Shaw *et al.* 1983). Neither of these helper plasmids is capable of replicating in *Agrobacterium*. These transfers are conveniently brought about by 'triparental' matings. In these matings three bacterial strains are mixed together. These are: (a) the *E. coli* carrying the conjugation-proficient helper plasmid, (b) the *E. coli* strain carrying the recombinant IV, and (c) the recipient *Agrobacterium*. During the course of the incubation the helper plasmid transfers to the *E. coli* strain carrying the recombinant IV which is then mobilized and transferred to *Agrobacterium*. The *Agrobacterium* recipient frequently receives both the IV and the helper plasmid.

Once the IV has been introduced into *Agrobacterium*, *in vivo* homologous recombination is exploited to insert the IV into a resident non-recombinant Ti-plasmid.

A single recombination event between two circular plasmids will produce a *co-integrate*. This is illustrated in Fig. 12.15 where it is apparent that a Ti-plasmid such as pGV3850 is particularly useful as an acceptor of the IV because the pBR322 sequences in its T-DNA region are homologous with most pBR322-based IV plasmids. In the example shown the IV includes a *neo* marker for selection of recombinant T-DNA in plant cells, and a *kan*[R]

240

*Section 3
Cloning in
Organisms
Other than
E. coli*

Intermediate vector transferred
into *Agrobacterium* by conjugation
Unable to replicate autonomously
in *Agrobacterium*

Disarmed Ti-vector pGV3850,
resident in *Agrobacterium*

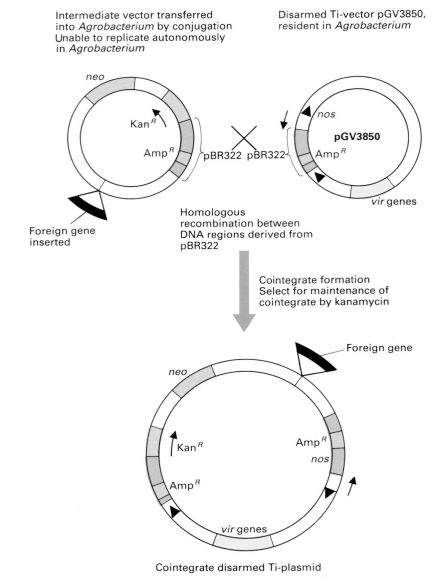

Foreign gene
inserted

Homologous
recombination between
DNA regions derived from
pBR322

Cointegrate formation
Select for maintenance of
cointegrate by kanamycin

Cointegrate disarmed Ti-plasmid

Fig. 12.15 Production of recombinant disarmed Ti-plasmid by cointegrate
formation.

determinant for selection of cointegrates in the *Agrobacterium*. (Neither the
IV nor the helper plasmid can replicate autonomously in *Agrobacterium*.)

Binary Ti-vectors

The *in vivo* recombination technique of co-integrate formation just described has been used widely. The procedure reconstructs a very large,
recombinant, disarmed Ti-plasmid. In fact this is unnecessary. We saw
earlier that the T-region is distinct from the *vir* region whose functions are
responsible for transfer and integration of T-DNA into the plant genome.

It is possible to take advantage of this by providing the manipulated disarmed T-DNA carrying foreign DNA and the *vir* functions on separate plasmids (de Framond *et al.* 1983). This is the principle of *binary* vectors (Hoekma *et al.* 1983, Bevan, 1984). Thus a modified T-DNA region carrying foreign DNA is constructed in a small plasmid which replicates in *E. coli* such as pRK252. This plasmid, called *mini*-Ti or *micro*-Ti can then be transferred conjugatively in a tri-parental mating into *A. tumefaciens* which contains a compatible plasmid carrying virulence genes. The *vir* functions are supplied in *trans*, causing transfer of the recombinant T-DNA into the plant genome. This binary system simplifies the transfer of foreign genes to plant cells (Fig. 12.16).

Binary vector systems without Ti sequences

In discussing the mechanism of T-DNA transfer from *Agrobacterium* to plant cells we noted the similarity of the process to normal bacterial conjugation, in which double-stranded plasmid DNA is nicked in one strand by the action of *mob* (mobilization) proteins at the *ori*T (origin of transfer) site, and a single strand of plasmid DNA is transferred in a polar manner from this site into the recipient bacterium. The apparent similarity is reinforced by the remarkable observations of Buchanan-Wollaston *et al.* (1987). In place of a T-DNA-containing mini-Ti vector they constructed pJP181 (Fig. 12.17), a plasmid based on the incQ group, wide host range, plasmid pRSF1010 (see p. 172). A *neo* gene was included for selection of transferred DNA in plant cells. pJP181 was conjugated into *Agrobacterium*

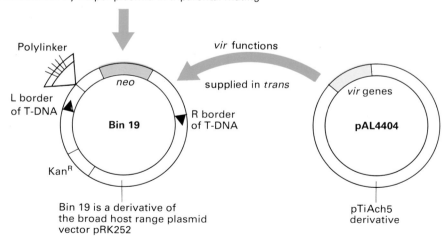

Fig. 12.16 Binary Ti-vector system. The binary vector (Hoekma *et al.* 1983, Bevan 1984) illustrated, Bin 19, is based upon the incP group, broad host range plasmid pRK252. It contains a *neo* marker for selection of transferred DNA in plant cells and includes a *lacα* peptide gene with inserted polylinker region for blue/white detection of insertion of foreign DNA. *Vir* functions are supplied in *Agrobacterium* by pAL4404, a pTiAch5 derivative from which the T-DNA has been deleted.

242

*Section 3
Cloning in
Organisms
Other than
E. coli*

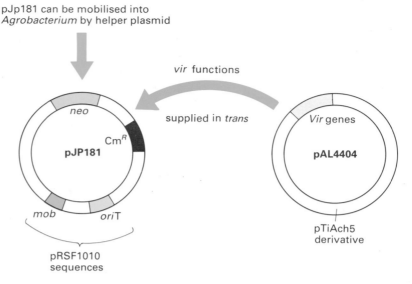

pJp181 can be mobilised into
Agrobacterium by helper plasmid

Fig. 12.17 Binary vector system without T-DNA sequences. The binary vector illustrated pJP181 (Buchanan-Wollaston *et al.* 1987) is based upon the incQ group broad host range plasmid pRSF1010, and includes the *mob* and *ori*T sequences. pJP181 includes a *neo* marker for selection in plant cells. *Vir* functions are supplied in *Agrobacterium* by pAL4404.

which contained pAL4404 and the *Agrobacterium* then used to infect tobacco leaf discs. It was found that kanamycin (or G418) resistant transformant plant cells were produced and that these contained pJP181 DNA integrated into the plant genome. These experiments showed that the bacterial *mob* and *ori*T functions required for normal bacterial conjugation can also promote plasmid transfer to plants and could form the basis of a variety of new binary vectors. It appears therefore that *Agrobacterium* infection provides a means by which bacteria become closely associated with plant cells, and that the actual DNA transfer mechanism to plant cells has evolved from the familiar bacterial conjugation mechanism. An important evolutionary implication of these findings is that plant genomes have access to the gene pool of Gram-negative bacteria.

A. rhizogenes and Ri plasmids

We noted previously that Ri plasmids induce hairy root disease in plants. The Ri plasmids are analogous to Ti plasmids. They have been of interest from the point of view of vector development because opine-producing root tissue induced by Ri-plasmids in a variety of dicots can be regenerated into whole plants by manipulation of phytohormones in the culture medium. Ri T-DNA is transmitted sexually by these plants and affects a variety of morphological and physiological traits but does not in general appear deleterious. The Ri plasmids therefore appear to be already equivalent to disarmed Ti-plasmids (Tepfer 1984). Many of the principles explained in the context of disarmed Ti-plasmids are applicable to Ri-plasmids. An intermediate vector co-integrate system has been developed

(Jensen *et al*. 1986) and applied to the study of nodulation in transgenic legumes, *Lotus corniculatus* (bird's-foot-trefoil).

243
Chapter 12
Gene Transfer
to Plants

Van Sluys *et al*. (1987) have exploited the fact that *Agrobacterium* containing both an Ri plasmid and a disarmed Ti plasmid can frequently co-transfer both plasmids. The Ri plasmid induced hairy root disease in recipient *Arabidopsis* and carrot cells, serving as a transformation marker for the co-transferred recombinant T-DNA, and allowing regeneration of intact plants. No drug resistance marker on the T-DNA was necessary with this plasmid combination.

A simple and general experimental procedure for transferring genes into plants with agrobacterium Ti-plasmid vectors

Once the principle of selectable, disarmed T-DNA vector regions was established, there followed an explosion of experiments in which foreign DNA was transferred into regenerated fertile plants. A precise upper size limit for foreign DNA acceptable by T-DNA has not been determined. It is greater than 50 kb (Herrera-Estrella *et al*. 1983a).

The simple general protocol of Horsch *et al*. (1985) has been widely adopted (Fig. 12.18). First small discs (a few mm diameter) are punched from leaves of petunia, tobacco, tomato, or other dicot plant, surface-sterilized, and inoculated in a medium containing *A. tumefaciens* carrying the recombinant disarmed T-DNA (as co-integrate or binary vector) in which the foreign DNA is accompanied by a chimaeric *neo* gene conferring kamamycin resistance on plant cells. The discs are cultured for 2 days and transferred to medium containing kanamycin to select the transferred *neo* gene, and carbenicillin to kill *Agrobacterium*. After 2−4 weeks shoots develop which are excised from the callus and transplanted to root-inducing medium. Rooted plantlets can subsequently be transplanted to soil, about 4−7 weeks after the inoculation step.

This method has the advantage of being simple and relatively rapid. It is superior to previous methods in which transformed plants were regenerated from callus which had itself been derived from protoplasts that had been transformed by co-cultivation with the *Agrobacterium* (De Block *et al*. 1984, Horsch *et al*. 1984). Interestingly, in one application of the protoplast co-cultivation method it was found that transmission of the foreign DNA in regenerated tobacco plants was maternal; it was not transmitted through pollen. Southern blotting chloroplast DNA showed directly that the foreign DNA had become integrated into the chloroplast genome. This was the first demonstration of the, presumably relatively rare, introduction of foreign DNA into the chloroplast DNA by Ti-plasmid vectors (De Block *et al*. 1985).

Agrobacterium and monocots

Agrobacterium normally causes crown gall disease on dicots. In the laboratory, tumours have been induced on the monocot *Asparagus officinalis*. Hooykaas-Van Slogteren *et al*. (1984) showed that *Agrobacterium* T-DNA can be transferred into the genomes of the monocots *Chlorophytum* and

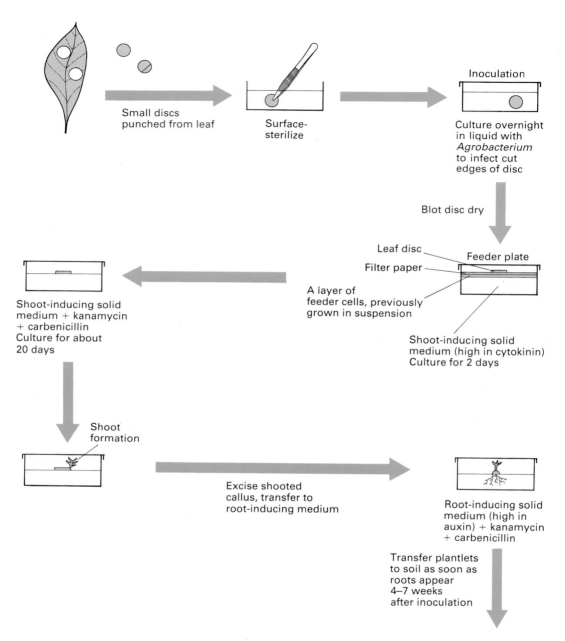

Fig. 12.18 Leaf-disc transformation by Ti-plasmid vectors.

Labels in figure:

Small discs punched from leaf

Surface-sterilize

Inoculation
Culture overnight in liquid with *Agrobacterium* to infect cut edges of disc

Blot disc dry

Leaf disc
Feeder plate
Filter paper
A layer of feeder cells, previously grown in suspension
Shoot-inducing solid medium (high in cytokinin)
Culture for 2 days

Shoot-inducing solid medium + kanamycin + carbenicillin
Culture for about 20 days

Shoot formation

Excise shooted callus, transfer to root-inducing medium

Root-inducing solid medium (high in auxin) + kanamycin + carbenicillin

Transfer plantlets to soil as soon as roots appear 4–7 weeks after inoculation

Narcissus. No tumours were induced but the plant cells synthesized opines. The T-DNA may therefore be naturally disarmed for these monocots.

It has been argued that *Agrobacterium* infection of monocots is inefficient because of the lack of production of *vir* gene activating substances by the wounded monocot tissue. By pretreating *Agrobacterium* with wound exudate from tubers of the potato, a susceptible dicot, Schafer *et al.* (1987) were able to induce crown gall tumours in the important monocot crop plant *Dioscorea bulbifera*, the yam. T-DNA was detected in tumour cell DNA. The experiments demonstrated the usefulness of pre-inducing the *Agrobacteria*, but to date little progress has been reported with the agriculturally important cereal crop plants.

DNA-mediated transfection of plant protoplasts

Removal of the wall from plant cells makes the resulting protoplasts amenable to transformation by DNA. The process has much in common with animal cell transformation. First, a method for actually getting DNA into the protoplasts is required. A variety of procedures have been used for this: electroporation (see p. 12) has become the favoured technique. Also, selectable genes are required to which foreign DNA can be ligated and which allow selection of transformants. The development of a selection system based on the bacterial genes conferring resistance to neomycin (G418, kanamycin), driven by plant expression signals such as those of *nos*, *ocs* and the CaMV gene VI, has been very useful in this context.

Transformants obtained by such procedures are placed on nutrient medium in which the protoplasts regenerate new cell walls and initiate cell division. Manipulation of the culture conditions then makes it possible, with a wide range of dicotyledonous plants, to induce shoot and root formation. Ultimately transformed plantlets may be grown into fertile transgenic plants.

In early demonstrations of the power of the protoplast transformation technique, Paszkowski, Potrykus and co-workers (Paszkowski *et al.* 1984, Hain *et al.* 1985, Potrykus *et al.* 1985a) showed that selectable foreign DNA became stably integrated into the nuclear genome of transgenic tobacco plants and was transmitted to progeny plants in a Mendelian fashion.

An example of this technology was provided by Meyer *et al.* (1987) who constructed a plasmid, in an *E. coli* vector, which contained a cDNA corresponding to the maize enzyme dihydroquercetin 4-reductase (an enzyme of anthocyanin pigment biosynthesis) linked to promoter and transcript termination signals of the 35S promoter of cauliflower mosaic virus (see later, p. 252). The plasmid also contained a *neo* gene, flanked by the *nos* promoter and the *ocs* polyadenylation site. Protoplasts of a mutant, white coloured, *Petunia* strain were transformed with the plasmid DNA using the mitotic-arrested-cell electroporation technique (p. 13). Kanamycin-resistant microcalli were derived from transfected cells and were induced to regenerate whole plants. The plants produced flowers of a new brick red colouration, owing to expression of the maize cDNA. Other work demonstrated that foreign DNA can be stably introduced into protoplasts from graminaceous plants, including wheat (Lorz *et al.* 1985) and Italian ryegrass, *Lolium multiflorum* (Potrykus *et al.* 1985b). However, by contrast with dicotyledonous plants, cereal species present a problem; with such cereals it has not proven feasible to regenerate whole plants from cultured cells of cereal species. (Rice is exceptional in having been shown to regenerate from cultured cells; Uchimaya *et al.* 1986, Shimamoto *et al.* 1989.) This aspect of plant cell culture is receiving much attention. At

246

*Section 3
Cloning in
Organisms
Other than
E. coli*

present it remains to be seen how amenable graminaceous plants will become to this technology.

Transgenic rye plants

Because of the apparent inability to regenerate whole plants from cultured cells of cereal species, effort has been directed towards ways of producing transgenic cereals which circumvent this requirement. Success has been achieved by directly injecting plasmid DNA, including a foreign selectable gene, into developing floral tillers of rye, *Secale cereale* (de la Pena *et al.* 1987) (Fig. 12.19). The selectable gene was once again a bacterial neomycin phosphotransferase gene (which confers kamamycin resistance) under the control of the *nos* promoter.

The injection of DNA solution into young floral tillers was carried out with a conventional syringe at a point above each tiller node, until several drops of the solution came out from the top of the young inflorescence (usually about 0.3 ml was injected). The timing of the injection is thought to be critical, occurring 14 days before meiosis. After injection the floral tillers were allowed to grow to maturity and seeds from the injected plants were obtained by cross-pollination with other injected plants. The seeds were then germinated in the presence of kamamycin. Out of over 3000 seedlings tested, just two were found to be genuinely kamamycin resistant and to have incorporated the foreign DNA. Southern blots of the DNA from these transgenic plants indicated that the foreign DNA had become integrated in the genome at more than one site or as more than one copy.

It is likely that after injection the vascular system of the plant transported DNA to the germ cells of the inflorescence. It is not known whether DNA uptake occurred in the male or female germ line, but the membrane characteristics of the male cells at the time of injection make them the likely target.

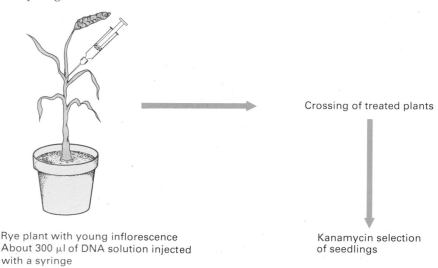

Rye plant with young inflorescence
About 300 µl of DNA solution injected
with a syringe

Crossing of treated plants

Kanamycin selection
of seedlings

Fig. 12.19 Transgenic rye plants obtained by injecting DNA solution into the plant, at a stage about 14 days before meiosis.

These results show that this injection procedure, although relatively inefficient, can result in the integration of new genetic information into the genome of a cereal species. The success demonstrates the power of the selection system and the importance of selectable markers constructed by previous manipulations. The applicability of direct injection procedures to other cereals is being investigated.

Microprojectiles for delivering nucleic acids into living cells

A completely novel way of delivering nucleic acids into plant cells involves the use of high velocity microprojectiles, which carry the RNA or DNA, and are literally shot into the cell (Klein *et al.* 1987). Figure 12.20 shows a gun for firing the microprojectiles. A gunpowder charge fires spherical tungsten particles (average diameter 4 μm) into intact cells. In the test system used by Klein *et al.* intact epidermal cells of the onion, *Allium cepa*, were bombarded with the particles. Most of the cells in the target area of about 1 cm² were found to contain the microprojectiles following bombardment.

To demonstrate the technique, genomic RNA of tobacco mosaic virus was adsorbed onto the tungsten particles before firing. Three days after bombardment it was evident that 30−40% of the *A. cepa* cells which contained the particles also showed signs of virus replication. It was also shown that the gun could deliver DNA into cells. The test DNA consisted of a CAT gene construct driven by the CaMV 35S promoter. Extracts of the *A. cepa* epidermal tissue which had been bombarded with microprojectiles coated with the DNA were found to transiently express very high levels of chloramphenicol acetyltransferase.

This remarkable technique does not require cell culture or pretreatment of recipient tissue. The technique may prove useful for the stable transformation of plant cells from a variety of species, including monocots, and should circumvent problems associated with regeneration of plants from protoplasts. To accomplish this, particle bombardment of regenerable tissues such as meristems or embryogenic callus with DNA-bearing microprojectiles will be necessary. The cells of these tissues are much smaller than those of *A. cepa* epidermal tissue, and will probably require correspondingly small microprojectiles.

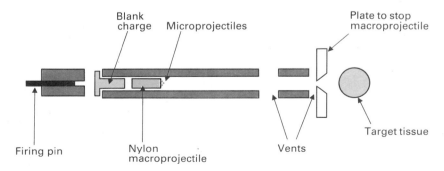

Fig. 12.20 Gun for delivering micro-projectiles, coated with DNA, to plant tissue.

248

*Section 3
Cloning in
Organisms
Other than
E. coli*

Regulated tissue-specific expression of transferred genes in plants

We have already seen how expression signals from constitutively expressed plant genes have been exploited in constructing generally useful selectable markers based on bacterial aminoglycoside phosphotransferases. The expression signals of the *nos*, *ocs* genes and CaMV gene VI have been most widely used. In addition to constitutive expression, it is necessary to be able to direct expression of manipulated genes in particular tissues of a plant. This ability is essential for many actual or foreseen applications of recombinant DNA technology in agriculture. Evidence from animal systems confirms that signals for differentiated expression lie associated with the promoter region and that these signals, together with tissue-specific enhancer and silencer elements, confer tissue specificity of transcription. Fortunately, in plants it also has proven feasible to identify promoter sequences and associated regulatory sequences which normally confer tissue specificity upon their natural, linked gene sequence or upon foreign genes linked to them artificially. One example, of many which illustrate this, follows. Stockhaus *et al.* (1987) used CAT as a reporter activity for assaying the functioning of manipulated promoter and upstream sequences of a light-inducible gene (named ST-LS1) from the potato. The ST-LS1/CAT fusions were transferred into tobacco using a pGV3850 co-integrate vector and whole plants were regenerated. It was found that sequences derived from the 5'-upstream region of the ST-LS1 gene, comprising nucleotides −334 to +11, were sufficient to confer leaf/stem-specific as well as light-inducible expression of the CAT activity. This upstream region was also shown in other constructs to have an orientation-independent, enhancer-like activity.

Concluding remarks

The extraordinary achievements in developing the *Agrobacterium*-based gene transfer systems are truly impressive. The outcome of this research has been the development of simple and efficient gene transfer procedures which have been widely used.

An additional application of these systems is *transposon tagging*, i.e. random insertion of transferred DNA into the plant genome can be used to insertionally inactivate plant genes, in a genetically amenable species such as *Arabidopsis*, which has a small genome and short life cycle. When bred into homozygous condition any gene whose inactivation gives an 'interesting' phenotype can be cloned by virtue of its being 'tagged' with the inserted DNA sequence (compare p. 315).

DNA-mediated transfection of protoplasts can do many of the things that *Agrobacterium*-based systems have done. Protoplast transfection may have greater potential for cereals if the problems of plant regeneration can be overcome for these species. In addition, as we shall discuss for animal cells, there is a need to be able to target transferred DNA to particular sites in the plant genome. Gene-targeting in plants will probably make use of protoplast transfection. This technology is just beginning to be developed but we can assume that rapid progress will be made (Paszkowski *et al.* 1988). The convenience of the modern, highly developed *Agrobacterium*-based systems will probably ensure their continued use in addition to protoplast transfection.

PLANT VIRUSES AS VECTORS

Plant viruses are attractive as vectors for several reasons.
1 Viruses absorb to and infect cells of intact plants.
2 Relatively large amounts of virus can be produced from infected plants, leading to the prospect of large amounts of foreign protein being expressed form recombinant viruses. This is an aspect of the inherent gene amplification which accompanies virus replication.
3 Some virus infections are *systematic.* They are spread throughout the whole plant. In some cases intact virions are transported through the vascular system of the plant.
4 Viruses are known which infect plants such as cereals, for which current alternative technology is limited.

It is possible to envisage the use of a plant virus vector in two distinct ways. One would be a massive infection which is deleterious to the plant and would perhaps ultimately kill it, but which meanwhile leads to the expression of a foreign gene product. The alternative would be a less harmful systematic infection which leads to more-or-less healthy plants in which foreign protein is expressed either to alter the plant to make it a better crop plant, or to express a foreign protein which is a desirable product when purified from the plant tissues. Crop plants can produce biomass cheaply without high technology. Ultimately the production of proteins for therapeutic or other uses may be undertaken not in expensive fermentors with microorganisms such as yeast or bacteria as the host, but in fields of genetically engineered plants. These plants could be engineered by the gene transfer systems discussed in previous sections, or using virus vectors.

At present, virus vectors have not been developed to the stage where they are widely used as vectors, but progress has been made with the only two groups of plant viruses which have DNA genomes; the caulimoviruses and geminiviruses. Most plant viruses have RNA genomes. Plant RNA viruses which have great potential as prototype vectors are Brome mosaic virus and tobacco mosaic virus.

Cauliflower mosaic virus (CaMV)

CaMV is the best studied of the caulimovirus group. This is a group of spherical viruses (Fig. 12.21) which contain a circular double-stranded DNA genome of about 8 kb. The caulimoviruses are responsible for a number of economically important diseases of cultivated crops. They have restricted host ranges individually but as a group infect a range of dicots.

One feature of CaMV which makes it attractive as a vector is that infection becomes systemic. In order for CaMV to be transmitted through

250

*Section 3
Cloning in
Organisms
Other than
E. coli*

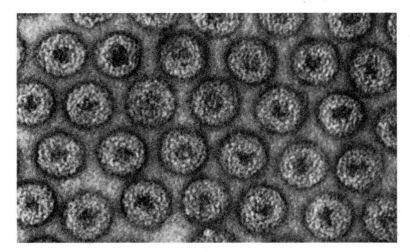

Fig. 12.21 Semi-crystalline array of cauliflower mosaic virus purified from turnip (approx. magnification ×200 000). The dark spots in the centres of the particles are typical and are the result of the outer protein shell being sucked into the hollow core during preparation for electron microscopy. (Photograph courtesy of M. Webb, National Vegetable Research Station.)

the vascular system of the plant the DNA must be assembled within virions. In infected cells, refractile, round inclusions form which consist of many virus particles embedded in a protein matrix. The matrix protein is virus-encoded and can account for up to 5% of the protein in infected cells. In nature the virus is transmitted on the stylets of aphids. The transmission depends upon a transmission factor which is also a virus-encoded protein present in infected cells, and is not part of the virion (Woolston *et al.* 1983).

CaMV DNA. Several different isolates of CaMV have been sequenced (Franck *et al.* 1980, Gardner *et al.* 1981, Balazs *et al.* 1982). The DNA of CaMV has an unusual structure. There are three discontinuities in the duplex, two in one strand, one in the other (Fig. 12.22). These are regions of sequence overlap (Frank *et al.* 1980, Richards *et al.* 1981).

CaMV DNA is infectious. Infections can be initiated by inoculating abraded surfaces of the host plant with DNA. The DNA has a single *Sal* I site, and *Sal* I-linearized DNA is infectious even in the absence of religation *in vitro*. Recircularization occurs in the plant cell. CaMV DNA has been cloned into an *E. coli* vector and propagated in *E. coli*. When the CaMV was released from the vector as a linear CaMV DNA molecule it was found to be infectious, despite the lack of the single-strand discontinuities in the inoculating DNA (Howell *et al.* 1980).

Another unusual feature of CaMV DNA is the presence of ribonucleotides covalently attached to the 5'-termini of the discontinuities. These and other observations have led Hull & Covey (1983 a,b) and Pfeiffer & Hohn (1983) to suggest that CaMV replication involves reverse transcription with an RNA genomic intermediate. The replication cycle resembles that of retroviruses and hepatitis B virus (Hohn *et al.* 1985).

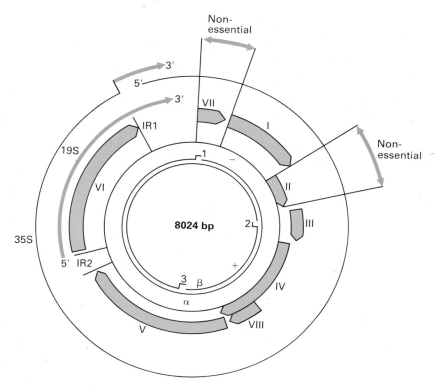

Fig. 12.22 Map of the cauliflower mosaic virus genome. The eight coding regions are shown by coloured boxes, and the different reading frames are indicated by the radial positions of the boxes. The thin lines in the centre indicate the (plus and minus) DNA strands with the three discontinuities. The major transcripts, 19S and 35S, are shown around the outside.

The sequence of CaMV DNA reveals eight closely packed reading frames. There are only two small intergenic regions, and the only non-essential genes are the two small genes II and VII. Thus, most attempts to alter the DNA by insertion or deletion have caused loss of infectivity (Delseny & Hull 1983). The absence of substantial non-essential DNA limits scope for substituting sequences with foreign DNA.

CaMV as a vector. The CaMV envelope does not allow the DNA which it encapsidates to be substantially larger than normal. This was evident when foreign DNA was inserted at the unique *Xho* I site which lies in the non-essential gene II (Gronenborn *et al.* 1981). If DNA longer than a few hundred nucleotides was inserted the infectivity was destroyed. This packaging limitation and the absence of long non-essential sequences which can be deleted in the genome severely limit the capacity for foreign DNA.

In Chapter 13 we demonstrate how similar constraints can be overcome in SV40 by exploiting a helper virus to complement a defective recombinant virus. This does not appear feasible in CaMV because homologous recombination between the helper and recombinant virus DNAs readily expels the foreign DNA with production of wild-type virus.

252

*Section 3
Cloning in
Organisms
Other than
E. coli*

It appears therefore that capacity is a major limitation of CaMV vectors. But a small foreign DNA, comprising the bacterial *dhfr* gene, replacing the gene II coding sequence, has been successfully expressed in plants (Brisson *et al.* 1984).

As a vector, CaMV has been overshadowed by the success of the *Agrobacterium*-based, and DNA-mediated transfection, systems. Knowledge of CaMV has been of value in providing very strong promoters for driving expression of other genes in plants. We met an example of the use of the 35S promoter (p. 245, Meyer *et al.* 1987). Other constructs employ the promoter of gene VI (Paszkowski *et al.* 1984, Balazs *et al.* 1985). This gene encodes the abundant matrix protein of the inclusion body and has a very strong promoter.

Agroinfection with a geminivirus

Geminiviruses are attractive for vector development because members of this group infect a wide range of monocot and dicot crop plants. Very little is known about the molecular biology of this group so that vector development has not progressed very far. The following experiments with maize streak virus, MSV, are interesting from several points of view.

MSV is a leaf-hopper-transmitted geminivirus. The intact virus can only infect if transmitted by the insect. The viral genome consists of a circle of single-stranded DNA. Replication of the viral DNA is thought to occur by DNA intermediates only, and the viral genome replicates to a high copy number in the nucleus of proliferating cells. In nature the virus causes stunting and yellow-streaked leaves of infected maize plants.

The MSV genome has never been introduced successfully into plants as native or cloned DNA. Grimsley *et al.* (1987) were able to introduce the DNA with the aid of *Agrobacterium* as follows. They constructed a plasmid in which MSV cloned DNA was inserted as a tandem dimer of the intact genome. These were inserted into an *Agrobacterium* T-DNA vector, and maize plants were infected with *A. tumefaciens* containing this recombinant T-DNA. Viral symptoms appeared within two weeks of plant inoculation. This process has been termed 'agroinfection'. It has in fact been demonstrated for a number of viruses and viroids that if the T-DNA contains partially or completely duplicated genomes, single copies of the genome can escape and initiate infections. Agroinfection is a very sensitive assay for transfer to the plant cell because of the inherent amplification of virus infection and resulting visible symptoms.

The successful agroinfection of maize plants with MSV leads to two important conclusions. Firstly, *Agrobacterium* can transfer DNA to maize, even if inefficiently by comparison with dicot plants. Secondly, cloned MSV DNA is biologically active and now amenable to manipulation and vector development.

Brome mosaic virus as a vector

Plant viruses with DNA genomes are in the minority; the great majority of plant viruses have RNA genomes. The range of potential vectors is very great. In addition, many of these RNA viruses have filamentous morphology and so it is expected that the length of the virus particle is determined by the length of the viral nucleic acid. Thus there should be no strict size limitation on the RNA to be packaged.

Another common feature of plant viruses is that they may be multi-component. An example of this is Brome mosaic virus. The virus infects a number of Graminae including barley. There are three separate RNA components of the genome, each of which is packaged into a separate particle. Each of these RNAs is also an mRNA and during infection a fourth mRNA is produced which is a subgenomic derivative of RNA 3. This RNA 4 is the mRNA encoding the very abundant viral coat protein. The cloned cDNAs corresponding to RNAs 1−3 can be transcribed *in vitro* to produce transcripts which are infectious when mixed and introduced into barley protoplasts. Only RNAs 1 and 2 are necessary for replication and expression of the genome, so that the coat protein sequence on RNA 3 is available for manipulation, although in the absence of coat protein no virus particles are produced (Ahlquist *et al.* 1987).

The amenability of this system has been demonstrated by French *et al.* (1986) who inserted a bacterial CAT coding sequence into the coat protein gene sequence of RNA 3.

When this construct was transcribed *in vitro*, and introduced into barley protoplasts together with RNAs 1 and 2, high levels of CAT activity were expressed. It is therefore likely that further manipulation of this promising system will be possible, and that the coat protein expression signals may be exploited for high level expression of foreign proteins.

Tobacco mosaic virus as a vector

The approach of using a cDNA copy of an RNA virus genome has also been employed with the tobacco mosaic virus (TMV) genome. The TMV genome is a single-component RNA which is also a messenger RNA. The genome encodes at least four polypeptides. The 130 kDal. and 180 kDal. proteins are translated directly from the same initiation codon on genomic RNA. The other two proteins, 30 kDal. and coat protein, are translated from processed subgenomic RNAs. The 130 kDal. and 180 kDal. proteins are involved in viral replication, whereas the 30 kDal. protein is necessary for cell-to-cell movement of virus. These three proteins are therefore probably essential for TMV propagation in whole plants. The coat protein is not essential for viral multiplication, but is necessary for long-distance spread of infection in the plant. Since the coat protein is synthesized in large amounts (up to several mg per g of infected tobacco leaf), and since it is non-essential, it is an attractive target site for introducing and expressing a foreign gene.

Takamatsu *et al.* (1987) have modified a full-length TMV cDNA clone from which infectious TMV RNA can be transcribed *in vitro*. A bacterial CAT gene sequence was placed just downstream of the initiation codon of the coat protein gene. When *in vitro* transcripts of the recombinant TMV cDNA were inoculated into tobacco plants, CAT activity was observed in

254

*Section 3
Cloning in
Organisms
Other than
E. coli*

the inoculated leaves, although the infection was unable to spread systemically throughout the plant. These initial experiments have therefore demonstrated that TMV can be used as an expression vector in higher plants.

13 Introducing Genes into Animal Cells

Analysis of the regulation of gene expression in animal cells is one of the central themes of current molecular biology. For this and other reasons procedures for introducing manipulated genes into animal cells, where their regulation can be assayed, have been the subject of intensive research. There has been a great diversity of approaches. By direct micro-injection of DNA into fertilized eggs of *Xenopus*, or the mouse, or into early *Drosophila* embryos, it has been possible to incorporate foreign DNA into the genomes of the resulting adult animals (see Chapter 14). Here we shall consider methods for introducing foreign DNA into animal cells in culture. Most of this research has made use of mammalian cell lines. Insect cells are also important as hosts for baculovirus vectors.

Integration of DNA into the genome of mammalian cells

The ability of mammalian cells to take up exogenously added DNA, and to express genes included in that DNA, has been known for many years. Szybalska & Szybalski (1962) were the first to report DNA-mediated transfer. They transfected* HGPRT⁻ mutant human cells to HGPRT⁺ using total, uncloned, human nuclear DNA as the source of the wild-type gene. The rare HGPRT⁺ transformants were selected by means of the HAT selection system which they had devised (Fig. 13.1). Much later, it was appreciated that successful DNA transfer in these experiments was dependent upon the formation of a co-precipitate of the DNA with calcium phosphate, which is insoluble, and must be formed freshly in the presence of the DNA when the transfection mixture is assembled. Apparently the calcium phosphate granules are phagocytosed by the cells and in a small proportion of the recipients some of the DNA becomes stably integrated into the nuclear genome. The technique became generally accepted after its application by Graham & Van der Eb (1973) to the analysis of infectivity of viral DNA.

---

* The term transformation commonly has two different meanings in the context of this chapter. (a) An inherited change in genotype due to the uptake of foreign DNA, analogous to bacterial transformation. (b) A change in properties of an animal cell possessing normal growth characteristics to one with many of the characteristics of a cancer cell, i.e. growth transformation. In order to avoid confusion, in this chapter the term *transfection* will be used for the first meaning although originally transfection applied to the uptake of viral DNA. The term transformation will be used only with meaning (b).

256

Section 3
Cloning in
Organisms
Other than
E. coli

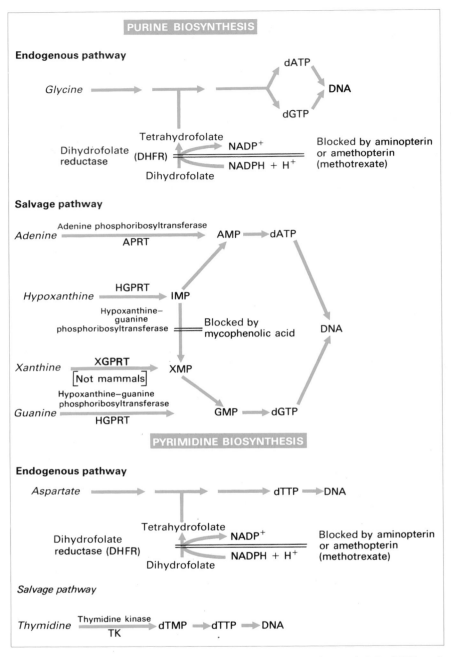

Fig. 13.1 Commonly used mutants and inhibitors in cell culture. dATP, dGTP and dTTP have two synthetic pathways. A loss of either pathway for any one nucleotide, therefore, is not lethal, so mutants can be isolated which lack one of these pathways. APRT⁻ cells can be isolated because they are *resistant* to toxic base analogues (2,6-diamino-purine, 2-fluoroadenine) but are *killed* in medium containing adenine and hypoxanthine plus azaserine, whereas wild-type cells survive. Azaserine inhibits several reactions in the endogenous pathway of purine biosynthesis. HGPRT⁻ cells are *resistant* to toxic base analogues (thioguanine, azaguanine), but are *killed* in medium containing aminopterin and hypoxanthine in which wild-type cells survive. TK⁻ cells are *resistant* to the thymidine analogue BUdR, but are *killed* in medium containing aminopterin and thymidine. HAT medium contains hypoxanthine, aminopterin and thymidine and is commonly used to select against TK⁻ and HGPRT⁻ cells.

XGPRT is an enzyme that is found in *E. coli* but not in mammalian cells.

Other transformation techniques

257

Chapter 13
Introducing
Genes into
Animal Cells

The calcium phosphate co-precipitate provides a general method for introducing any DNA into mammalian cells. It can be applied to relatively large numbers of cells in a culture dish but, as originally described, it is limited by the variable and usually rather low (1–2%) proportion of cells that take up exogenous DNA. Only a subfraction of these cells will be stably transfected. Procedures have been designed to increase to about 20% the proportion of cells that take up the DNA (Chu & Sharp 1981). A related procedure involves fusion of cultured cells with bacterial protoplasts containing the exogenous DNA (Schaffner 1980, Rassoulzadegan *et al.* 1982). Here, the proportion of cells taking up exogenous DNA can approach 100%. An alternative approach is to microinject DNA into the nucleus of cultured cells (see, for example, Kondoh *et al.* 1983), a procedure which has the advantage that 'hits' are almost certain but which cannot be applied to large numbers of cells.

Recently other general and convenient methods for introducing DNA into cultured mammalian and other vertebrate cells have gained popularity. One problem with the calcium phosphate co-precipitate is that many cell lines do not like having the solid precipitate adhering to them and the surface of their culture vessel. An alternative transfection system employs DEAE-dextran (diethylaminoethyl-dextran) in the transfection medium, in which the DNA is also present. The DEAE-dextran is soluble and no precipitate is involved. It is polycationic and probably acts by mediating, in some unknown way, the productive interaction between negatively charged DNA and components of the cell surface in endocytosis. The DEAE-dextran procedure is particularly convenient for transient assays in COS cells (see below). It does not appear to be efficient for the production of *stable* transfectants (Sussman & Milman 1984). *Electroporation* and *lipofection* have been gaining favour for many cell types (Chapter 1), for the production of stable transfectants.

Direct application of calcium phosphate-mediated transfection: transient assays without eukaryotic vector sequences. An example of the direct application of calcium phosphate-mediated transfection is the work of Rutter's group on DNA sequences controlling cell-specific expression of insulin and chymotrypsin genes (Walker *et al.* 1983). This work illustrates how the activity of manipulated gene sequences can be assayed after they have been introduced into mammalian cells. No mammalian vector is involved — the activity of the introduced sequences was assayed after only a short time period following transfection.

The genes coding for human and rat insulin, and the rat chymotrypsin gene had been cloned and characterized. These genes are expressed at a high level only in the pancreas. Each is expressed in clearly distinct cell types: insulin is synthesized in endocrine β-cells and chymotrypsin in exocrine cells. DNA sequences containing the promoter and 5'-flanking sequences of these genes were linked to the coding sequence of bacterial chloramphenicol acetyltransferase (CAT). This enzyme activity can be assayed very sensitively (see p. 262) and was used here as a 'reporter

258

Section 3
Cloning in
Organisms
Other than
E. coli

enzyme' whose activity was taken to be a measure of the transcription of the insulin or chymotrypsin genes. Recombinant plasmids containing such genes were constructed and grown in E. coli. These plasmids were not designed to be replicating eukaryotic vectors. Although it is possible that some exogenous DNA replication may occur when they are introduced into mammalian cells (this was not tested), they were expected to persist just long enough in the cell for their *transient* expression to be assayed. Therefore, plasmid DNA was introduced into either pancreatic endocrine or pancreatic exocrine cell lines in culture, and after a subsequent 44-hour incubation cell extracts were assayed for CAT activity. It was found that the constructs retained their preferential expression in the appropriate cell type. The insulin 5'-flanking DNA conferred a high level of CAT expression in the endocrine, but not the exocrine cell line, with the converse being the case for the chymotrypsin 5'-flanking DNA.

The analysis was extended by creating deletions in the 5'-flanking sequences and testing their effects on expression. From such experiments it could be concluded that there are sequences located upstream of the promoter, between 150 and 300 base pairs of the transcription start site, which are essential for appropriate cell-specific gene transcription.

Co-transfection (co-transformation)

Following the general acceptance of the calcium phosphate method, subsequent experiments showed that the thymidine kinase gene of herpes simplex virus was effective in transfecting Tk$^-$ mammalian cells (see Fig. 13.1) to a stable Tk$^+$ phenotype which can be selected in HAT medium (Wigler et al. 1977). However, the isolation of cells transfected by other genes which do not encode selectable markers remained problematic. A breakthrough was made when it was discovered that cells can be simultaneously co-transfected by a mixture of two *physically unlinked* DNAs in the calcium phosphate precipitate (Wigler et al. 1979). To obtain co-transfectants, cultured cells were exposed to the thymidine kinase gene in the presence of a vast excess of a well-defined DNA, such as pBR322 or bacteriophage ϕX174 DNA, for which hybridization probes could be readily prepared. In order to achieve a suitably high DNA concentration for the formation of an effective co-precipitate, 'carrier' DNA was also included. This often consisted of total cellular DNA isolated from salmon sperm. Tk$^+$ transfectants were selected and scored by molecular hybridization for the co-transfer of the *unselected*, pBR322 or ϕX174, DNA.

Wigler et al. (1979) demonstrated the co-transfection of mouse Tk$^-$ cells with pBR322, bacteriophage ϕX174 DNA and rabbit β-globin gene sequences and we shall use their ϕX experiments for illustrative purposes. ϕX replicative form DNA was cleaved with *Pst* I which recognizes a single site in the circular genome. Five hundred picograms of the purified thymidine kinase gene were mixed with 10 µg of *Pst* I-cleaved ϕX replicative form DNA. This DNA mixture was added to Tk$^-$ mouse cells and 25 Tk$^+$ transfectants were observed per 10^6 cells after 2 weeks in HAT medium. To determine if these Tk$^+$ transfectants also contained ϕX DNA sequences, high mol. wt DNA from the transfectants was cleaved with *Eco* RI which recognizes no sites in the φX genome. The cleaved DNA was fractionated by agarose gel electrophoresis, transferred to nitrocellulose filters by 'Southern blotting' and annealed with labelled φX DNA. These annealing experiments demonstrated that 14 out of 16 Tk$^+$ transfectants had acquired one or more φX sequences. Subsequent studies of the integrated DNA tell us something about the molecular events taking place during co-transfection. Although the added DNAs in the calcium phosphate co-precipitate are not physically linked, Southern blot analysis of the DNA integrated into the host genome reveals large concatemeric structures, up to 2000 kb long. These concatemers include copies of the selectable marker, the co-transfecting DNA and fragments of carrier DNA (Perucho *et al.* 1980b). Therefore at some stage during the co-transfection process the DNAs must be physically ligated. The foreign DNA can probably integrate virtually anywhere in the genome (Robbins *et al.* 1981, Scangos *et al.* 1981).

The co-transfection phenomenon allows the stable introduction into cultured mammalian cells of any cloned gene. There is no requirement for a vector capable of replication in the host cell. Originally, co-transfection was discovered using two unlinked DNAs. However, this is an unimportant feature; analogous results can be obtained if the selected and unselected genes are ligated in prior manipulations. Originally, herpes simplex virus DNA provided pure tk DNA. Later this tk gene was cloned in *E. coli* plasmids, hence providing a very convenient source.

It has been found subsequently that the HSV tk gene fragment cloned in *E. coli* plasmids as a 3.4 kb *Bam* HI fragment bears a promoter which is rather weak in mammalian cells. It requires the addition of an enhancer for full activity. Since it is now apparent that efficient expression of selectable markers is required for high transfection efficiencies, the weakness of the tk promoter may explain the rather low transfection frequencies obtained in early experiments.

The requirement for a tk$^-$ recipient cell line was a serious limitation of the use of tk selection. This was overcome by the development of so-called *dominant* selectable markers which could be used with non-mutant cell lines.

Selectable markers other than tk

These markers include:
1 dihydrofolate reductase and associated methotrexate resistance;
2 rodent CAD;
3 bacterial XGPRT, and
4 bacterial neomycin phosphotransferase.

Dihydrofolate reductase: methotrexate resistance: amplicons. Dihydrofolate reductase (Fig. 13.1) is sensitive to the inhibitor methotrexate (Mtx). Cultured wild-type cells are sensitive to concentrations of the drug at about 0.1 μg/ml. Mtx-resistant cell lines have been selected and have been found to fall into three categories:
1 cells with decreased cellular uptake of Mtx,
2 cells overproducing DHFR,

260

*Section 3
Cloning in
Organisms
Other than
E. coli*

3   cells having structural alterations in DHFR, lowering its affinity for Mtx.

It has been found that cells overproducing DHFR contain increased copy numbers of the gene (Schimke *et al.* 1978). The DNA sequence that is amplified can be large, about 100 kb, and it can be amplified to a copy number of up to 1000. Often it is present as small extrachromosonal elements called double minute chromosomes, alternatively the amplified DNA can remain at its original chromosomal site or be transposed elsewhere (Schimke 1984).

The Chinese hamster ovary 'A29' cell line is notable because it is extremely resistant to Mtx and has been shown to synthesize *increased amounts* of an *altered* DHFR (Flintoff *et al.* 1976). Wigler *et al.* (1980) have used genomic DNA of the A29 cell line as a donor of the DHFR gene in co-transfection experiments with Mtx-sensitive cells. This system has the advantage that the high gene copy number in donor DNA gives efficient transfection. Additionally, the selection system is powerful and can be applied to *non-mutant* Mtx-sensitive cell lines.

There is one further advantage of the Mtx system: highly resistant variants of the transfected cell line can be selected so as to give concomitant amplification of the unselected DNA. In order to explore this possibility, Wigler *et al.* (1980) first showed that *Sal* I digestion of A29 DNA did not destroy its ability to transfect cells to Mtx-resistance. The *Sal* I-cleaved A29 DNA was then ligated to *Sal* I-linearized pBR322 and used to transfect a Mtx-sensitive mouse cell line. Resistant colonies were picked, grown, and exposed to increasing concentrations of the drug so as to select highly resistant variants. DNAs of certain variants resistant to 40 µg/ml were analysed by Southern blot hybridization with pBR322 DNA as the probe. This analysis showed that the pBR322 sequence had undergone a substantial amplification of at least 50-fold.

This illustrates the fact that the unit of amplification, i.e. the *amplicon*, can be much larger than the selected DHFR gene. DNA that is covalently linked to the DHFR gene is co-amplified. This has important biotechnological implications because increasing the copy number of a foreign gene creates the opportunity for obtaining expression of the gene product at a high level.

There are several variations in the use of amplicons. One approach is simply to use co-transfection to integrate and link the foreign gene to a DHFR gene. For example, Christman *et al.* (1982) used a procedure very similar to that already described, co-transfecting a cloned hepatitis B virus genome with A29 genomic DNA into mouse 3T3 cells and selecting a cell line expressing HBV surface antigen. Other approaches involve pSV-dhfr (see p. 263) or related vectors expressing mouse dhfr cDNA in co-transfection with genes encoding useful products such as tissue plasminogen activator (Kaufman *et al.* 1985), human interferon-gamma (Scahill *et al.* 1983) or hepatitis B virus surface antigen (Patzer *et al.* 1986). The mouse DHFR gene is rather long, 31 kb (Crouse *et al.* 1982) so that DHFR cDNA is conveniently linked to expression cassettes in such constructs (Kaufman & Sharp 1982).

A variation on the use of DHFR depends upon the fact that bacterial DHFR is intrinsically resistant to Mtx (although sensitive to trimethoprim;

see Chapter 8). O'Hare *et al.* (1981) have constructed plasmids in which the bacterial gene is transcribed from an SV40 promoter. Such plasmids transfected mouse cells to a Mtx-resistant phenotype by integration into the mouse genome.

CAD: PALA resistance. The CAD protein is a multifunctional enzyme catalysing the first three steps of *de-novo* uridine biosynthesis: carbamyl phosphate synthetase; aspartate transcarbamylase; and dihydroorotase (Swyryd *et al.* 1974). One of these activities, aspartate transcarbamylase, is inhibited by N-phosphonacetyl-L-aspartate (PALA). PALA-resistant mammalian cells overproduce CAD from highly amplified copies of the CAD gene, in a manner analogous to Mtx-resistance (Wahl *et al.* 1984). The CAD gene of the Syrian hamster has been cloned on cosmid vectors in *E. coli* and has been shown to provide a dominant, amplifiable genetic marker that can be selected in non-mutant cells on the basis of resistance to high concentrations of PALA (De Saint Vincent *et al.* 1981, Wahl *et al.* 1984).

XGPRT: mycophenolic acid resistance. The *E. coli* enzyme xanthine−guanine phosphoribosyltransferase, XGPRT, is a bacterial analogue of mammalian HGPRT. However, by contrast with HGPRT, it has the additional ability to convert xanthine to XMP and hence ultimately to GMP (Fig. 13.1). Only hypoxanthine and guanine are substrates of HGPRT. Mulligan & Berg (1980, 1981a,b) have cloned the XGPRT gene and incorporated it into a variety of vectors in which transcription of the bacterial gene is directed by a SV40 promoter (see p. 263). Such constructs are capable of transfecting HGPRT⁻ cells to HGPRT⁺, but, more importantly, provide a selectable marker for non-mutant cells in medium containing adenine, mycophenolic acid and xanthine (Fig. 13.1). This selection can be made more effective by adding aminopterin, which blocks endogenous purine biosynthesis.

Neomycin phosphotransferase: G418 resistance. Bacterial transposons Tn5 and Tn601 encode distinct neomycin phosphotransferases, whose expression confers resistance to aminoglycoside antibiotics (kanamycin, neomycin and G418), which are protein synthesis inhibitors, active in bacterial or eukaryotic cells. Berg (1981) has incorporated a neomycin phosphotransferase gene into constructs analogous to those containing XGPRT. Other constructs have linked the neomycin phosphotransferase gene to the Herpes simplex virus tk promoter in an *E. coli* plasmid (Colbère-Garapin *et al.* 1981). Transfectants of non-mutant mammalian cells which contain such constructs can be selected by antibiotic resistance. Colbère-Garapin *et al.* (1981) demonstrated the application of their construct to the co-transfection of a variety of cell lines from different mammalian species. Grosveld *et al.* (1982) have also constructed cosmid cloning vectors which include selective markers for growth in the host bacterium (β-lactamase) and animal cells. The markers for animal cells were neomycin phosphotransferase, HSV thymidine kinase, or XGPRT. Such cosmids can be used to construct libraries of eukaryotic genes from which a particular recombinant can be isolated. The recombinant cosmid DNA can then be transfected into animal cells at high efficiency where transfectants in

262

*Section 3
Cloning in
Organisms
Other than
E. coli*

which the DNA has integrated into the nuclear genome can be readily selected. The power of aminoglycoside antibiotic resistance as a selective system in eukaryotes is now very evident. It has application in yeast (Jimenez & Davis 1980; see Chapter 11) and plants (Chapter 12).

The pSV and pRSV plasmids and pSV-CAT

In previous sections we have mentioned the use of mouse DHFR cDNA, and the bacterial genes encoding XGPRT and neomycin phosphotransferase, in order to provide dominant selectable markers. We stated that expression of these genes was obtained from an SV40 promoter. Here we explain these constructions in a little more detail.

In a subsequent section we shall describe the molecular biology of SV40, but for our purposes here it is only necessary to note that the small *Pvu* II−*Hin*d III fragment of SV40 DNA (Fig. 13.2) contains the promoter for early transcription and the transcriptional start point but no AUG translational initiation codon. The plasmid pSV contains an expression cassette consisting of this promoter fragment, followed by a sequence containing the small t intron, followed by the transcript polyadenylation signal. These SV40 sequences are inserted into the pBR322 vector. The genes for neomycin phosphotransferase, or XGPRT or mouse DHFR were inserted as *Hin*d III−Bg III fragments into the expression site to produce pSV2-neo (Southern & Berg 1982), pSV2-gpt (Mulligan & Berg 1980, 1981a,b) and pSV2-dhfr (Subramani *et al.* 1981). The SV40 expression signals function in a wide variety of mammalian cells, hence providing a generally useful set of dominant selectable markers. However, it has been observed that the promoter from the retrovirus Rous sarcoma virus (RSV) is more powerful than the SV40 promoter in various cell types (Gorman *et al.* 1982a). The promoter elements of retroviruses are located in their long terminal repeat (LTR) sequences (see p. 292). A 524 nucleotide pair fragment of one LTR of RSV has been isolated and incorporated in place of the *Pvu* II−*Hin*d III SV40 promoter fragment in the pSV series of plasmids. This has given rise to the pRSV series of expression plasmids, including pRSV-neo and pRSV-gpt (Gorman *et al.* 1983). These plasmids appear to give a relatively high frequency of DNA-mediated transfection, presumably because of efficient expression of the selected marker gene.

The expression cassettes of the pSV and pRSV types have also been used to drive expression of the bacterial chloramphenicol acetyltransferase (CAT) gene. This gene is derived from Tn9 of *E. coli*, where it confers resistance to the antibiotic chloramphenicol. We have already discussed a particular application of CAT as a reporter enzyme earlier in this chapter (p. 257). The pSV-CAT and pRSV-CAT constructs (Gorman *et al.* 1982b) have been very widely used as tools for analysing transient expression of transfected DNAs where a promoter (SV40 early, or RSV) which is expressed in many cell types is convenient. The CAT enzyme activity can be assayed very sensitively and rapidly in simple homogenates of cells or tissues, and there is no endogenous eukaryotic activity (Fig. 13.3). The pSV-CAT construct has been the source of a DNA fragment bearing a CAT coding sequence, linked to a polyadenylation signal, which has subsequently been included in a huge range of gene constructs for particular applications.

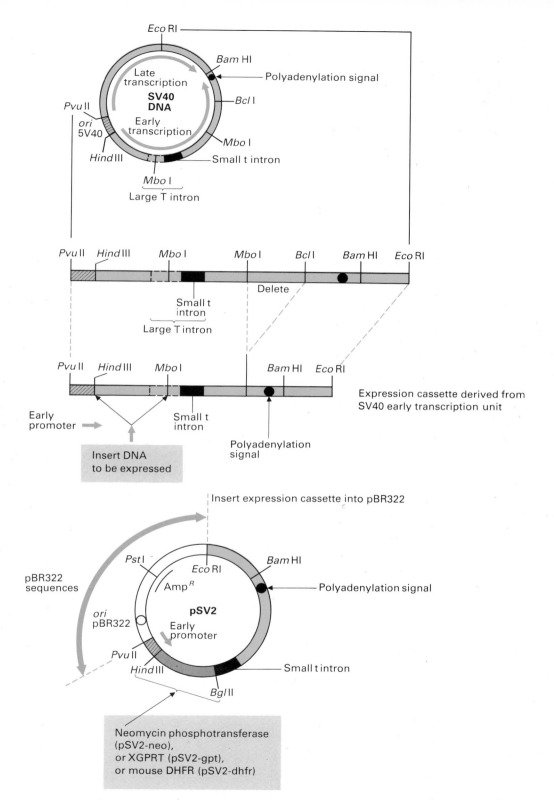

Fig. 13.2 Construction of pSV2-neo, pSV2-gpt and pSV2-dhfr.

264

*Section 3
Cloning in
Organisms
Other than
E. coli*

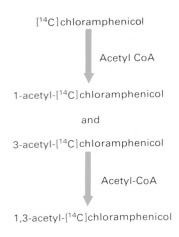

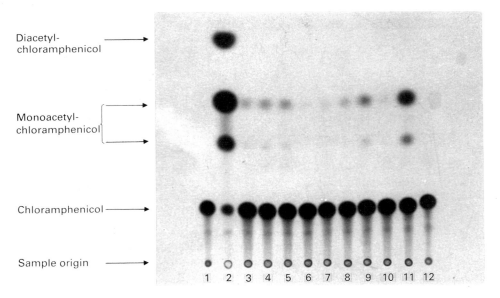

Fig. 13.3 The enzyme chloramphenicol acetyltransferase (CAT) catalyses the transfer of acetyl groups from acetyl-CoA to chloramphenicol. A mixture of monoacetylated forms of chloramphenicol is the first product of the reaction. Further reaction produces 1,3-acetyl-chloramphenicol. The activity is conveniently assayed with ^{14}C-labelled chloramphenicol as substrate. Thin-layer chromatography separates the products, which can be detected by autoradiographing the chromatogram. An autoradiograph of a typical set of assays is shown. Sample 1 is a negative control with no enzyme activity. Sample 2 is a positive control showing extensive activity. Samples 3–12 exhibit intermediate CAT activity.

The fate of transfecting DNA

In discussing the fate of transfecting DNA it is necessary to distinguish between procedures in which a high concentration of high mol. wt 'carrier DNA' (usually total genomic DNA from a cheap, convenient source such as salmon or herring sperm, or chicken red blood cells) has been used, and those where it has not been used. Carrier DNA was used in the co-transfection procedure of Wigler *et al.* (1979) discussed above. We noted there that the selectable marker DNA, the unselected DNA and the carrier DNA are eventually found in large concatemeric structures. It appears that during the transformation the DNAs become fragmented and then joined in random combinations. Recombination events also take place (Miller & Temin 1983). The large concatemer of selected and non-selected DNA interspersed with carrier DNA is called a *transgenome*. It forms early in the transformation process and at a later stage integrates at an apparently 'random site' into the genome, and thereupon the cells become stably transformed. In stable transformations there is usually a single transgenome (Robbins *et al.* 1981). The transgenome is susceptible to partial or more-or-less complete deletion (studied for example by reversion from tk$^+$ to a tk$^-$ phenotype) at a relatively high frequency (Perucho *et al.* 1980b).

The apparently random integration of the transfecting DNA is turned to advantage in gene isolation procedures which exploit the 'tagging' approach described in the following section.

Where stable transfection has taken place with pure plasmids in the absence of carrier DNA using the calcium phosphate technique, the plasmids integrate into the genome as single (or, rarely, a few) copies found at one to five separate chromosomal locations. This is the case, for example, with pSV2-neo (Southern & Berg 1982) or pSV2-gpt (Mulligan & Berg 1981a). Using transfection with pure plasmid by the electroporation technique a similar small number of copies are integrated. Where the DNA is microinjected into the nucleus the DNA also becomes integrated at random sites. Multiple head-to-tail concatemers are integrated if a large amount of DNA is microinjected; single copies are integrated upon micro-injection of smaller amounts. Whether the transfection technique employs calcium phosphate precipitation, electroporation or microinjection, linearization of the plasmid seems to increase the efficiency of integration and the possibility of concatemer formation (Huttner *et al.* 1981, Folger *et al.* 1982, Potter *et al.* 1984)

Isolation of genes transferred to animal cells in culture

The calcium phosphate method for transferring DNA into cultured animal cells has been described. Techniques have been devised for screening libraries specifically for the transferred sequences. Thus any gene which can be selected or recognized by its phenotypic effects in tissue culture can be isolated. There are several variants of this approach. The simplest is 'plasmid rescue' (Hanahan *et al.* 1980, Perucho *et al.* 1980a). In the example shown in Fig. 13.4, plasmid rescue is used to isolate a chicken tk$^+$ gene. First *Hin*d III fragments of non-mutant, total nuclear chicken DNA are ligated into pBR322. This recombinant DNA is used directly to transform mouse tk$^-$ cells, from which tk$^+$ transformants are selected in HAT medium. DNA from these transformants is prepared and used in a second round of transformation. This second round eliminates most of the non-selected recombinant plasmid DNA which becomes integrated in the first round. DNA from these secondary tk$^+$ transformants is cut with an ap-

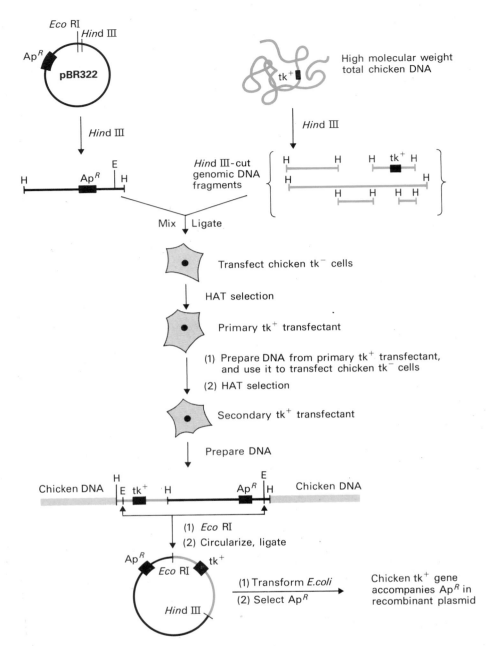

Fig. 13.4 Plasmid rescue, applied to the isolation of the chicken thymidine kinase gene (Perucho *et al.* 1980a). (See text for details.)

propriate restriction enzyme, circularized and selected in E. coli on the basis of plasmid-borne drug resistance. The tk⁺ gene accompanies the drug resistance marker in the plasmid which has been rescued. Thus in the plasmid rescue approach the transferred DNA is 'tagged' with the drug resistance marker, which can be selected in E. coli. In a related approach the transferred DNA is tagged with an amber suppressor gene.

This is illustrated in Fig. 13.5, in which the cloning of a human oncogene is used as an example (Goldfarb *et al*. 1982).

In other variants of this approach, the transferred DNA is 'tagged' with DNA sequences that are not selected genetically but which can be detected by molecular hybridization probes. Figure 13.6 shows the procedure of Lowy *et al*. (1980) in which this tag is simply the plasmid pBR322 to which the transferred DNA has been ligated. There is, therefore, a formal resemblance to the plasmid rescue approach, but an advantage of this procedure is the incorporation of the phage λ cloning step. As we saw in Chapter 4, genomic libraries can be prepared in phage λ very efficiently when combined with *in-vitro* packaging. In an alternative variant the 'tag' for the transferred DNA is not provided artificially, but is provided by 'Alu' sequences (Rubin *et al*. 1983). These are members of a highly repetitive family of sequences. Copies are dispersed throughout the human genome such that any substantial fragment of human DNA several kb in size is likely to contain at least one copy of an Alu sequence. It has been demonstrated that under stringent hybridization conditions the presence of Alu sequences can be used to establish the presence of human DNA in mouse or hamster cells (Gusella *et al*. 1980, Murray *et al*. 1981, Perucho *et al*. 1981).

Gene targeting in animal cells: homologous recombination and allele replacement

The armoury of techniques available to the gene manipulator allows any cloned gene sequence to be altered as desired *in vitro*. It would be a great advance if such alterations could be engineered into copies of a chosen gene *in situ* within the chromosomes of a living animal cell. The strategy for achieving this desirable aim is to bring about the change in the endogenous gene through homologous recombination between it and incoming mutated copies of the gene introduced by a DNA transfection procedure. If this capability were available for mouse cells it would, for example, be possible to introduce a mutation into a chosen gene within embryo-derived stem (ES) cells in culture. These cells can then be incorporated into mouse embryos at the blastocyst stage.

ES cells have pluripotential developmental capacity, and can give rise to many cell types in the resulting adult mouse, including germ cells. Interbreeding of heterozygous sibling progeny produced from such adults would yield animals homozygous for the desired mutation (Fig. 13.7). Thus the functioning of any gene could be studied in whole animals, providing only that it has been previously cloned.

In Chapter 11 we discussed how, in yeast, homologous recombination is relatively easily detected. Homologous recombination between transforming DNA and the endogenous gene is stimulated by the free DNA ends of the transforming DNA. The situation in animal cells seems to be similar to that in yeast, with the important difference that random integration of the transfecting DNA occurs relatively very frequently, hence making it difficult to detect homologous recombination (or gene conversion) events. One solution to this problem in studying homologous re-

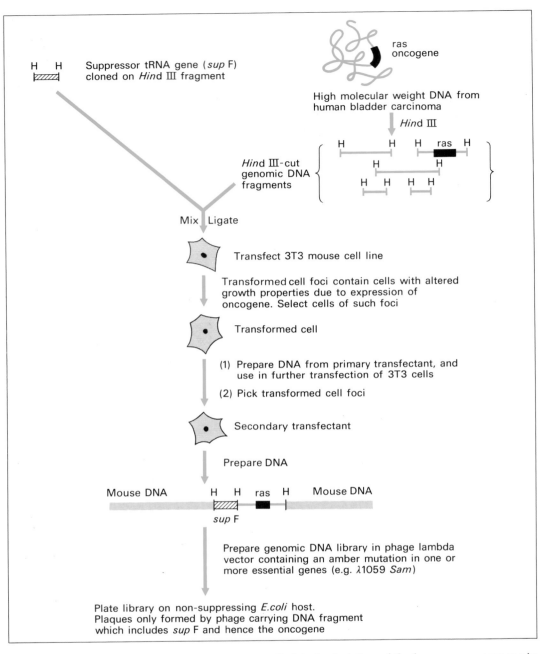

Fig. 13.5 Suppressor rescue, applied to the isolation of the human, *ras* oncogene in the T24 bladder carcinoma (Goldfarb *et al.* 1981).

combination is to use test systems in which selectable inactivation of a test gene (HGPRT) or activation (Thomas *et al.* 1986) of a mutated test gene (neo) occurs by homologous recombination (or gene conversion) but not by random integration. These experiments have defined some parameters of homologous recombination in animal cells, and have also revealed

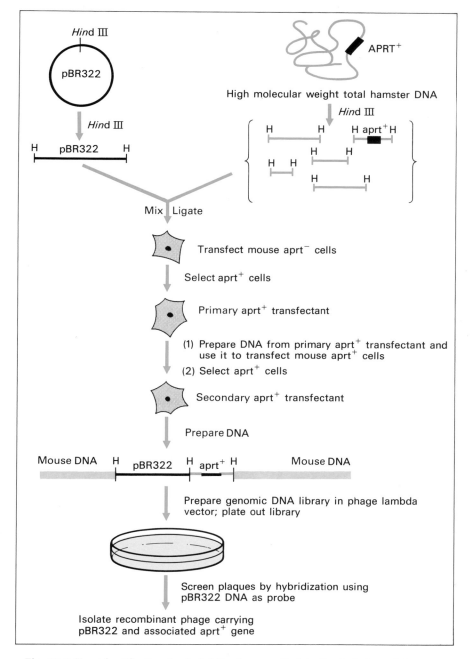

Fig. 13.6 Tag identification by hybridization, applied to the isolation of the hamster adenine phosphoribosyl transferase (aprt) gene (Lowy *et al.* 1980).

evidence for stimulation of unplanned mutations in the test gene as a result of the introduction of homologous DNA (Thomas & Cappechi 1986). However, these test systems are artificial in that a suitable selectable phenotype is not available for the great majority of genes for which allele replacement might be desired.

270

*Section 3
Cloning in
Organisms
Other than
E. coli*

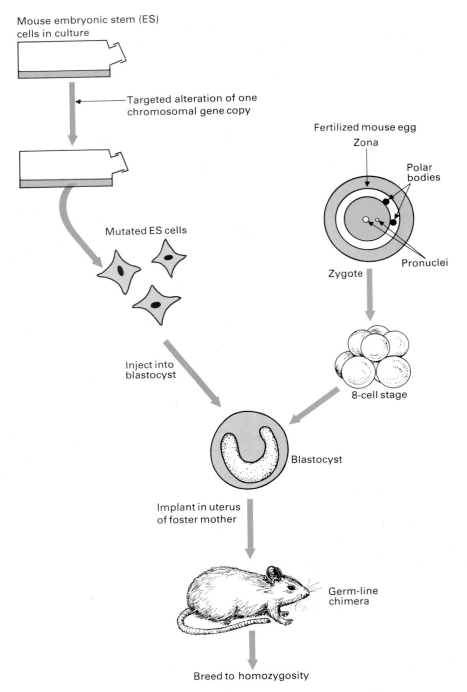

Fig. 13.7 Gene targeting in ES cells to give homozygous mutant mice.

An alternative and general approach to the problem of detecting rare homologous recombination events is simply to use brute force and examine very large numbers of individual transfectants. This has been the approach used for studying homologous recombination in the human β-

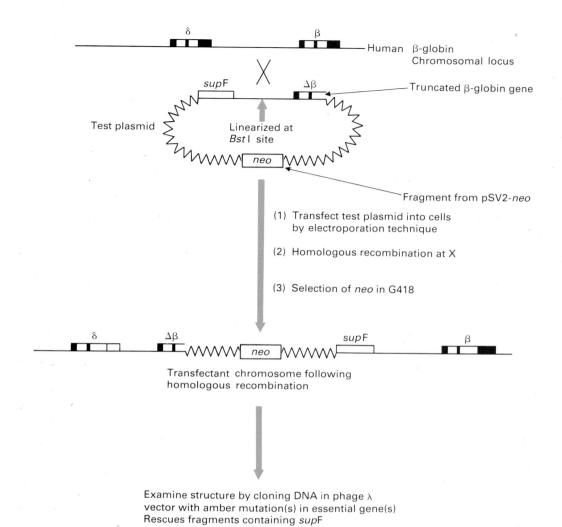

Fig. 13.8 Homologous recombination between test plasmid and human β-globin locus.

globin locus of a human fibroblast × mouse erythroleukaemia hybrid somatic cell line (Smithies *et al.* 1985). A modified β-globin sequence was 'tagged' with a *sup*F gene (Fig. 13.8). This tag was used to isolate cloned DNA from transformed cells into which the plasmid had been introduced by electroporation. These experiments were designed to study integration of a test plasmid by a single recombination event stimulated by free DNA ends, i.e. the whole plasmid DNA became integrated by a single recombination event as though it were circular, although it had been linearized to provide free ends to stimulate recombination. The frequency of homologous recombination was found to be between 10^{-2} and 10^{-3} that of random integration. Other workers have used the polymerase chain reaction (Chapter 15) as a technique for examining large numbers of different transfectants (see Hogan & Lyons 1988). Screening can be carried out

272

*Section 3
Cloning in
Organisms
Other than
E. coli*

relatively quickly without the need for cloning. In summary, these experiments show that with either electroporation or nuclear microinjection the frequency of homologous recombination can be in the range of $10^{-2}-10^{-3}$ transformants or injected cells.

Note that as shown in Fig. 13.8 a single homologous recombination event leads to duplication of sequences and not allele replacement: this requires either a gene conversion event or two homologous recombination events. A practicable strategy for achieving a particular allele replacement event, leading to disruption of any chosen gene, with a generally applicable selection system for enriching cells in which the rare event has occurred, has recently been developed (Mansour *et al.* 1988). The strategy is illustrated in Fig. 13.9 and involves both positive and negative selection. The transfecting DNA contains a disrupted copy of any gene, depicted as gene *A*. The disruption is due to the insertion of a *neo* gene which confers G418 resistance. The DNA also contains the tk gene of herpes simplex virus. When two homologous recombination events occur between the targeting vector and a chromosomal copy of gene *A*, the disrupted gene replaces the chromosomal copy without including the HSV−tk gene. Random integration events elsewhere in the genome will usually lead to insertion of the entire targeting vector including the HSV−tk gene. Transfection by both kinds of process is positively selected by virtue of G418 resistance, but random integration is excluded by selection for the absence of the HSV−tk gene. (HSV−tk confers sensitivity to the nucleoside analogue gancyclovir.) This combination of positive and negative selection leads to a 2000-fold enrichment for ES cells that contain a targeted mutation. Although this strategy is expected to be general in that gene *A* could be any gene, in its present form it is limited to disruption of targeted genes by *neo*. This is quite adequate for the envisaged application of disrupting genes in ES cells with a view to creating mutant mice. But an important application of homologous recombination is in the field of gene therapy for human inherited diseases. Here the requirement is to introduce functional, wild-type alleles into cells with a mutant allele (Chapter 15).

There are two reasons for the desire to introduce the functional gene at the precise chromosonal location of the mutated gene. First, random integration may have deleterious effects due to disruption of important gene functions at the unplanned site of integration. Second, it is clear that transferred genes are often expressed much less efficiently than endogenous genes, and hence insufficient gene product may not correct the deficiency. In the case of transfected globin genes, sequences which normally flank the β-globin locus at considerable distances have been found to have an important effect, possibly by influencing chromatin structure so as to place it in an 'open' configuration. Inclusion of these sequences with transfected β-globin genes allows normal, relatively high level expression in transgenic mice (Grosveld *et al.* 1987). For homologous recombination in the gene therapy field it may be that brute force methods like those described above may prove to be the most useful.

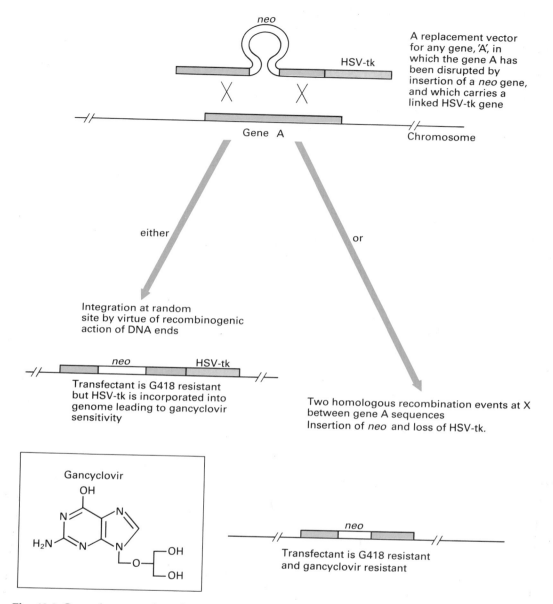

Fig. 13.9 General strategy for selecting ES cell transfectants in which gene 'A' has been disrupted by insertion of a *neo* gene. Gancyclovir is a nucleoside analogue which selectively kills cells expressing HSV-tk because the substrate requirement of HSV-tk is less stringent than that of cellular tk.

VIRAL VECTORS

Many animal viruses have been subjugated as vectors. Virtually every virus that has been studied in any detail and that has a DNA genome or a DNA stage in its replication cycle has been manipulated in this way. As we have seen in bacterial systems, viruses provide an efficient way of

274

Section 3
Cloning in
Organisms
Other than
E. coli

introducing foreign gene sequences into a cell: the gene manipulator can exploit their natural ability to adsorb to cells and infect them. Also as in bacterial systems, animal viruses often contain powerful promoters which can be of general applicability in driving gene expression, and in many cases they have the ability to replicate their genomes to high intracellular copy numbers, hence providing a route to high level expression of foreign genes. Certain animal viruses, of which retroviruses are the most prominent example, naturally integrate their DNA into the host chromosome as part of their replication cycle: another feature open to exploitation by the gene manipulator. Viral vectors can provide alternative and distinctive means for transferring genes into animal cells in addition to the transfection approaches discussed so far.

Examples of vectors derived from several types of animal virus will follow in this chapter, but we begin with a close look at some early SV40 manipulation. Development of SV40 derivatives preceded others because SV40 molecular biology was the first to be worked out in detail. Its small genome was the first animal virus genome to be completely sequenced. Ramifications of SV40 'vectorology' are to be found widespread in animal cell molecular biology of the past decade.

Basic properties of SV40

The virus particle contains the circular duplex DNA genome, of about 5.2 kb (Fig. 13.10), associated with the four histones H4, H2a, H2b and H3 in a mini-chromosome. H1 is absent. The capsid is composed of 420 subunits of the 47 K polypeptide VP1. Two minor polypeptides VP2 and VP3, which consist largely of identical amino acid sequences, are also present.

SV40 can enter two types of life cycle depending upon the host cell. In *permissive* cells, which are usually permanent cell lines derived from the African green monkey, virus replication occurs in a normal infection. In

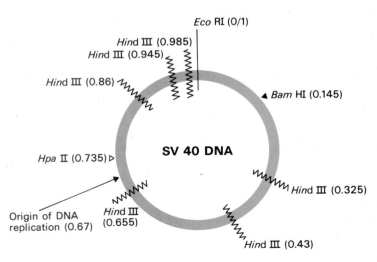

Fig. 13.10 Restriction endonuclease sites on SV40 DNA. The map co-ordinates of each site are shown in parentheses.

non-permissive cells, usually mouse or hamster cell lines, there is no lytic

275

*Chapter 13
Introducing
Genes into
Animal Cells*

infection because the virus is unable to complete DNA replication. However, growth transformation of non-permissive cells can occur. In such cells SV40 DNA sequences are integrated into the host genome. The SV40 sequences are often amplified and rearranged in such transformed cell DNA. Because of the unpredictability in this integration and the subsequent rearrangements, recombinant SV40 virus vectors are not commonly used as integrative vectors.

The lytic infection of monkey cells by SV40 can be divided into three distinct phases. During the first 8 hours the virus particles are uncoated and the DNA moves to the host cell nucleus. In the following 4 hours, the *early* phase, synthesis of early mRNA and early protein occur and there is a virus-induced stimulation of host cell DNA synthesis. The *late* phase occupies the next 36 hours and during this period there is synthesis of viral DNA, late mRNA, and late protein, and the phase culminates in virus assembly and cellular disintegration.

A functional map of the SV40 genome is given in Fig. 13.11. The region of about 400 base pairs around the origin of DNA replication is extremely interesting. Closely associated with the origin are control signals regulating the initiation of early and late primary transcripts. The early transcription is stimulated by two tandem, 72 base pairs, enhancer sequences located within this region.

An important feature of SV40 gene expression is the complex pattern of RNA splicing (Fig 13.11). Alternative splicing pathways process the early primary transcript into two different early mRNAs. The late primary transcript can be processed into three different late mRNAs.

The two early mRNAs encode the large T and small t proteins (*Tumour* proteins or antigens). The late mRNAs, designated according to their sedimentation coefficients as the 16s, 18s and 19s mRNAs, encode VP1, VP3 and VP2 respectively. The 18s and 19s mRNAs have coding sequences in common and direct the synthesis of VP3 and VP2 such that they contain amino acid sequences in common. More striking, however, is the finding that the VP1 coding region overlaps VP2 and VP3 in a different translational reading frame.

SV40 vectors: strategies

The assembly of SV40 virions imposes a strict size limitation on the amount of recombinant DNA which can be packaged. In view of this, two strategies have been employed in developing SV40 vectors. The first is to replace a region of the viral genome with an equivalently sized fragment of foreign DNA, and hence produce a recombinant DNA that can replicate and be packaged into virions in permissive cells. In order to supply genetic functions lost by replacement of virus sequences, a helper SV40 virus must genetically complement the recombinant.

In the alternative strategy, the size limitation of the virion is avoided. Recombinants are constructed which are never packaged into virions and give no lytic infection. These are maintained in host cells transiently as high copy number, unintegrated, plasmid-like DNA molecules.

276

*Section 3
Cloning in
Organisms
Other than
E. coli*

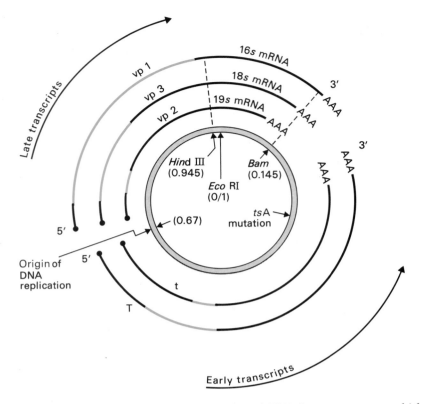

Fig. 13.11 Transcripts and transcript processing of SV40. Intron sequences which are spliced out of the transcripts are shown by red lines.

First experiments with late region replacement

SV40 lacking the entire late region functions can be propagated in *mixed infections* with a temperature-sensitive helper virus that can complement the late region defect. The mixed infection is maintained since the *ts* defect in the helper's early region must be complemented. A suitable helper is a *ts*A mutant, which produces a temperature-sensitive T protein (Mertz & Berg 1974).

Based on this observation Goff & Berg (1976) prepared an SV40 vector by excising virtually the entire late region of the viral DNA by cleavage with *Hpa* II and *Bam* HI restriction endonucleases (Fig. 13.12). Cleavage with these two enzymes produces fragments approximately 0.6 and 0.4 of the genome length. These two fragments were separated by electrophoresis and the smaller fragment discarded. The large fragment, called SVGT-1, was then modified by the addition of 5′ poly (dA) tails using deoxynucleotidyl terminal transferase.

A suitable insert was prepared by cleaving λ DNA with *Eco* RI and *Hin*d III restriction endonucleases. This treatment yielded 12 DNA fragments which were separated by electrophoresis in agarose. Fragment 8 is 1.48 kb in length and contains *ori*, the origin of λ DNA replication, two structural genes, *cII* and *cro*, as well as four transcriptional promoters.

After elution of fragment 8 from the agarose gel it was modified by the addition of poly (dT) tails (Fig. 13.12). SVGT-1 containing the poly (dA) termini and the poly-(dT)-tailed fragment 8 were mixed and annealed. The annealed DNA was then used to transfect monkey kidney cells in the presence or absence of *ts*A helper virus DNA. At the restrictive temperature (41°C) *ts*A DNA alone gave no plaques, the annealed DNA alone gave no plaques, but transfection with the two DNAs together produced 2.5×10^3 p.f.u. μg of annealed DNA.

Thirty-four plaques from the mixedly infected cultures were extracted, mixed with more *ts*A mutant virus and used to infect fresh monkey cells at 41°C. DNA was extracted from the progeny virus and the presence of the λ phage DNA segment was detected by measuring the reassociation kinetics with ^{3}H-labelled λ fragment 8 DNA. DNA from infections with nine of the 34 plaques caused a striking increase in the reassociation rate of the labelled λ DNA fragment indicating the presence of high levels of the λ-SGVT hybrid. Heteroduplex analysis and restriction endonuclease mapping of

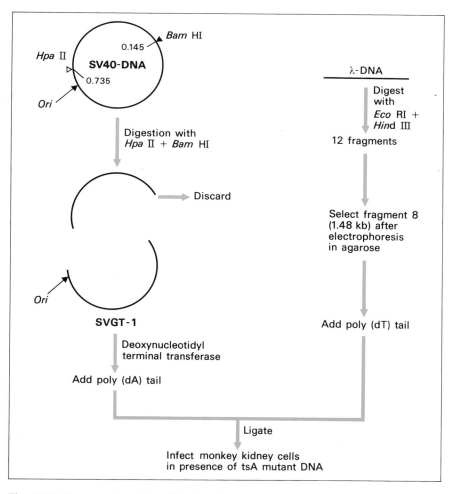

Fig. 13.12 Construction of an SV40-λ DNA hybrid as described in the text.

278

Section 3
Cloning in
Organisms
Other than
E. coli

the viral DNA from these nine infected cultures confirmed that it was the expected mixture of *ts*A DNA and recombinant DNA.

Although the SVGT-λ hybrid DNA replicated in monkey cells in the presence of *ts*A helper DNA, the λ DNA sequences were not transcribed. At a time after infection when substantial amounts of SV40-specific RNA could be detected no λ-specific RNA was found. The search for new polypeptides in the SVGT-hybrid infected cells was also fruitless. These disappointingly negative results were still obtained when hybrids with the λ insert in opposite orientations were used.

Hamer (1977) synthesized SV40-*E. coli* Su$^+$ III recombinant DNA by a method similar to that of Goff & Berg (1976). In contrast to the results of the latter workers, Hamer (1977) found that monkey cells infected with the SV40-Su$^+$ III recombinant DNA synthesized surprisingly large amounts of RNA complementary to the insert. The normal product of the *E. coli* Su$^+$ III gene is a suppressor RNA but all attempts to isolate a free or charged tRNA were unsuccessful.

It is now clear that the lack of proper transcription of the foreign DNA was due to the disruption of the late region post-transcriptional processing in these constructs.

Construction of an improved SV40 vector

From the information presented above on transcription of SV40 it is clear that the ideal vector would retain all the regions implicated in transcriptional initiation and termination splicing and polyadenylation. Inspection of the SV40 map (Fig. 13.11) reveals two restriction endonuclease sites that could be used to generate a suitable vector. First, there is a *Hin*d III site at map position 0.945 which is six nucleotides *proximal* to the initiation codon for VP1 (Fig. 13.13) and 50 nucleotides *distal* to the site at which the leader sequence is joined to the body of 16s mRNA. Second, the *Bam* HI site at map position 0.145 is 50 nucleotides proximal to the termination codon for VP1 translation and 150 nucleotides before the poly A sequence at the 3' end of 16s RNA (Fig. 13.13). If the DNA between co-ordinates 0.945 and 0.145 were removed, the remaining molecule could be used as a vector for it would retain:

1 the origin of replication;
2 the regions at which splicing and polyadenylation occur;
3 the entire early region, and hence could be complemented by a *ts*A mutant.

Such a vector (SVGT-5) has been constructed and used successfully to clone the rabbit β-globin gene in monkey kidney cells (Mulligan *et al.* 1979).

The first step in constructing SVGT-5 was partially to digest SV40 DNA with restriction endonuclease *Hin*d III. The digestion products were separated by electrophoresis and full-length molecules, i.e. those with a single cut, selected. Clearly these full-length molecules could have a cut at any one of the six *Hin*d III sites on SV40 DNA. The mixture of different full-length molecules was then digested with a mixture of restriction endonucleases *Bam* HI and *Eco* RI and fragments of 4.2 kb (SVGT-5) selected by

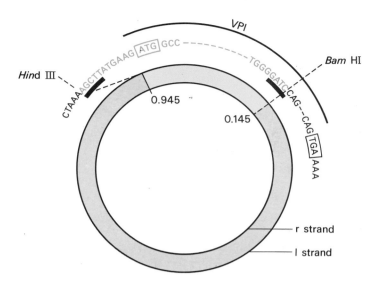

Fig. 13.13 Location of the *Hin*d III and *Bam* HI cleavage sites in relation to the coding sequence for VP1. The triplets enclosed in boxes show the initiation and termination signals for translation of VP1. The sequences underlined are the recognition sites for the *Hin*d III and *Bam* HI restriction endonucleases.

electrophoresis on an agarose gel. Examination of the SV40 restriction map (Fig. 13.14) shows that cleavage of the *Hin*d III site at position 0.945 and at the *Bam* HI site produces the desired SVGT-5 fragment which is 4.2 kb long. However, cleavage at the *Hin*d III site at position 0.325 and at the *Bam* HI site would produce a fragment of similar size. Since only the latter fragment contains an *Eco* RI site, it can be selectively eliminated by including endonuclease *Eco* RI in the digestion mixture.

Example use of SVGT-5: cloning of rabbit β-globin gene in monkey kidney cells

The starting point for this experiment was the recombinant plasmid pβG1 which contains the rabbit β-globin coding sequence. pβG1 was constructed by Maniatis *et al.* (1976) by inserting a cDNA copy of purified β-globin mRNA at the *Eco* RI site of plasmid pMB9 using the homopolymer tailing method. Nucleotide sequence analysis confirmed that pβG1 contained the entire β-globin coding sequence, all of the 3' and 80% of the 5' non-coding sequences in β-globin mRNA (Efstratiadis *et al.* 1976). Basically, the experiment was conducted in two stages:
1 excision of the β-globin cDNA from cDNA from pβG1 and its modification to produce *Hin*d III cohesive sites at either end;
2 insertion of the altered cDNA into SVGT-5.

Alteration of β-globin cDNA (Fig. 13.15)

pβG1 was incubated at 50°C in the presence of S1 nuclease. At this temperature the poly (dA.dT) joints melt and the resulting single strands

280

*Section 3
Cloning in
Organisms
Other than
E. coli*

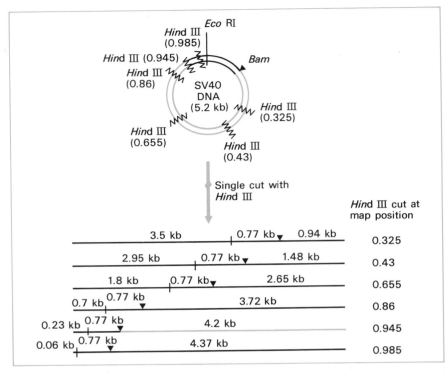

Fig. 13.14 Molecules produced by cutting SV40 DNA once with restriction endonuclease *Hin*d III. Note that when these six different molecules are cleaved with a mixture of *Eco* RI and *Bam* HI restriction endonucleases only one of them will produce a fragment 4.2 kb in length. For convenience only one strand of the DNA is shown.

are digested with nuclease S1. The two fragments were separated by electrophoresis and the β-globin cDNA fragment isolated and treated with DNA polymerase I to convert the 'ragged' single-stranded ends to 'blunt' ends. A synthetic decanucleotide was then attached to either end of the cDNA with the aid of T4 DNA ligase. The resultant molecules were digested with endonuclease *Hin*d III and cloned in the E. *coli* vector pBR322. One of the recombinants between pBR322 and the modified β-globin cDNA was selected and designated pBR322-βG2.

Construction of SVGT-5-RaβG (Fig. 13.16)

The function of the cloning step in pBR322 was the enrichment of the altered cDNA prior to its insertion into SVGT-5. This altered cDNA was excised from pBR322-βG2 by sequential digestion with *Hin*d III and *Bgl* II endonucleases. This yielded a fragment having the codon for initiating translation of β-globin 37 nucleotides downstream from the *Hin*d III endonuclease-generated cohesive end and the translation termination codon just proximal to the *Bgl* II endonuclease-generated end. Even though the *Bam* HI (GGATCC) and *Bgl* II (AGATCT) endonuclease recognition sites differ, the cohesive ends generated by the two endonucleases are identical

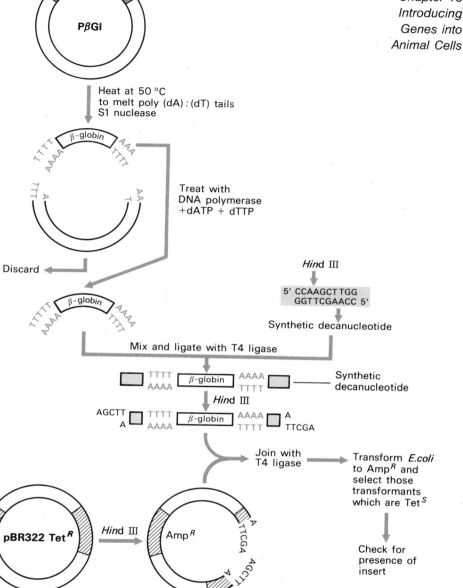

Fig. 13.15 Cloning of the β-globin cDNA in pBR322. (See text for details.)

(GATC). Accordingly, the 0.485 kb β-globin cDNA fragment was ligated to SVGT-5 to produce SVGT-5-RaβG.

To propagate the SVGT-5-RaβG genome, monkey kidney cells were transfected at 41°C with a mixture of SVGT-5-RaβG DNA and the DNA from a *ts*A mutant of SV40.

282

*Section 3
Cloning in
Organisms
Other than
E. coli*

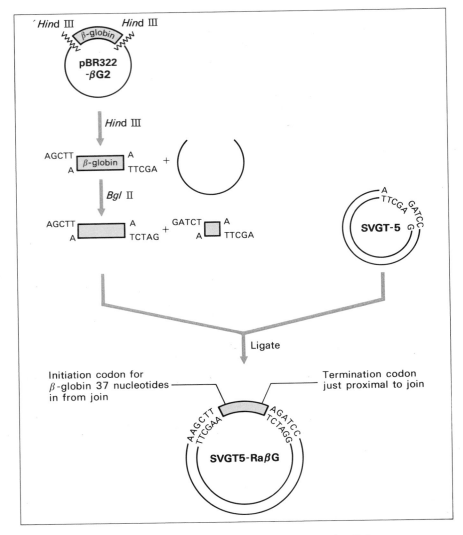

Fig. 13.16 The construction of SVGT-5-RaβG. (See text for details.)

Expression of SVGT-5-RaβG

In SVGT-5-RaβG the coding sequence of VP1 has been precisely replaced by a cDNA copy of rabbit β-globin mRNA SV40 late region leader and splice junctions are linked to the cDNA sequence, and the downstream polyadenylation signal is specified by viral sequences. Following infection of monkey cells with the mixed recombinant/helper virus stock, the synthesis of the appropriate recombinant late mRNA was detected. The hybrid derivative of 16s late mRNA (it is actually 0.5 kb smaller than the normal 16s late mRNA) directed the synthesis of a protein indistinguishable from authentic rabbit β-globin. The protein was synthesized in large amounts, comparable to normal amounts of VP1, but appeared to be unstable, presumably because it was not assembled into haemoglobin.

Application of late-region replacement vectors

Vectors of this type have been effective in obtaining expression of many foreign genes in monkey cells: mouse β-globin (Hamer & Leder 1979a); rat preproinsulin (Gruss & Khoury 1981, Gruss *et al*. 1981a); the p21 protein of Harvey murine sarcoma virus (Gruss *et al*. 1981b); influenza virus hae-magglutinin (Gething & Sambrook 1981; Sveda & Lai 1981), and hepatitis B virus surface antigen (Moriarty *et al*. 1981). Late-region replacement vectors have been used to study post-translational RNA processing and stability. Some of these studies have demonstrated a requirement for an intron in the late-region primary transcript if a cytoplasmic mRNA is to be produced (Gruss *et al*. 1979, Hamer & Leder 1979b, Lai & Khoury 1979). However, this has not proved consistently to be the case for all constructs (Gething & Sambrook 1981, Gruss *et al*. 1981a, Sveda & Lai 1981) so that the reasons for the requirement remain unclear.

Early region replacement vectors; COS cells

When the early region of SV40 is replaced, the essential T function is lost and must be provided by complementation. A major breakthrough in the development of SV40 vectors came when it was found that the T protein can be complemented in the COS cell line (this must not be confused with *cos*, the cohesive end site of phage λ). This cell line is a derivative of the permissive CV-1 monkey cell line. When CV-1 cells are infected with SV40, the normal lytic cycle ensues and no growth transformants are recovered. Gluzman (1981) found that CV-1 cells could be transformed by a segment of SV40 early-region DNA (cloned in an *E. coli* plasmid vector) in which the SV40 origin of DNA replication had been inactivated by a 6-base-pair deletion. Had the origin of replication been functional, the plasmid DNA would have replicated many times under the influence of T protein and made the host cell inviable. The resulting COS cell line (CV-1, origin of SV40) express T protein from integrated SV40 sequences and does so in a cellular background which is permissive for SV40 DNA replication. Infection of these cells with SV40 lacking a functional early region therefore leads to a normal lytic cycle. Since mixed infections are not necessary there is no contaminating helper virus. Therefore SV40 recombinants in which early DNA sequences have been replaced with foreign DNA can be constructed and propagated. Gething & Sambrook (1981) have made recombinants of this type which express influenza virus haemagglutinin in COS cells.

Overview of recombinant SV40 viruses

Both late- and early-region replacement vectors have the advantage of producing recombinant virion particles which introduce the foreign DNA into cells without the need for DNA-mediated transfection. However, advances in transfection technology are reducing the importance of this advantage. The replication of the recombinant DNA to a high copy number is a considerable advantage. For many biotechnological production pro-cesses a limitation on the use of recombinant SV40 viruses is the fact that

284

*Section 3
Cloning in
Organisms
Other than
E. coli*

expression of foreign genes is transient in infected cells; this is often not as convenient as a stable transfected cell line. The maximum capacity of about 2.5 kb of foreign DNA is another potential limitation.

COS cells and SV40 replicons: transient expression: the SV40 enhancer

In COS cells *any* circular DNA, with no definite size limitation, containing a functional SV40 origin of replication should be replicated independently of the cellular DNA as a plasmid-like episome. Many laboratories have constructed vectors based on this principle (Myers & Tjian 1980, Lusky & Botchan 1981, Mellon *et al.* 1981). In general, these vectors consist of a small SV40 DNA fragment containing the SV40 origin cloned in an *E. coli* plasmid vector. Recombinant derivatives can be constructed and grown in *E. coli*. These are then transfected into COS cells where very high copy numbers of the recombinants are obtained. It is important that the *E. coli* plasmid sequence does not contain the so-called 'poison' sequence. This has been identified as a small region of pBR322 which in simian cells causes inefficient replication of any DNA in which it is included (Lusky & Botchan 1981). Plasmid pAT153 (see Chapter 3) and other related deletion derivatives of pBR322 have lost this site and are, therefore, suitable for constructing shuttle vectors for mammalian cells. The poison sequence is coincident with, or very near to, the *nic/bom* site necessary for conjugative mobilization of pBR322 but it is not known if this coincidence is significant.

Permanent cell lines are not established when transient vectors are transfected into COS cells because the massive vector replication makes the cells inviable, but even though only a low proportion of cells are transfected, the high copy number is compensatory and the *transient* expression of cloned genes can be analysed. The major application of these vectors is in providing a rapid means of screening the effects of *in-vitro* manipulations upon transcriptional and post-transcriptional control sequences. In such studies the important transcriptional stimulation by the SV40 enhancer sequences was discovered (Banerji *et al.* 1981, Khoury & Gruss 1983).

The enhancer sequences are a pair of 72-nucleotide, tandemly repeated sequences located near the origin of DNA replication and which are necessary for efficient early transcription. By linking the enhancer sequences to other genes it was found that they stimulate or *enhance* transcription from virtually any promoter placed near them. This effect operates on both sides of the enhancer, i.e. it is independent of enhancer orientation and extends to a promoter placed several kilobases away.

Subsequent to the discovery of the SV40 enhancer, virtually every cellular gene transcribed by RNA polymerase II has been found to have one or more associated enhancers. Some enhancers are cell-type specific. It is not entirely clear how enhancers work; it is possible that several mechanisms are involved. Enhancers and their relationship to transcription factors and other proteins which influence transcription are the subject of the most intensive research, which is beyond the scope of this book (Maniatis *et al.* 1987, Wasylyk 1988). The SV40 enhancer sequences are important to gene manipulators because of their stimulatory effect on the SV40 early transcription unit to which foreign genes may be fused and hence expressed.

Alternatively, transcription from *cellular* promoters may be enhanced by the enhancer. The SV40 enhancer has the advantageous property of stimulating transcription in a wide variety of cell types. It is very complex. Detailed analysis has revealed many distinct sequence motifs within the enhancer region, with different cellular specificities, which in combination give the rather non-specific generalized enhancement.

In the original COS cell lines the SV40 T protein was expressed constitutively from integrated SV40 sequences. Subsequent COS-type cell lines have been produced in which the SV40 T protein activity is regulated. This has been achieved either by using a temperature-sensitive T protein (Rio *et al.* 1985) or by placing the T protein gene under the control of a human metallothionein gene promoter which is inducible by heavy metals (Gerard & Gluzman 1985). Using such cell lines it is possible to establish *stable* transfectants in which the SV40 replicon is maintained episomally at a low copy number. The T protein activity can be increased when expression of foreign genes carried on the vector is required.

SV40 and polyoma virus mini-replicons

A series of vectors has been constructed which comprise an *E. coli* plasmid replicon derived from pBR322, an SV40 origin of DNA replication, a functional early region providing T protein, and an SV40 transcription unit into which foreign genes can be inserted and expressed. These vectors are grown and manipulated in *E. coli* and then transferred into permissive monkey cells by standard transfection procedures (Berg 1981, Mulligan & Berg 1981b).

Analogous vectors contain a virus origin of DNA replication and early region derived from polyoma virus, whose structural organization closely resembles that of SV40. Mouse cells, but not monkey cells, are permissive for polyoma virus and therefore are the host for such polyoma-derived vectors.

Because of the intact T function, these mini-replicon vectors replicate in permissive cells and attain very high copy numbers over a period of a few days. This eventually kills the transfected host cells, but massive transient expression of a foreign gene is possible. If appropriate, the foreign gene can be expressed from its own promoter rather than a viral one. Deans *et al.* (1984) used such a polyoma-based mini-replicon vector to express an immunoglobulin heavy-chain gene in transfected mouse lymphocytes (Fig. 13.17).

Vectors based upon bovine papillomavirus (BPV) DNA

Bovine papillomavirus causes warts in cattle. It is a member of a group of viruses which induce warts and papillomas in a range of mammals. Certain of the human papillomaviruses are implicated in causing cervical cancer.

286

*Section 3
Cloning in
Organisms
Other than
E. coli*

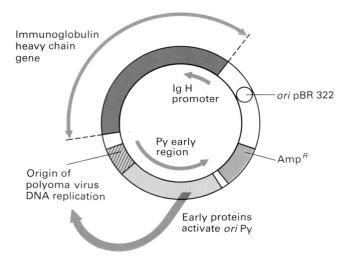

Fig. 13.17 Polyoma-based mini-replicon for expression of a mouse immuno-globulin heavy chain gene in mouse lymphocytes.

Like other papillomaviruses, BPV normally infects terminally differentiated squamous epithelial cells, but it can cause growth transformation in dividing cultured mouse cell lines. In such transformed mouse cells the viral DNA is found as a multicopy plasmid. This property has been the impetus for developing vector molecules (reviewed in Campo 1985) based on the BPV replicon.

BPV has a circular double-stranded DNA genome of about 7.9 kb, which has been completely sequenced (Chen *et al.* 1982). The molecular biology of the virus (Giri & Danos 1986) is considerably more complex than that of SV40, to which it appears to be distantly related. The early region of the genome is contained on a large subgenomic fragment of about 5.5 kb. This fragment makes up 69% of the entire genome and has been found to contain all the functions necessary for growth transformation of mouse cells, including the origin of BPV DNA replication. This fragment is called BPV_{69T}. Simple vectors have been constructed which consist of this fragment cloned in pBR322. Such a vector has been used to express a rat preproinsulin genomic DNA in mouse cells, using growth transformation as the basis for selecting transfected cells (Sarver *et al.* 1981a,b). The pBR322 sequences appeared to be inhibitory and so were excised from the plasmid before DNA-mediated transfection into the mouse cells. The remaining BPT_{69T} rat gene fragment cyclized spontaneously in the host cell and replicated as a plasmid at about 100 copies per cell. Expression of the rat gene from its own promoter was readily detected. Other workers have constructed similar vectors and found that the *E. coli* plasmid sequence could be retained. For example, a recombinant carrying human β-globin gene sequences was maintained at 10−30 copies per cell (Di Maio *et al.* 1982). In similar experiments with a simple BPV_{69T}-based vector, transformed cells contained as many as 200 copies of the recombinant DNA (Ostrowski *et al.* 1983). There are examples of transfectants that have maintained the recombinant plasmid as an episome, without integration into the chromosome, for long periods (Fukunaga *et al.* 1984), but in other cases the DNA has been found to integrate into the host genome (Ostrowski *et al.* 1983, Sambrook *et al.* 1985). The simple vectors use growth transformation as the means of selecting transfectants. This is not always convenient and limits the range of mouse cell lines available as hosts. By inserting a neomycin phosphotransferase gene into the vector plasmids, transfectants can then be selected for resistance to G418 (Law *et al.* 1983).

BPV-derived vectors are useful because *permanent* cell lines can be obtained carrying the recombinant DNA either episomally or integrated at relatively high copy numbers. An additional advantage is the ability to carry large DNA inserts.

Recombinant vaccinia viruses, and other large DNA viruses

Vaccinia virus is closely related to variola virus, which causes smallpox, and inoculation with vaccinia virus provides a high degree of immunity to smallpox. There has been a proposal to construct vaccinia virus recombinants which express antigens of unrelated pathogens and use them as live vaccines against those pathogens. Smith *et al.* (1983a/b) adopted a clever strategy for expressing the hepatitis B virus surface antigen (HBsAg) in vaccinia which took into account:

1 the large size (187 kb) of vaccinia DNA,
2 the lack of infectivity of isolated viral DNA,
3 the packaging of viral enzymes necessary for transcription within the virion, and
4 the probability that vaccinia virus has evolved its own transcriptional regulatory sequences operative in the cytoplasm where viral transcription and replication occur.

Briefly, fragments of vaccinia DNA were cloned in an *E. coli* plasmid vector that contained a non-functional vaccinia thymidine kinase gene. This gene had been rendered inactive owing to the insertion of a vaccinia DNA fragment containing a promoter derived from another early vaccinia gene. The HBsAg gene was inserted next to the vaccinia promoter in the correct orientation. This chimaeric HBsAg gene was then inserted into vaccinia DNA by homologous recombination as follows. Monkey cells were infected with wild-type vaccinia and simultaneously transfected with the recombinant *E. coli* plasmid. Homologous recombination could then replace the functional thymidine kinase gene of the wild-type virus with the non-functional tk gene sequence which included the HBsAg chimaeric gene. Such virus would be TK⁻ and would be selectable on the basis of resistance to BUdR (Fig. 13.1). When cells were infected with such TK⁻ virus, th found to synthesize HBsAg and secrete it into the culture medium nated rabbits rapidly produced high-titre antibodies to HBsAg. A strategy (Fig. 13.18) was then used to construct an infectious vaccini recombinant which expresses the influenza haemagglutinin gene and i resistance to influenza virus infection in hamsters (Smith *et al.* 1983

Subsequently recombinant vaccinia viruses expressing other imp genes have been constructed, including an AIDS virus envelope gene

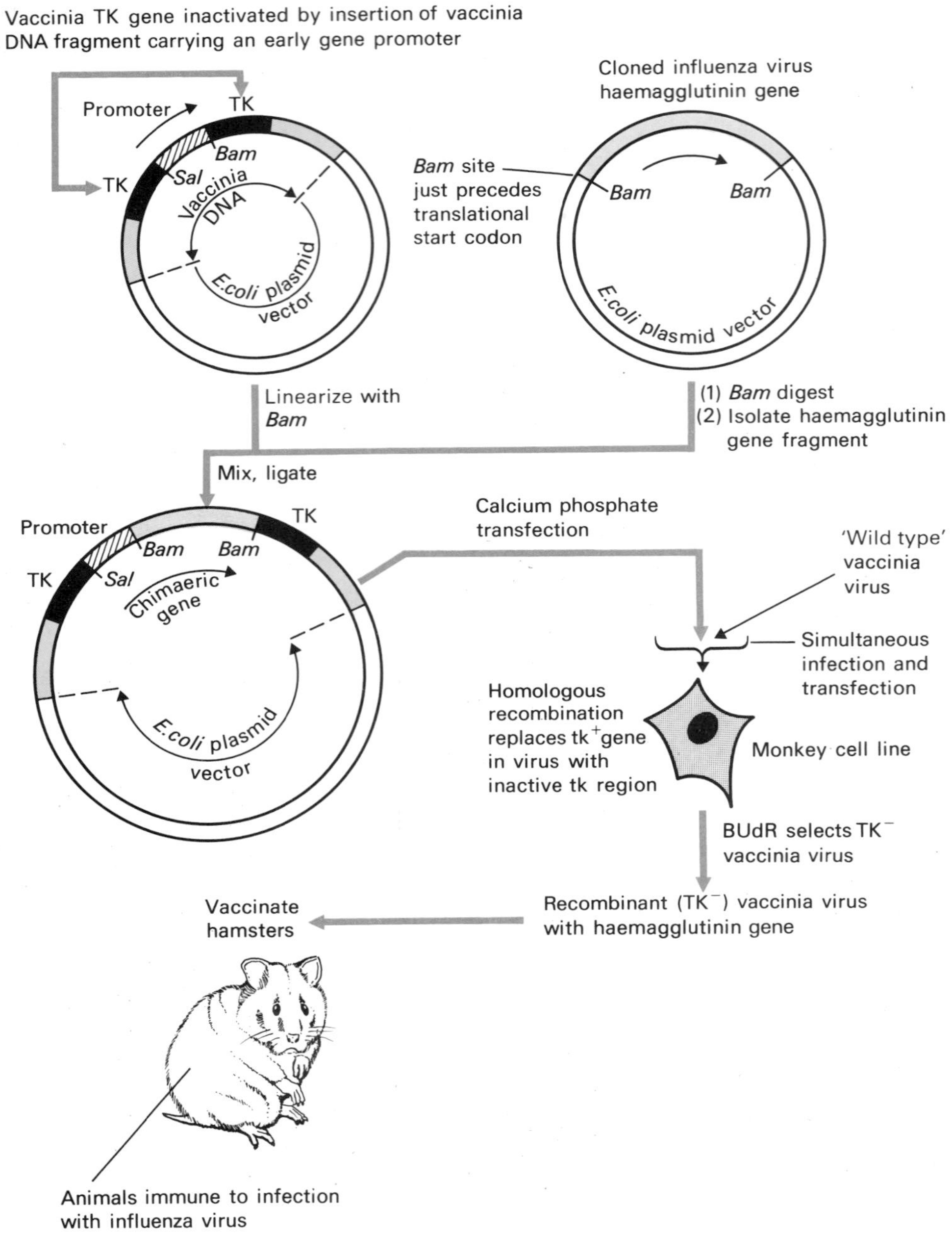

Vaccinia TK gene inactivated by insertion of vaccinia
DNA fragment carrying an early gene promoter

Cloned influenza virus
haemagglutinin gene

Promoter — TK
Bam
Sal
Vaccinia DNA
TK
E.coli plasmid vector

Bam site
just precedes
translational
start codon

Bam   Bam

E.coli plasmid vector

Linearize with
Bam

(1) Bam digest
(2) Isolate haemagglutinin
    gene fragment

Mix, ligate

Promoter   TK
Bam   Bam
TK   Sal
Chimaeric
gene
E.coli plasmid
vector

Calcium phosphate
transfection

'Wild type'
vaccinia
virus

Simultaneous
infection and
transfection

Homologous
recombination
replaces tk⁺gene
in virus with
inactive tk region

Monkey cell line

BUdR selects TK⁻
vaccinia virus

Vaccinate
hamsters

Recombinant (TK⁻) vaccinia virus
with haemagglutinin gene

Animals immune to infection
with influenza virus

Fig. 13.18 Construction of an infectious vaccinia virus recombinant expressing
influenza virus haemagglutinin.

et al. 1986), HTLV-III envelope gene (Chakrabarti et al. 1986) and hepatitis B
virus surface antigen gene (Moss et al. 1984, Smith et al. 1983).

These experiments showing the potential of vaccinia vectors for immunization have been followed by the actual immunization of wild foxes
against rabies. A recombinant vaccinia virus that expresses the rabies antigen
was administered to the wild population of foxes in North Eastern France

by providing 'bait' consisting of chicken heads spiked with the virus. A substantial proportion of the wild foxes were shown to have acquired immunity to rabies by this means (Blancou *et al.* 1986).

Polyvalent vaccines

The prospect of using recombinant vaccinia virus for immunization of human populations has received considerable attention. Experience with vaccinia virus in large-scale immunization is extensive following the global programme leading to the eradication of smallpox. The last reported case of smallpox was in 1977 (reviewed by Bebehani 1983). Large-scale vaccination of human populations must take into account production and delivery costs as well as efficacy and safety. A heat-stable polyvalent vaccine, simultaneously providing immunity to several diseases, that requires only one inoculation, may be both feasible and economical, using a live vector such as vaccinia. It has been possible to construct an infectious vaccinia recombinant which expresses multiple foreign antigens (Perkus *et al.* 1985). Other vector systems based on mycobacteria (Jacobs *et al.* 1987), *Salmonella typhimurium* (Hosieth & Stocker 1981), adenoviruses (Davis *et al.* 1985) and herpes viruses, including varicella-zoster (Shih *et al.* 1984, Lowe *et al.* 1987) have been proposed as live, possibly polyvalent, vaccines. A possible limitation of such live polyvalent vaccines is existing immunity within the target population, not only to a potential vector, but to any of the expressed antigens. This might restrict replication of the vector and reduce the overall immune response to the vaccine. In order to test this possibility a recombinant vaccinia virus was constructed which expresses the surface antigens from two distinct human pathogens, influenza A virus haemagglutinin and herpes simplex virus type 1 glycoprotein D (Flexner *et al.* 1988). This recombinant protected mice against influenza A virus and HSV-1 infection, and importantly, was even able to immunize mice that had recently recovered from infection with either influenza A virus or HSV-1. These results are encouraging for the further development of polyvalent recombinant vaccines.

Whether recombinant vaccinia viruses will ever be used for human vaccination remains to be seen. The following arguments have been put forward against them: (a) recombinant vaccinia viruses are often less virulent than wild-type vaccinia, but too little is known about the mechanism of attenuation, (b) vaccinia virus has a broad host range and transmission might occur from man to animals and back again. Therefore the deliberate spreading of a recombinant virus could be viewed as particularly undesirable, and (c) the fact that the present human population contains individuals who have been vaccinated against smallpox may render recombinant vaccinia ineffective.

Interestingly in this context, an accidental trial of recombinant vaccinia in man has been reported (Jones *et al.* 1986). A researcher accidentally inoculated a small cut on his finger with a drop of recombinant vaccinia which expresses the vesicular stomatitis virus nucleoprotein (N). VSV causes a highly contagious vesicular disease of cattle, pigs and horses. In man VSV causes flu-like symptoms. It was evident that the researcher had

290

*Section 3
Cloning in
Organisms
Other than
E. coli*

become infected with the recombinant vaccinia, but he did not experience fever or malaise. The mildness of the infection was notable because a non-vaccine strain of the virus was involved. This could be related to the fact that the researcher had been vaccinated some 30 years previously, or to attenuation of the virus by the insertional inactivation of the viral tk gene. Importantly, serum from the researcher showed a high antibody titre against VSV N-protein. Therefore this event showed that despite previous vaccination, a new immune response was elicited to non-vaccinia protein in man. It is clear that recombinant vaccinia, even if not ultimately used as a vector for human immunization, is a prototype which has provided a great stimulus for the development of other recombinant viruses for immunization.

Baculovirus vectors for insect cells and insects

Baculoviruses infect insects. They have large double-stranded, circular DNA genomes within a rod-shaped capsid. The baculoviruses which have been exploited as vectors are the *Autographa california* nuclear polyhedrosis virus (AcNPV) (Smith *et al.* 1983, Miller *et al.* 1987) and the silkworm virus, *Bombyx mori* nuclear polyhedrosis virus (BmNPV) (Maeda *et al.* 1985). During normal infection the viruses produce nuclear inclusion bodies which consist of virus particles embedded in a protein matrix, the major component of which is a virus-encoded protein called polyhedrin. Large amounts of virus and polyhedrin are produced. Transcription of the polyhedrin gene is driven by an extremely active promoter, which is therefore ideally suited as a promoter for driving expression of foreign genes. This is all the more attractive because the polyhedrin gene product is not essential for viral replication. Construction of expression vectors has therefore consisted of inserting a foreign coding sequence just downstream of the polyhedrin promoter. This cannot be achieved directly because the large size (about 130 kb) of the viral DNA precludes simple *in vitro* manipulation. The strategy for inserting the foreign DNA into the virus has many similarities to that already described for vaccinia virus.

Consider the insertion of a human interferon (IFN)-α gene into BmPNV (Maeda *et al.* 1985) as an example. Essentially, the same strategy has been used for expressing foreign genes in AcPNV. First, a fragment of DNA from the polyhedrin gene region was cloned into the *E. coli* vector pUC9 so as to place a polylinker site just downstream from the start point of transcription directed by the polyhedrin promoter. On the other side of the polylinker, opposite from the promoter fragment, was inserted a DNA fragment containing downstream sequences of the polyhedrin gene. A human IFN-α coding sequence was then ligated into the polylinker site so that it was placed between the polyhedrin promoter and downstream sequences in the correct orientation so as to give an expression construct. Homologous recombination *in vivo* was then used to replace the wild-type polyhedrin gene in viral DNA with the disrupted polyhedrin gene of the expression construct. This was achieved by co-transfecting *Bombyx mori* cultured cells with expression plasmid DNA and BmPNV DNA. Recombinant viruses gave characteristic plaques because of their failure to form inclusion bodies, and so could be isolated. These recombinant viruses replicated in silkworm caterpillars, producing up to 50 μg per larva.

The silkworm host has advantages as a production system employing recombinant BmPNV because silkworms can be cultured easily and at low cost, as in the long-established silk industry. Expression of recombinant AcPNV is usually carried out by infecting cultured cells of the insect *Spodoptera frugiperda*. The foreign gene is expressed during the infection and very high yields of protein can be achieved by the time the cells lyse, about 3 days post-infection (e.g. Miyamoto *et al*. 1985, Kuroda *et al*. 1986).

Retroviruses

Retroviruses have useful properties for exploitation as vectors: (a) they cover a wide host range including avian, mammalian and other animal hosts, (b) infection does not lead to cell death, infected cells produce virus over an indefinite period, (c) viral gene expression is driven by strong promoters — these can be harnessed to foreign genes, and (d) in the case of murine mammary tumour virus the promoter function can be switched on and off experimentally. Transcription is induced by glucocorticoid hormones (Lee *et al*. 1981, Scheidereit *et al*. 1983).

Retroviruses contain RNA genomes. The virus particle actually contains two copies of viral RNA. Each RNA genome has many features similar to a eukaryotic mRNA in that it has a poly (A) sequence of approximately 200 residues at the 3′ terminus and a typical cap structure at the 5′ terminus. Figure 13.19 shows a simple map of a typical retrovirus, Moloney-MuLV.

A very abbreviated scheme of the retroviral replication cycle is given in Fig. 13.20. This illustrates the main points necessary for an appreciation of vector development. (For a complete account of retroviral biology see Weiss *et al*. 1985.) When the viral RNA enters the cell it is accompanied by reverse transcriptase and integrase which are packaged into the virion. The reverse transcriptase (and associated RNaseH activity) then engages in a complex series of cDNA synthesis reactions which lead to the production of a double-stranded DNA copy of the viral RNA. This DNA copy, which is called the proviral DNA, is slightly longer than the RNA from which it

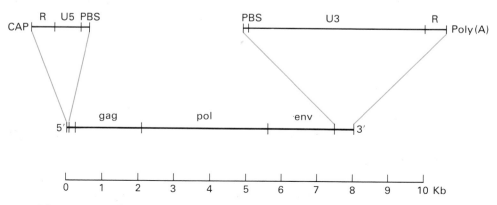

Fig. 13.19 The genome of murine leukaemia virus.

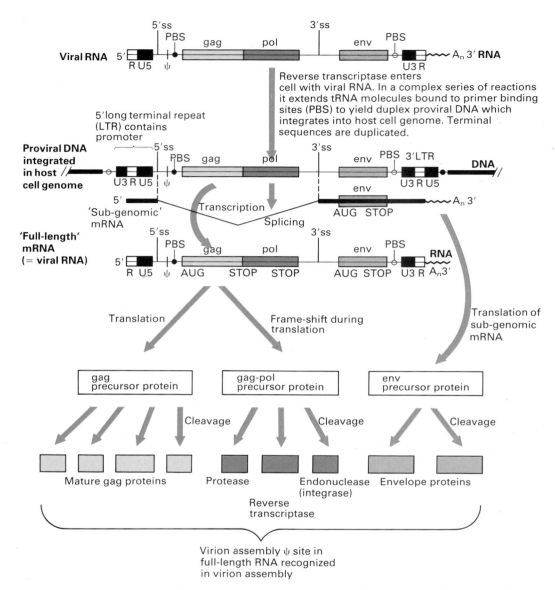

Fig. 13.20 Abbreviated scheme of retroviral replication cycle. (See text for details.)

was derived because terminal sequences are duplicated in the process of converting it to double-stranded form. The proviral DNA circularizes and, through the action of the integrase protein, inserts into the host genome. There is usually only a single copy (or a small number of copies) of proviral DNA integrated per cell, the site of integration in the genome is 'random' (there may be a preference for transcriptionally active regions) and the proviral DNA integrates such that it is bounded by the long terminal repeat, LTR, sequences which include a strong promoter for RNA polymerase II. The proviral genome contains three genes: *gag*, *pol* and *env*. These are transcribed and translated into precursor proteins which are subject to proteolytic cleavages to produce mature proteins. Full-length transcripts of the proviral DNA constitute the RNA genome which is packaged into virus particles. The *psi* site in full-length RNA is important for the interaction of viral RNA with proteins in the assembly of virions. Assembly takes place at the cell membrane with virus budding off from the cell. There is no cell lysis. Figure 13.21 is a schematic representation of the structural proteins within the virion.

Strategies of vector construction

Certain retroviruses are acutely oncogenic. Most of these have genome structures similar to that of Moloney MuLV shown in Fig. 13.19 but differ in that they contain an oncogene sequence. These viral oncogenes are derived from cellular genes and are often the result of obvious gene fusions with viral genes. As a result such oncogenic viruses have lost essential viral gene functions. These viruses are therefore defective. They can replicate in a mixed infection with a helper virus which provides the lost functions. Such oncogenic retroviruses demonstrate a natural ability of retroviruses to act as vectors.

For the deliberate construction of retrovirus vectors the starting point has been cloned DNA of the retrovirus. It is necessary to know which *cis*-acting sequences in the genome are essential for its replication and packaging. The LTRs are essential for efficient transcription of proviral DNA and for generating the 3' end of the full-length transcripts. The LTRs are also essential for integration of the proviral DNA into the host cell DNA. Other sequences important for viral replication are located between the 5' LTR and the *gag* gene, and between the *env* gene and the 3' LTR. These include a minus-strand tRNA primer-binding site (PBS) (necessary for initiation of the first strand of DNA synthesis during reverse transcription), a plus-strand tRNA primer-binding site (required for initiation of the second strand of DNA during reverse transcription), and the *psi* site. The *gag*, *pol* and *env* genes are dispensable because their functions can be provided from another proviral DNA.

Generation of infectious recombinant retroviruses

Several groups of researchers developed retroviral vectors in which regions of the proviral DNA were replaced by foreign DNA (Tabin *et al.* 1982, Wei *et al.* 1981, Shimotohno & Temin 1981, 1982). Subsequently a great variety of replacement vectors have been produced (see Weiss *et al.* 1985 for review). It is common to insert a selectable marker gene, such as *neo*, along with the non-selectable gene of interest. A variety of arrangements can be used to obtain expression of inserted genes. If the LTR is to be used to drive expression then the inserted gene is most simply expressed if its ATG is in a similar location to that of the *gag* gene which it replaces. Alternatively, the foreign gene may be driven by its own accompanying promoter. There are examples where selection for expression of one gene appears to interfere with high expression of the second inserted gene in integrated retrovirus recombinants (Emerman & Temin 1986).

The construction of such recombinants takes place in *E. coli* using

294

*Section 3
Cloning in
Organisms
Other than
E. coli*

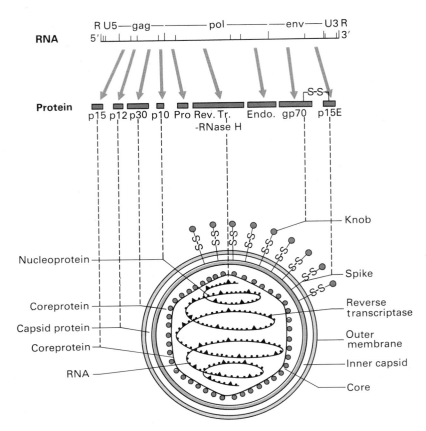

Fig. 13.21 Gene map, polypeptide products and schematized virion particle of a typical retrovirus, murine leukaemia virus.

proviral DNA sequences carried by an *E. coli* plasmid vector. Infectious recombinant virus can then be generated by using standard calcium phosphate-mediated transfection of a suitable animal cell line. The missing functions of the recombinant retrovirus are most simply provided either by co-infecting the transfected cell line with non-defective helper virus, or by co-transfection with non-defective helper virus DNA. In the latter case the transfected DNAs integrate into the genome in a non-specific way with respect to their own sequences. Transcription of intact, non-permuted copies leads to virus expression. The mixed virus preparation can then be used to infect cells in the normal way. These infected cells will continue to produce virus indefinitely. Since the packaging of the recombinant genome is dependent upon the helper virus, it will have the host-range properties of the helper virus (and the helper will also provide the proteins necessary for the initial stages of infection, e.g. reverse transcriptase). The helper virus is said to *pseudotype* the recombinant.

The use of the helper virus just described leads to the production of a mixed virus preparation containing defective recombinant virus and non-defective helper. An important alternative strategy is to use a *helper cell line* to provide the missing functions. Such helper cell lines contain mutant proviruses which have non-functional packaging signals. These mutant

proviruses have deletions of the *psi* site. The first of these helper cell lines·
to be constructed, *psi*-2, is a NIH 3T3 mouse cell line with a transfected
provirus derived from Moloney-MuLV (Mann *et al.* 1983). When such a
helper cell line is transfected with defective, recombinant proviral DNA,
fully infectious recombinant virus particles are produced. These viruses
can be applied to other normal cell lines where recombinant provirus
formation, integration, and expression occur efficiently, but where no
infectious viruses are produced.

The host range of the recombinant retrovirus produced with the aid of
the helper cell line will depend upon its pseudotye. The Moloney-MuLV
provirus in *psi*-2 cells confers a narrow host range. Therefore in order to
broaden the range of hosts available, helper cell lines have now been
produced which are based upon amphotropic strains of murine leukaemia
virus. Amphotropic MuLVs replicate in cells from a wide variety of mammalian and even avian species. These new helper cell lines allow the
transfer of retroviral recombinant DNAs to human cells (Cone & Mulligan
1984, Sorge *et al.* 1984, Miller *et al.* 1985).

In addition to the helper-virus or helper cell line approaches already
mentioned, there is a final possible strategy, that is to produce non-
defective recombinant virus which contains a complete set of essential
genes. This has been achieved with derivatives of the avian virus, Rous
sarcoma virus (e.g. Hughes & Kosick 1984). This approach has not been
extensively explored for mammalian retroviruses, but in view of the fact
that mammalian retroviruses can accommodate an increase in overall
genome size it should be possible to include foreign DNA *in addition* to
the essential gene sequences. In fact it appears that retroviruses have
considerable flexibility in the size of the genome that can be packaged,
ranging from considerably smaller than, up to a few kb larger than, the
normal genome size.

Retroviruses can be used to be reduce the size of gene inserts containing
introns. Should the original insert in the DNA construct of a recombinant
retrovirus contain introns, then the retroviral life cycle which follows
transfection into a susceptible cell leads to the removal of intron sequences
by RNA splicing. A shortened, intronless RNA copy is packaged into
virions and is later represented in proviral DNA copies.

14 Transferring Genes into Animal Oocytes, Eggs and Embryos

The successful introduction of exogenous DNA into cultured mammalian cells has led to new experimentation on gene regulation in animals. We saw in Chapter 13 that by using differentiated cell types in culture it has been possible to analyse DNA sequences essential for cell-specific gene expression. A drawback of this approach is that cultured cell lines are incapable of development into organisms *in vitro*. However, the microinjection systems described in this chapter allow the study of transferred gene sequences in the context of normal embryonic development. These are based upon direct microinjection of DNA into fertilized eggs or embryos. The experimental systems which are most advanced in this context involve the frog *Xenopus laevis*, the mouse and the fly *Drosophila melanogaster*. In addition, the oocytes (which are precursor cells of the mature egg cells) of *Xenopus* have been exploited very usefully as an assay for gene expression quite apart from the developmental context.

Each of the experimental systems has its particular advantages and limitations. Frog oocytes are readily available in large numbers, as are the eggs and embryos. The molecular biology and biochemistry of early development of this vertebrate are, therefore, relatively well studied since material is not limiting. The fact that amphibian embryogenesis occurs outside the mother means that access is not a problem. However, there is very little genetics of *Xenopus*, and breeding experiments requiring the raising of more than one generation to adulthood are tedious. The mouse has a much shorter generation time, but by comparison, access to and manipulation of the mammalian embryo is more complicated. *Drosophila* has all the advantages of an extremely powerful genetic system. Particularly useful is the ability to locate gene sequences simply and precisely by hybridization *in situ* to polytene chromosomes. The relatively small genome size (1.8×10^5 kb, compared with 3×10^6 kb and 2.3×10^6 kb for *Xenopus* and mouse) is also a distinct advantage for molecular genetics.

EXPRESSION OF DNA MICROINJECTED INTO *XENOPUS* OOCYTES AND UNFERTILIZED EGGS

Oocytes can be obtained in large numbers by removal of the ovary of adult female *Xenopus*. Each fully grown oocyte is a large cell (0.8–1.2 mm in diameter) arrested at first meiotic prophase (Fig. 14.1). This large cell has a correspondingly large nucleus (also called the germinal vesicle) which is located in the darkly pigmented, animal hemisphere of the oocyte; because of the large size, microinjection of DNA into the nucleus is technically easy. Typically 20–40 nl is injected through a finely drawn

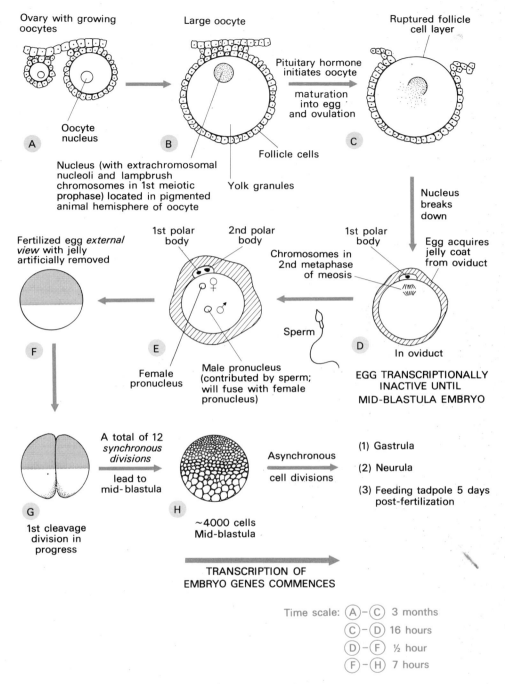

Fig. 14.1 Events in oogenesis and early embryogenesis of *Xenopus*.

glass capillary. The oocyte nucleus contains a store of the three eukaryotic RNA polymerases; enough to furnish the needs of the developing embryo at least until the 60 000 cell stage (Roeder 1974). This store is available for transcription of the exogenous DNA. The oocyte also contains large stores

298

Section 3
Cloning in
Organisms
Other than
E. coli

of histones and other essential components for early embryonic development (reviewed by Davidson 1986). The exogenous DNA is assembled into chromatin (Gargiulo & Worcel 1983, Ryoji & Worcel 1984), but there is no replication of injected duplex DNA in the oocyte. Single-stranded DNA, such as that obtained by cloning in M13 vectors, is converted to duplex form efficiently in the oocyte nucleus.

Expression of microinjected genes

Since the ability of oocytes to transcribe exogenous DNAs was first demonstrated by Mertz & Gurdon (1977) a wide variety of cloned genes from diverse sources has been microinjected. Genes transcribed by each of the three RNA polymerases, I, II and III, have been expressed in this system. When protein-coding genes transcribed by RNA polymerase II have been injected correct transcriptional initiation has often been observed, giving rise ultimately to stable transcripts which are exported to the cytoplasm and translated.

This ready ability to introduce a manipulated DNA into a living nucleus where its expression can be monitored has proved very powerful. The effects of directed changes in the DNA sequence, changes in the RNA transcribed from it, and changes in proteins encoded by it can all be assayed. This approach has been given the name 'surrogate genetics' by Birnstiel (Birnstiel & Chipchase 1977). He and his co-workers originally applied the surrogate genetic approach to the analysis of promoter elements, and related upstream regions of sea-urchin histone genes, which affect their transcription in *Xenopus* oocyte nuclei (Grosschedl & Birnstiel 1980a, b, Grosschedl *et al.* 1983).

Surrogate genetics in the oocyte can be applied not only to analysing promoter function but also to other cis-acting sequences affecting gene function. Thus Williams *et al.* (1984) analysed the DNA sequences necessary for correct polyadenylation at the 3'-terminus of globin mRNA. Other workers have examined the effects of DNA conformation on transcription in oocyte nuclei. Harland *et al.* (1983) found that circular molecules are transcribed more efficiently than linear molecules.

An especially noteworthy application of the oocyte as an assay has exploited the *failure* of the oocyte to produce a correct transcript from an injected gene. It was observed that a sea-urchin histone H3 gene was transcribed from the correct initiation site but failed to generate substantial amounts of mRNA because transcripts extended beyond the correct 3'-terminus. Birnstiel and his co-workers reasoned that the *Xenopus* oocyte may be deficient in an activity necessary for the generation of the correct 3'-terminus of the transcript. In accord with this idea they made extracts of sea-urchin cells and co-injected them with the H3 gene. They were able to identify and purify a component of the sea-urchin cells which complemented the *Xenopus* deficiency (Stunnenberg & Birnstiel 1982). This component was a nuclear RNA-protein complex. Even the pure 60 nucleotide nuclear RNA alone would stimulate correct 3'-terminus formation when co-injected with the H3 gene (Galli *et al.* 1983).

299

Chapter 14
Gene Transfer
to Oocytes,
Eggs and
Embryos

Biochemical complementation in *Xenopus* oocytes has been used as an assay for specific transcription factors which the oocyte appears to lack, and which therefore must be provided exogenously for efficient transcription of certain injected genes. For example Sweeney & Old (1988) found that a promoter containing an immunoglobulin gene octamer sequence was poorly transcribed in the oocyte nucleus. However if mRNA from myeloma cells (which normally express immunoglobulin genes) was first injected into the cytoplasm of the oocyte, then transcription of the immunoglobulin-like promoter, injected into the nucleus, was activated. The stimulation of transcription was probably caused by translation of the myeloma cell mRNA into a specific transcription factor necessary for immunoglobulin gene expression.

The surrogate genetic approach can be applied to RNA splicing. Thus Green *et al*. (1983) synthesized a human β-globin gene transcript *in vitro*, using the phage SP6 RNA polymerase (Chapter 4), and injected the transcript into the oocyte nucleus. The transcript was accurately spliced in the *Xenopus* oocyte nucleus. *In vitro* synthesis of transcripts with 3'-extensions beyond the 3'-terminus of mature mRNA have similarly been used to investigate 3'-end formation and polyadenylation (Krieg & Melton 1984, Chambers & Old 1988).

A further example of the power of the oocyte system follows from the observation that secretory proteins encoded by DNAs injected into the nucleus, or mRNAs injected cytoplasmically, are secreted by the oocyte into the surrounding incubation medium. It has been found that the oocyte recognizes the hydrophobic signal sequence of secretory proteins from a variety of phylogenetically diverse organisms, making this process amenable to surrogate genetics (Lane *et al*. 1980).

The oocyte translation system as an assay for isolating cDNA clones

The *Xenopus* oocyte is, as we have seen, a large cell. It also efficiently translates mRNAs injected into its cytoplasm. The large size makes it easy to measure electrophysiological responses to neurotransmitters and related substances, because electrodes for measuring the response can be inserted into the oocyte without difficulty (Fig. 14.2). It has been found, for example, that oocytes can be made responsive to the mammalian tachykinin neuropeptide, substance K, by injecting a mRNA preparation from bovine stomach into the oocyte cytoplasm. The mRNA preparation contains mRNA encoding the substance K receptor protein, which is evidently expressed and inserted into the oocyte membrane in functional form. Masu *et al*. (1987) exploited this property of oocytes to isolate a cDNA clone encoding the receptor. The principle was to make a cDNA library from stomach mRNA, using a vector in which the cDNA was flanked by a promoter for the SP6 or T7 RNA polymerase. This allowed *in-vitro* synthesis of mRNA from the mixture of cloned cDNAs in the library. The receptor clone was identified by testing for receptor expression following injection of synthetic mRNA into the oocyte cytoplasm. Repeated subdivision of the mixture of

300

*Section 3
Cloning in
Organisms
Other than
E. coli*

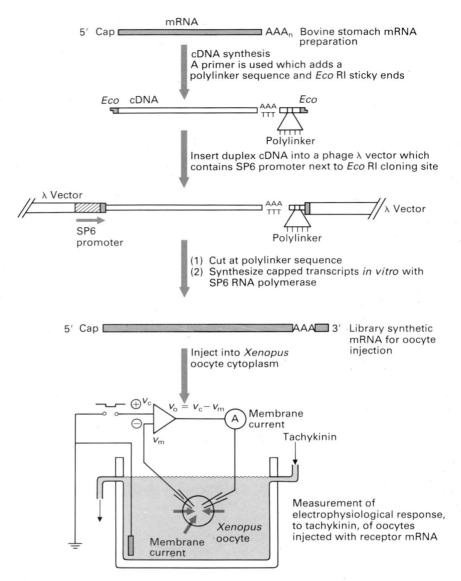

Fig. 14.2 Cloning isolation of bovine substance–K receptor through oocyte expression system. (See text for details.)

cDNAs in the library led to the isolation of a single cloned cDNA. Because oocytes injected with exogenous mRNAs can express a variety of specific receptors and channels, this strategy is very powerful.

Regulation of genes injected into *Xenopus* oocytes

While the oocyte has been a useful system for transcribing genes injected into the nucleus and translating mRNAs injected into the cytoplasm, it has been of limited value in analysing the *regulation* of transcription of injected genes by RNA polymerase II. Two problems occur: (a) The amount of expression observed can vary greatly between different batches of oocytes and can be very sensitive to the amount of DNA deposited in the nucleus (Bendig & Williams 1984, Walmsey & Patient 1987), and (b) certain genes which would be expected to be transcriptionally inactive in oocytes if they mimicked their endogenous counterparts, have been found to be transcribed relatively efficiently. For example, a *Xenopus* β-globin gene and a *Xenopus* albumin gene promoter, which are normally active in adult blood and liver respectively, are both active when injected into the oocyte nucleus (Walmsey & Patient 1987; Old *et al*. 1988).

301

Chapter 14
Gene Transfer
to Oocytes,
Eggs and
Embryos

Injection of DNA into fertilized eggs of *Xenopus*

DNA may be injected into de-jellied fertilized eggs at the one- or two-cells stage. The fate of this DNA has been the subject of recent experiments. In experiments with unfertilized eggs, nucleus-like bodies were formed in which the injected DNA is surrounded by an envelope similar to the normal nuclear envelope (Forbes *et al*. 1983). The DNA is also replicated. There is presently some controversy about whether specific origins of DNA replication are operative in *Xenopus* eggs, or in higher eukaryotes in general for that matter. However, it is clear that all circular, duplex DNAs tested so far are replicated in unfertilized *Xenopus* eggs (reviewed in Laskey *et al*. 1983). In eggs the injected DNA is immediately subject to replication, and thereafter is replicated further only in synchronization with the cell cycle of the activated egg (Harland & Laskey 1980). As an embryo develops, exogenous DNA persists, and at least some of it continues to be replicated. In a typical experiment in which a recombinant plasmid carrying *Xenopus* globin genes was injected, the amount of plasmid DNA increased 50- to 100-fold by the gastrula stage. At subsequent stages, the amount of DNA per embryo decreased and most of the persisting DNA co-migrated with high molecular weight chromosomal DNA (Bendig & Williams 1983). Etkin *et al*. (1987) have analysed the replication of a variety of DNAs injected into *Xenopus* embryos. It was found that various plasmids increase to different extents. This was not simply related to the size of the plasmid. Some sequences inhibited replication. Replication has also been found to depend upon the concatemerization conformation and number of molecules injected (Marini *et al*. 1989).

Some of the injected DNA becomes integrated into the genome (Rusconi & Schaffner 1981) and may be transmitted to the next generation, but the ability of frogs to transmit such DNA to their progeny is not readily investigated because the time scale of breeding experiments in *Xenopus* (at least 6 months from egg to adult) is long.

Transcription of DNA introduced into *Xenopus* embryos by injection of fertilized eggs

Transcription of a variety of exogenous genes has been observed following injection into fertilized eggs. These include: a yeast tRNA$_{leu}$ gene (Newport & Kirschner 1982b); rabbit β-globin gene (Rusconi & Schaffner 1981) and *Xenopus* adult globin genes (Bendig & Williams 1983). Transcription is low or absent immediately following injection, but subsequently becomes

302

Section 3
Cloning in
Organisms
Other than
E. coli

activated in accord with the mid-blastula transition in the developmental programme of *Xenopus* embryos (Newport & Kirschner 1982a,b).

These experiments show that expression of DNA introduced into *Xenopus* is feasible. Several studies have demonstrated apparently correct tissue-specific expression, or developmental regulation of *Xenopus* gene constructs, during early embryogenesis (Krieg & Melton 1985, Wilson *et al*. 1986, Giebelhaus *et al*. 1988). However it is becoming apparent that the embryos exhibit a mosaic pattern of expression of the introduced DNA. To what extent this mosaic pattern of expression is due to mosaicism in the persistence and integration of DNA in different cells of the embryo, is presently unclear.

TRANSGENIC MAMMALS

The ability to introduce genes into the germ line of mammals is one of the greatest technical advances in recent biology. The results of gene manipulation are inherited by the offspring of these animals. All cells of these offspring inherit the introduced gene as part of their genetic make-up. Such animals are said to be *transgenic*. Transgenic mammals have provided a means for studying gene regulation during embryogenesis and in differentiation, for studying the action of oncogenes, and for studying the intricate interactions of cells in the immune system. The whole animal is the ultimate assay system for manipulated genes which direct complex biological processes. In addition, transgenic animals provide exciting possibilities for expressing useful recombinant proteins and for generating precise animal models of human genetic disorders.

Methods for producing transgenic mammals

Several methods for introducing foreign DNA into the germ line of mammals have been developed (Fig. 14.3). A fundamental requirement was the availability of techniques for removing fertilized eggs or early embryos, culturing them briefly *in vitro*, and then returning them to foster mothers where further embryogenesis could proceed. This opened the way for the mixing of cells from different embryos, i.e. chimaera production, for introducing pluripotent cells such as ES cells into developing embryos, for microinjecting DNA, and for infection by retroviruses.

Microinjection of DNA. In early experiments on microinjecting DNA into mouse embryos, SV40 DNA was deposited in embryos at the pre-implantation blastocyst (four to 30 cell) stage (Jaenisch & Mintz 1974). These embryos were implanted into the uteri of foster mothers and allowed to develop. Some cells of the embryo incorporated DNA into their chromosomes, but the adult animals which resulted were mosaics, with only a proportion of the cells in a tissue containing integrated DNA. However, integration into some germ line cells did occur and genetically defined substrains could be obtained in the next generation. In later experiments viral DNA (cloned proviral Moloney murine leukaemia virus DNA) was

303

*Chapter 14
Gene Transfer
to Oocytes,
Eggs and
Embryos*

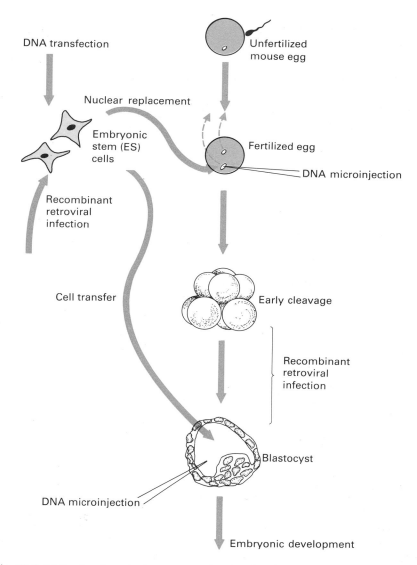

DNA transfection

Unfertilized
mouse egg

Nuclear replacement

Embryonic
stem (ES)
cells

Fertilized egg

DNA microinjection

Recombinant
retroviral
infection

Cell transfer

Early cleavage

Recombinant
retroviral
infection

Blastocyst

DNA microinjection

Embryonic development

Fig. 14.3 Methods of producing transgenic mice that have been devised or proposed. (After Palmiter & Brinster 1986.)

injected into the cytoplasm of one-cell embryos (zygotes). Such embryos developed into adults carrying a single inserted copy of the viral DNA in every cell (Harbers *et al.* 1981).

The procedure which has revolutionized transgenic mouse production is the direct microinjection of DNA into one of the *pronuclei* of the newly fertilized egg (reviewed by Palmiter & Brinster 1986). The process is shown in Fig. 14.4.

The male pronucleus is contributed by the sperm and, being larger than the female pronucleus, is the one usually chosen for microinjection. Typically about 2 pl of DNA-containing solution is introduced. The two pronuclei subsequently fuse to form the diploid, zygote nucleus of the

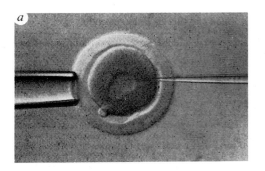
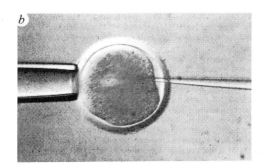
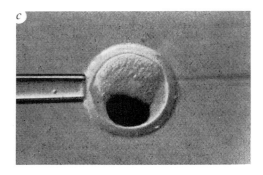

Fig. 14.4 Injection of DNA into the pronucleus of a newly fertilized mouse egg.

fertilized egg. The injected embryos are cultured *in vitro* to morulae or blastocysts and then transferred to pseudopregnant foster mothers (Gordon & Ruddle 1981). In practice, between 3% and 40% of the animals developing from these embryos contain copies of the exogenous DNA (Lacy *et al.* 1983). In such *transgenic* mice the foreign DNA must have been integrated into one of the host chromosomes at an early stage of embryo development. There is usually no mosaicism, and so the foreign DNA is transmitted through the germ line. In different transgenic animals the copy number of integrated plasmid sequences differs, ranging from one copy to several hundred in a head-to-tail array, and the chromosomal location differs (Palmiter *et al.* 1982a, Lacy *et al.* 1983).

In general, the foreign DNA is stably transmitted for generations.

Other methods: retroviruses, ES cells. As we discussed in Chapter 13, proviral DNA of retroviruses integrates by a precise mechanism into the genome of infected cells. Only a single proviral copy is integrated at a given chromosomal site, the chromosomal site is 'random' but the junctions of the proviral DNA are precise with respect to the viral sequences. Infection of pre-implantation embryos by natural or recombinant retroviruses can lead to germ-line integration and hence transgenic animals. An advantage of this method is its technical simplicity; eight-cell embryos with the zona pellucida removed are exposed to concentrated virus stock and transferred to foster mothers. However it has the limitations that additional steps are required for construction of recombinant virus, size limitations of foreign DNA, mosaicism of the founder animals because infection occurs after cell division begins, and possible interference of the proviral LTR sequences with the expression of foreign genes. Therefore this approach is promising for some applications but limited in general utility (reviewed by Jaenisch 1988).

A further method for producing transgenic mice was discussed in Chapter 13. This involved the transfection of ES cells, and their subsequent incorporation into mouse embryos. ES cells can also be transduced by recombinant retrovirus, as an alternative to transfection.

305

Chapter 14
Gene Transfer
to Oocytes,
Eggs and
Embryos

Expression of foreign DNA in transgenic mice

The mouse metallothionein gene promoter. The mouse metallothionein-1 (MMT) gene encodes a small cystein-rich polypeptide that binds heavy metals and is thought to be involved in zinc homeostasis and detoxification of heavy metals. The protein is present in many tissues of the mouse, but is most abundant in the liver. Synthesis of the protein is induced by heavy metals and glucocorticoid hormones. This regulation occurs at the transcriptional level (Durnam & Palmiter 1981).

Brinster *et al.* (1981) constructed plasmids in which the MMT gene promoter and upstream sequences had been fused to the coding region of the herpes simplex virus *tk* gene. The thymidine kinase enzyme can be assayed readily and provides a convenient 'reporter' of MMT promoter function. The fused MK (metallothionein-thymidine kinase) gene was injected into the male pronucleus of newly fertilized eggs which were then incubated *in vitro* in the presence or absence of cadmium ions (Brinster *et al.* 1982). The thymidine kinase activity was found to be induced by the heavy metal. By making a range of deletions of mouse sequences upstream of the MMT promoter sequences, the minimum region necessary for inducibility was localized to a stretch of DNA 40–180 nucleotides upstream of the transcription initiation site. Additional sequences that potentiate both basal and induced activities extended to at least 600 base pairs upstream of the transcription initiation site.

The same MK fusion gene was injected into embryos which were raised to transgenic adults (Brinster *et al.* 1981). Most of these mice expressed the MK gene and in such mice there were from one to 150 copies of the gene. The reporter activity was inducible by cadmium ions and showed a tissue distribution very similar to that of metallothionein itself (Palmiter *et al.* 1982b). Therefore these experiments showed that DNA sequences necessary for heavy metal induction and tissue specific expression can be functionally dissected in eggs and transgenic mice. For unknown reasons, there was no response to glucocorticoids in either the egg or transgenic mouse experiments.

As expected, the transgenic mice transmitted the MK gene to their progeny. The genes were inherited as though they were integrated into a single chromosome. When reporter activity was assayed in these offspring the amount of expression could be very different from that in the parent. Examples of increased, decreased, or even totally extinguished expression were found. In some, but not all, cases the changes in expression correlated

306

*Section 3
Cloning in
Organisms
Other than
E. coli*

with changes in methylation of the gene sequences (Palmiter *et al.* 1982b).

In a dramatic series of experiments, Palmiter *et al.* (1982a) fused the MMT promoter to a rat growth hormone genomic DNA. This hybrid gene (MGH) was constructed using the same principles as the MK fusion. Of 21 mice that developed from microinjected fertilized eggs, seven carried the MGH fusion gene and six of these grew significantly larger than their littermates. The mice were fed zinc to induce transcription of the MGH gene, but this did not appear to be absolutely necessary since they showed an accelerated growth rate before being placed on the zinc diet. Mice containing high copy numbers of the MGH gene (20–40 copies per cell) had very high concentrations of growth hormone in their serum, some 100–800 times above normal. Such mice grew to almost double the weight of littermates at 74 days old (Fig. 14.5).

These experiments have subsequently been repeated with the MMT promoter linked to a human growth hormone gene (Palmiter *et al.* 1983). Synthesis of growth hormone was found to be inducible by heavy metals. The gene was expressed in all tissues examined, but the ratio of human growth hormone mRNA to endogenous metallothionein-1 mRNA varied among different tissues and animals, suggesting that the expression of the fused gene was affected by factors such as cell type and integration site.

Fig. 14.5 Transgenic mouse containing the mouse metallothionein promoter fused to the rat growth hormone gene. The photograph shows two male mice at about 10 weeks old. The mouse on the left contains the MGH gene and weighs 44 g; his sibling without the gene weighs 29 g. In general, mice that express the gene grow 2–3 times as fast as controls and reach a size up to twice the normal. (Photograph by courtesy of Dr R. L. Brinster.)

307

*Chapter 14
Gene Transfer
to Oocytes,
Eggs and
Embryos*

These experiments showed the power of the MMT promoter in obtaining high levels of expression of any gene to which it is fused. The application of this technology to the production of important polypeptides in farm animals is discussed by Palmiter and his co-authors (1982a). The concentration of growth hormone in the transgenic mice was impressively high, much greater than bacterial or cell cultures genetically engineered for growth hormone production. The genetic farming concept is comparable to the practice of raising valuable antisera in animals, except that a single injection of a gene into a fertilized egg would substitute for multiple somatic injections. An added advantage is the heritable nature of genes, but the variability in expression already encountered in the progeny may be problematical here.

Gene regulation in transgenic mice. The similarities between the tissue distribution of MK expression and normal MMT expression encouraged the hope that transgenic mice would provide a general assay for functionally dissecting DNA sequences responsible for tissue-specific or developmental regulation of a variety of genes. But in early experiments correct tissue-specific expression of genes was not often observed. It was soon realised that the presence of bacterial vector sequences is detrimental to correct expression of certain genes including β-globin and α-fetoprotein (Chada *et al.* 1986, Kollias *et al.* 1986, Hammer *et al.* 1987). Other genes appeared to be less sensitive to this effect, and to chromosomal position effects, e.g. immunoglobulin and elastase genes (Storb *et al.* 1984, Swift *et al.* 1984, Davis & MacDonald 1988). Most researchers routinely remove prokaryotic vector sequences to avoid possible adverse effects. A large number of regulated genes have been assayed in transgenic mice (Table 14.1). In order to discriminate the products of the injected gene from the endogenous counterpart the gene must be 'marked' in some way such as by fusion to a reporter enzyme or by other modifications. This research allows the dissection of *cis*-acting sequences which are required for correctly regulated expression (Palmiter & Brinster 1986).

In certain cases, variable expression has been observed among offspring in a transgenic line. There are several possible explanations of this, including extreme sensitivity to variables of chromatin structure such as nucleosomal phasing. Variable methylation of the genes is also a likely explanation of differences in expression.

Transgenic mice and cancer. Many oncogene constructs have been introduced into transgenic mice, and have been found to elicit tumorigenic responses. For example, mice that carry the SV40 enhancer and region coding for the large T-antigen, reproducibly develop tumours of the choroid plexus which are derived from the cells lining the ventricles of the brain (Palmiter *et al.* 1985). In similar experiments the *c-myc* oncogene was fused to the LTR of mouse mammary tumour virus. In one transgenic line females characteristically developed mammary carcinomas during their second or third pregnancy (Stewart *et al.* 1984). In other experiments the *c-myc* oncogene was driven by immunoglobulin enhancers and gave rise to malignant lymphoid tumours in transgenic mice (Adams *et al.* 1985). The evidence

308

*Section 3
Cloning in
Organisms
Other than
E. coli*

Table 14.1 Expression of transgenes in specific tissues

Tissue	Gene or promoter
brain	MBP, *Thy-1*, NFP, GRH, VP
lens	crystallin
mammary epithelial cells	β-lactoglobulin, WAP
spermatids	protamine
pancreas	insulin, elastase
kidney	ren-2
liver	alb, AGP-A, CRP, α2u-G, AAT, HBV
yolk sac	α-fetoprotein
hemopoietic tissues	
erythroid cells	β-globin
B-cells	κ Ig, μ Ig
T-cells	T-cell receptor
macrophages	M-MuLV LTR
connective tissue	MSV LTR, collagen, vimentin
muscle	α-actin, myosin light chain
many tissues	H-2 (HLA), β2-m, CuZn SOD

AAT, α1-antitrypsin; AGP-A, α1-acid glycoprotein; alb, albumin; α2u-G, α2u globulin; β2-m, β2-microglobulin chain; CRP, C-reactive protein; CuZn SOD, Cu/Zn-superoxide dismutase; GRH, gonadotropin-releasing hormone; HBV, hepatitis B virus; HLA, histocompatibility antigen class; MBP, myelin basic protein; MSV, murine sarcoma virus; NFP, neurofilament protein; ren-2, renin-2; VP, vasopressin; WAP, whey acidic protein.

References can be found in Palmiter & Brunster (1986).

from these experiments suggests that the tumours were clonal in origin, i.e. derived from a single transformed cell. In contrast with this, when an oncogenic human *ras* gene was fused to an elastase promoter/enhancer construct, transgenic mice were born with pancreatic neoplasms that appeared to be due to transformation of *all* of the differentiating pancreatic acinar cells (Palmiter & Brinster 1986).

Such transgenic lines as those described, which have a predisposition to specific tumours, are of immense value in the investigation of the events leading to malignant transformation.

Correction of genetic defects in mice

The transgenic mouse technology associated with direct injection of DNA, in its present form, makes it possible to introduce a functional gene into an animal with a deficient gene pair (usually two genes, since the disorders are often recessive). It is not possible to replace genes unless strategies for gene targeting are employed, such as those described in Chapter 13. If the functional gene is inserted at random, the chances are that it will not be linked to the mutant gene, and hence will segregate from it at meiosis.

Despite these limitations, pioneering experiments have been reported. In the first experiments an MT-rat growth hormone gene was introduced into dwarf mice. These mice carry the mutation *little*, which results in the failure to produce adequate amounts of growth hormone, and hence their small size and male infertility. Several transgenic lines produced rat growth hormone in the liver and grew to about three times the size of *little* littermates. Male fertility was restored, but other, undesirable, effects were observed including impaired female fertility (Hammer *et al.* 1984, 1985). Several other genetic disorders have been corrected using introduced genes; for example *shiverer* (Redhead *et al.* 1987). It is clear from such experiments that transgenic correction of these defects is a powerful tool in analysing the disease, and also that unexpected side-effects can result from this state-of-the-art technology.

309

*Chapter 14
Gene Transfer
to Oocytes,
Eggs and
Embryos*

Insertional mutagenesis in transgenic mice

The insertion of foreign DNA into the mouse genome can disrupt the functioning of an endogenous gene. Such insertional mutagenesis can potentially lead to the inactivation of a gene, or by providing a promoter/enhancer, to the inappropriate activation of a gene. Many examples of insertional mutagenesis have been observed in transgenic mice (reviewed by Jaenisch 1988, and by Palmiter & Brinster 1986). The generation of mutants by insertional mutagenesis is attractive because the introduced DNA can act as a tag for the affected gene, and hence provide a means for isolating cloned DNA sequences.

INTRODUCTION OF CLONED GENES INTO THE GERM LINE OF *DROSOPHILA* BY MICROINJECTION

P elements of *Drosophila*

P elements are transposable DNA elements which, in certain circumstances, can be highly mobile in the germ line of *Drosophila melanogaster*. The subjugation of these sequences as specialized vector molecules in *Drosophila* represents a landmark in modern *Drosophila* genetics. Through the use of P element vectors any DNA sequence can be introduced into the genome of the fly.

P elements are the primary cause of a syndrome of related genetic phenomena called P−M hybrid dysgenesis (Bingham *et al.* 1982, Rubin *et al.* 1982). Dysgenesis occurs when males of a P (paternally contributing) strain are mated with females of an M (maternally contributing) strain, but usually not when the reciprocal cross is made. The syndrome is confined mainly to effects of the germ line and includes a high rate of mutation, frequent chromosomal aberrations and, in extreme cases, failure to produce any gametes at all.

P strains contain multiple genetic elements, the P elements, which may be dispersed throughout the genome. These P elements do not produce dysgenesis within P strains because transposition is repressed, probably due to the presence of a P-encoded repressor of a P element-specific *transposase* which is also encoded by the P element. However, when a sperm carrying chromosomes harbouring P elements fertilizes an egg of a

310

*Section 3
Cloning in
Organisms
Other than
E. coli*

strain that does not harbour P elements (i.e. an M strain), the P element transposase is temporarily derepressed owing to the absence of repressor. P element transposition occurs at a high frequency and this leads to the dysgenesis syndrome, the high rate of mutation results from the insertion into and consequent disruption of genetic loci.

Several members of the P transposable element family have been cloned and characterized (O'Hare & Rubin 1983). It appears that the prototype is a 2.9 kb element and that other members of the family have arisen by different internal deletion events within this DNA. The elements are characterized by a perfect 31 base-pair inverted terminal repeat. It is likely that this repeat is the site of action of the putative transposase. Three long open-reading frames have been identified in the prototype DNA sequences. One of these open reading frames encodes the transposase protein. The messenger RNA for this protein is produced by removal of introns from its primary transcript. The germ line specificity of transposition is due to the fact that splicing out of one of these introns occurs in germ cells but not somatic cells. Laski *et al.* (1986) showed this clearly by making a P element construct in which the intron had been precisely removed. This element, which lacked the intron, showed a high level of somatic transposition activity. Some naturally occurring short P elements are defective. They cannot encode functional transposase but are transposable in *trans* in the presence of a non-defective P element within the same nucleus.

Spradling & Rubin (1982) have devised an approach for introducing the P element DNA into *Drosophila* chromosomes which mimics events taking place during a dysgenic cross. Essentially, a recombinant plasmid which consisted of a 2.9 kb P element together with some flanking *Drosophila* DNA sequences, cloned in the pBR322 vector, was microinjected into the posterior pole of embryos from an M-type strain. The embryos were injected at the syncytial blastoderm stage. This is a stage of insect development in which the cytoplasm of the multinucleate embryo has not yet become partitioned into individual cells (Fig. 14.6). The posterior pole was chosen because it is the site at which the cytoplasm is first partitioned, resulting in cells that will form the germ line. P element DNA introduced in this way became integrated into the genome of one or more posterior pole cells. Because of the multiplicity of such germ line precursor cells the integrated P element DNA was expected to be inherited by only some of the progeny of the resulting adult fly. Therefore the progeny of injected embryos were used to set up genetic lines which could be genetically tested for the presence of incorporated P elements.

A substantial proportion of progeny lines were indeed found to contain P elements integrated at a variety of sites in each of the five major chromosomal arms, as revealed by *in-situ* hybridization to polytene chromosomes. It may be asked whether integration really does mimic normal P element transposition or whether it is simply some non-specific integration of the microinjected plasmid. The answer is that integration occurs by a mechanism analogous to transposition. By probing Southern blots of restricted DNA it was found that the integrated P element was not accompanied by the flanking *Drosophila* or pBR322 DNA sequences present in the recombinant plasmid that was microinjected (Spradling & Rubin 1982). Injected

311
Chapter 14
Gene Transfer
to Oocytes,
Eggs and
Embryos

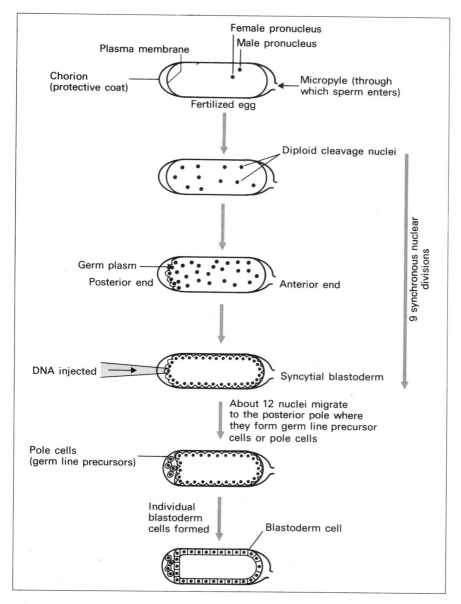

Fig. 14.6 Early embryogenesis of *Drosophila*. DNA injected at the posterior end of the embryo just prior to pole cell formation is incorporated into germ-line cells.

plasmid DNA must presumably have been expressed at some level *before* integration so as to provide transposase activity for integration by the transposition mechanism.

These experiments, therefore, showed that P elements can transpose with a high efficiency from injected plasmid into diverse sites in chromosomes of germ line cells. At least one of the integrated P elements in each progeny line remained functional, as evidenced by the hypermutability it caused in subsequent crosses to M strain eggs.

312

*Section 3
Cloning in
Organisms
Other than
E. coli*

Development of P element as a vector

Rubin & Spradling (1982) exploited their finding that P elements can be artificially introduced in the *Drosophila* genome. A possible strategy for using the P element as a vector would be to attempt to identify a suitable site in the 2.9 kb P element sequence where insertion of foreign DNA could be made without disrupting genes essential for transposition. However, an alternative strategy was favoured. A recombinant plasmid was isolated which comprised a short (1.2 kb), internally deleted member of the P element family together with flanking *Drosophila* sequences, cloned in pBR322. This naturally defective P element cannot encode any of the putative protein products of the 2.9 kb prototype element (O'Hare & Rubin 1983). Target DNA was ligated into the defective P element. The aim was to integrate this recombinant P element into the germ line of injected embryos by providing transposase function in *trans*. Two approaches for doing this were tested. In one approach a plasmid carrying the recombinant P element was injected into embryos derived from a P–M dysgenic cross in which transposase activity was therefore expected to be high. This approach does have the disadvantage that frequent mutations and chromosomal aberrations would also be expected. In the other approach the plasmid carrying recombinant P element was co-injected with a plasmid carrying the non-defective 2.9 kb element. This approach is formally similar to the application of bacterial transposons discussed in Chapter 10.

In the first experiments of this kind, embryos homozygous for a *rosy* mutation were microinjected with the P element vector containing a wild-type *rosy* gene. Both methods for providing complementing transposase were effective. Rosy$^+$ progeny, recognized by their wild-type eye colour, were obtained from 20 to 50% of injected embryos. The chromosomes of these flies contained one or two copies of the integrated *rosy*$^+$ DNA.

The *rosy* gene is a particularly useful genetic marker. It produces a clearly visible phenotype: Rosy$^-$ flies have brown eyes instead of the characteristic red colour of Rosy$^+$ flies. The *rosy* gene encodes the enzyme xanthine dehydrogenase, which is involved in the production of a precursor of eye pigments. The *rosy* gene is not cell autonomous: expression of *rosy*$^+$ anywhere in the fly, for example in a genetically mosaic fly developing from an injected larva, results in a wild-type eye colour.

A simple modern P element vector is shown in Fig. 14.7 (Rubin & Spradling 1983). It consists of a P element cloned in a bacterial vector, pUC8. Most of the P element has been replaced by a *rosy*$^+$ gene, but the terminal repeats essential for transposition have been retained. The vector includes a polylinker site for inserting foreign sequences. The capacity of such a vector is large: a definite upper limit has not been determined, but inserts of over 40 kb have been successfully introduced into flies by a P element vector (Haenlin *et al.* 1985). Transposition of the recombinant vector into the genome of injected larvae is brought about by co-injecting a helper P element which provides transposase in *trans*, but which cannot transpose itself because of a deletion in one of its terminal inverted repeats. Such an element is referred to as a *wings-clipped* element.

313

Chapter 14
Gene Transfer
to Oocytes,
Eggs and
Embryos

Structure of P-element derivative : Carnegie 20

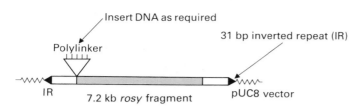

Structure of helper P-element : pπ 25.7 wings clipped

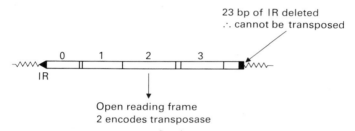

Fig. 14.7 P element derivatives as vector system. (See text for details.)

Pπ25.7 wc is an example of a *wings-clipped* helper element (Fig. 14.7) (Karess & Rubin 1984). The *rosy* marker has many advantages. But it is not selectable. Therefore P-element vectors have been constructed which include selectable marker genes such as alcohol dehydrogenase, *Adh* (Adh$^+$ flies can be selected on food containing 6% ethanol), or *neo*, which confers resistance to food containing G418 (Goldberg *et al*. 1983, Steller & Pirotta 1985). These markers have the disadvantage that they do not confer an immediately visible phenotype, hence complicating the maintenance of fly stocks. However they can be combined in a vector with *rosy*$^+$ to give both a visible and a selectable marker.

Application of the P element vector

The initial successful demonstration of the P element vector system was rapidly followed by the simultaneous publication of three reports which described the reintroduction of three cloned *Drosophila* genes (Dopa decarboxylase, xanthine dehydrogenase and alcohol dehydrogenase) into the chromosomes of the fly. The genes were accompanied by substantial flanking sequences. These experiments were particularly exciting because in each case the regulation of the gene in question was correct.

314

*Section 3
Cloning in
Organisms
Other than
E. coli*

Dopa decarboxylase is a gene that is subject to both temporal and tissue-specific regulation during development. Expression of the reintegrated copies of the gene showed the expected pattern of regulation (Scholnick *et al.* 1983). Xanthine dehydrogenase is the product of the *rosy* gene. Again, when the gene was reintroduced into the chromosomes of the fly, the tissue distribution of xanthine dehydrogenase was normal (Spradling & Rubin 1983). Finally reintroduction of a cloned alcohol dehydrogenase gene resulted in normal expression according to several criteria: quantitative amounts of enzyme in larvae and adults; tissue specificity; and a previously recognized and characteristic developmental switch in transcription initiation site which is a feature of this gene, which has two sets of promoter elements (Goldberg *et al.* 1983).

The above successes with a variety of genes promise rapid progress in understanding DNA control signals which are required for the developmental and tissue-specific regulation of gene expression in *Drosophila*. A further aspect of gene regulation raised by these experiments concerns dosage compensation. *Drosophila* females are XX and males are XY, as in mammals. Both the fly and mammals face the problem of controlling the level of expression of genes on their X chromosome, since females have a double dose and the males a single dose of each. In mammals this is achieved by total inactivation and heterochromatinization of almost all of an X chromosome in each somatic cell of the female. *Drosophila* has a different mechanism; transcription of each copy of the X-linked genes in the female is less active than in the male. How this is brought about is not known. An insight is given by the finding that when the autosomal genes encoding dopa decarboxylase or xanthine dehydrogenase were integrated into the X chromosome, expression was at least partially dosage compensated. This should be amenable to further analysis using the P element vector.

A final example of the application of this vector involves the heat shock response of *Drosophila*. High temperatures and other stress-inducing treatments evoke a dramatic change in the pattern of gene expression in *Drosophila*. Among many effects, transcription of certain genes, the heat shock genes, is greatly increased. This can be examined in many ways (see, for example, Bienz & Pelham 1982) but is most striking in the polytene chromosomes where the heat shock rapidly induces large puffs. Lis *et al.* (1983) have constructed a hybrid gene comprising the 5' flanking region of a *Drosophila* heat shock gene and some accompanying aminoterminal heat shock protein codons, fused in phase with the *E. coli* β-galactosidase gene. The fused gene and a selectable *rosy*[+] gene were cloned into the P element vector and then inserted into the *Drosophila* genome. Three lines of flies were established; two contained a single inserted DNA and one contained two copies of the DNA at separate sites. On heat shocking such flies large chromosomal puffs occurred in the polytene chromosomes at the sites of insertion. By dissecting out organs of the larvae and adult flies and then incubating them in the chromogenic β-galactosidase substrate, X-gal, it was evident that β-galactosidase activity was inducible by heat shock and showed the expected widespread distribution throughout the animal's tissues.

Transposon tagging using P element derivatives

315

*Chapter 14
Gene Transfer
to Oocytes,
Eggs and
Embryos*

P elements can be mobilized to insert into genes and cause insertional mutagenesis. The mutated gene can then be cloned by virtue of having been tagged with P element sequences. Systems for controlled transposon mutagenesis have been developed (reviewed by Cooley *et al.* 1988). Two fly strains are involved. One strain contains the 'mutator' a defective P element carrying useful marker genes, which can be mobilized when provided with a source of transposase. A second strain contains a *wings-clipped* type of element which provides transposase in *trans*. These are called 'jumpstarter' elements. During a controlled mutagenesis screen, a single jumpstarter element is crossed into a mutator-containing strain. Transposition occurs. In subsequent generations the mutator element is stabilized when the chromosome bearing the jumpstarter element segregates from the target chromosome into which the mutator element has been inserted.

Recovery of the P element and associated gene sequences from the inactivated gene is most easily accomplished by a plasmid rescue technique (see Chapter 13). For this approach, an *E. coli* plasmid origin of replication must be included in the mutator element.

Section 4
Applications of Recombinant DNA Technology

15 Nucleic Acid Probes and their Applications

Many of the techniques of gene manipulation depend on the hybridization of a nucleic acid probe to a target DNA or RNA sequence. In previous chapters we have described many applications of the technology and some examples are given in Table 15.1. Of equal importance are the diagnostic applications of probe technology and, since these have not been dealt with elsewhere, they will be covered in the discussion that follows. These applications are not just of interest for their commercial value: the principles behind them have much to offer the research scientist. Before describing these applications it is necessary to consider the methods available for labelling nucleic acids, for without a label hybridization cannot be detected easily!

Methods for labelling nucleic acids

There are five basic methods for labelling nucleic acids. These are:
1 nick translation,
2 primer extension,
3 methods based on RNA polymerase,
4 end-labelling methods, and
5 direct labelling methods.

A comparison of the different methods for uniformly labelling nucleic acids is given in Table 15.2.

Nick translation. Although predating many of the methods described below, nick translation remains the commonest means of labelling nucleic acids

Table 15.1 Applications for labelled hybridization probes.

southern blots	detection of gel-fractionated DNA molecules following transfer to a membrane
northern blots	detection of gel-fractionated RNA molecules following transfer to a membrane
dot blots	detection of unfractionated DNA or RNA molecules immobilized on a membrane
colony/plaque blots	detection of DNA released from lysed bacteria or phage and immobilized on a membrane
S1/RNase mapping	positional mapping of termini of target molecules
In-situ **hybridization**	detection of DNA or RNA molecules in cytological preparations

320

*Section 4
Applications of
Recombinant
DNA
Technology*

Table 15.2 Comparison of methods for uniform labelling of nucleic acids.

	Nick translation	Random primer	Phage polymerase
Template	dsDNA	ssDNA (and denatured ds DNA)	dsDNA
Optimal form	linear or circle	linear	linear
Amount of template	0·5–1 μg	25 ng	1–2 μg
Reaction time	~2 hours	~3 hours	~1 hour
Efficiency of incorporation	~60%	~75%	~75%
Nature of probe	DNA	DNA	RNA
Competing strand	yes	yes	no
Amount of probe	0·5–1 μg	~50 ng	~250 ng
Potential specific activity of probe (dpm/μg)	2×10^9	5×10^9	5×10^9
Probe length	~500 bp (medium variance)	~200 bp (high variance)	defined
Insert specific	no	no	yes
Requirement for subcloning	no	no	yes
Labelling in agarose	(yes)*	yes	no

Values quoted are for commonly used protocols.
* Although effective labelling in agarose has been observed, its efficiency is variable.

for hybridization. This somewhat confusing term (Rigby *et al.* 1977) does not relate to protein synthesis (translation), but rather to the translation or movement of a nick along a duplex DNA molecule (Fig. 15.1). Nicks can be introduced at widely separated sites in DNA by very limited treatment with DNase I. At such a nick, which exposes a free 3′-OH group, DNA polymerase I of *E. coli* will incorporate nucleotides successively. Concomitant hydrolysis of the 5′-terminus by the 5′→3′ exonucleolytic activity of polymerase I releases 5′-mononucleotides. If the four deoxynucleoside triphosphates are radiolabelled (e.g. α^{32}P-dNTPs), the reaction progressively incorporates the label into a duplex that is unchanged except for translation of the nick along the molecule. Because the original nicking occurs virtually at random, a DNA preparation is effectively labelled uniformly to a degree depending upon the extent of nucleotide replacement and specific radioactivity of the labelled precursors. Often the reaction is performed with only one of the four deoxynucleoside triphosphates in labelled form. Nick translation can be used with a variety of labels to generate probes suitable for most hybridization applications. For example, using [α-^{32}P] deoxynucleoside triphosphates (dNTPs) it is relatively simple to generate probes of a specific activity high enough to detect single copy genes on Southern blots of mammalian DNA. Nick translation also is appropriate for the generation of biotinylated probes (see p. 327).

A major advantage of nick translation over other uniform labelling methods is its flexibility with respect to probe size, specific activity and concentration. It is particularly suitable for the production of large quantities

321
Chapter 15
Nucleic Acid
Probes and
their
Applications

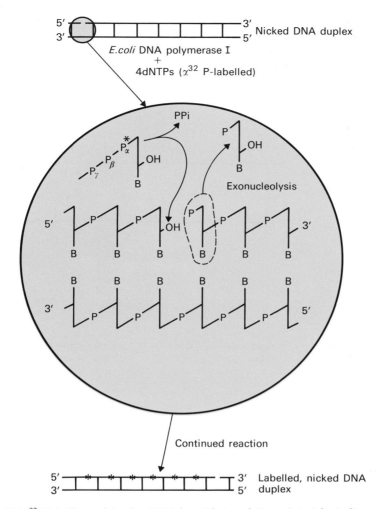

Fig. 15.1 ^{32}P-labelling of duplex DNA by nick translation. Asterisks indicate radiolabelled phosphate groups.

of probe for use in multiple hybridization reactions and/or where a high probe concentration is required.

Primer extension methods. In common with nick translation, primer extension methods utilize the ability of DNA polymerases to synthesize a new DNA strand complementary to a template strand, starting from a free 3'-hydroxyl group. In this case the latter is provided by a short oligonucleotide primer annealed to the template (Fig. 15.2). Two general approaches are possible. In the first, a mixture of primers of random sequence is used in order to produce a uniformly labelled DNA copy of any sequence (Feinberg & Vogelstein 1983). The second method uses a unique primer to restrict labelling to a particular sequence of interest. In the primer extension method it is essential to use a polymerase lacking a 5' → 3' exonuclease activity, otherwise degradation of the primer will occur. Both the Klenow

322

*Section 4
Applications of
Recombinant
DNA
Technology*

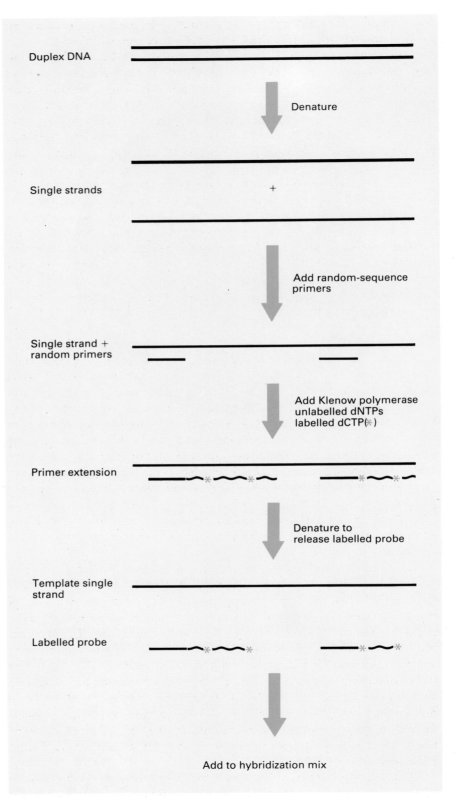

Fig. 15.2 The random primer method for preparing labelled DNA.

323

*Chapter 15
Nucleic Acid
Probes and
their
Applications*

fragment of *E. coli* DNA polymerase I, which lacks the $5' \rightarrow 3'$ exonuclease activity, and reverse transcriptase have been used successfully.

An advantage of the random priming method is that it may be used to label relatively impure DNA. For example, it is often necessary to label specific DNA fragments which have been purified by agarose gel electrophoresis. This procedure is used frequently to separate a cloned insert from a vector sequence which may show some hybridization with the target DNA. Feinberg & Vogelstein (1984) have demonstrated that it is possible to label DNA fragments in the presence of low gelling temperature agarose without prior purification.

Because net synthesis occurs in primer extension methods it is possible to obtain good incorporation of labelled nucleotide added at a concentration higher than that of equivalent nucleotide in the template. As the level of input DNA is low, the labelled nucleotide can be of high specific activity, but the amount of probe produced is substantially lower than in a nick translation reaction. Thus probes produced by primer extension are ideal for situations such as single-copy gene detection on a Southern blot where conditions of high specific activity and low probe concentration are frequently employed.

Preparation of RNA probes. RNA polymerases catalyse the synthesis of RNA from ribonucleoside triphosphates using a DNA template. Thus they have the ability to incorporate labelled ribonucleotides into RNA molecules which can be used in all the applications appropriate for uniformly labelled DNA probes. Originally RNA probes were prepared using RNA polymerase from *E. coli*. This enzyme when used *in vitro* shows very little template and promoter specificity and therefore produces transcripts which have been initiated and terminated at random (Schleif & Wensink 1981). This approach has been superseded by the more elegant use of bacteriophage RNA polymerases. The RNA polymerases from a number of phages possess a high degree of specificity for their own promoters *in vitro* (Butler & Chamberlain 1982, Davanloo *et al.* 1984, Morris *et al.* 1986). Thus transcription can be limited to sequences cloned downstream of an appropriate promoter.

The need for a cloning step has led to the development of a variety of vectors in which a phage promoter is present upstream from a multiple cloning site, thus allowing transcription from any cloned insert. One of the most frequently used types of vector contains two phage promoters in opposite orientation separated by a multiple cloning site. Transcription from the vector sequence is avoided by linearization of the vector downstream of the insert so that run-off transcripts are produced (Fig. 15.3). The use of two promoters allows transcription from either strand to be chosen so that sense or anti-sense RNA can be produced (Melton *et al.* 1984). In the presence of high concentrations of precursor nucleotides it is possible to produce large amounts of transcript, up to 10 μg from 1 μg of template, useful for a variety of applications such as splicing studies and *in-vitro* translation (Kreig & Melton 1984).

It should be noted that, in contrast to nick translation or primer extension methods, the specific activity of the probe produced is unrelated to the

324

*Section 4
Applications of
Recombinant
DNA
Technology*

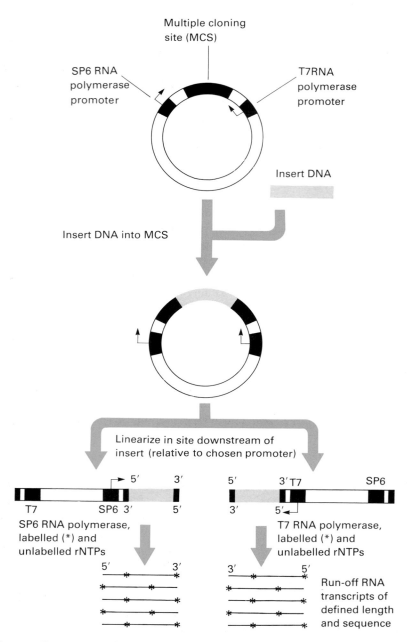

Fig. 15.3 Preparation of RNA probes using a plasmid containing SP6 and T7 polymerase promoters.

efficiency of incorporation of label. This is because the RNA probe does not need to be denatured prior to hybridization. The excess of double-stranded DNA does not participate in the hybridization and there is no dilution with unlabelled template. Thus probe specific activity is determined solely by the specific activity of the labelled nucleotide, the efficiency of incorporation giving a measure of the yield of probe from the reaction.

325

Chapter 15
Nucleic Acid
Probes and
their
Applications

RNA probes have been reported to show higher sensitivity than equivalent nick-translated probes in both Northern blots (Melton *et al.* 1984) and *in-situ* hybridization (Cox *et al.* 1984). Most probably this is due to the avoidance of probe reannealing during hybridization and because RNA−RNA hybrids are more stable than DNA−DNA hybrids. A further advantage of RNA probes is that non-specifically bound probe can be removed by treatment with RNase A which is very specific for single-stranded RNA.

Probably the most significant disadvantage of RNA polymerase based labelling is the requirement to subclone an insert into a vector containing an appropriate promoter. However, once an insert has been cloned into such a vector, DNA probes may still be prepared using nick translation or random primer labelling. A variety of transcription vectors has been developed that can be used as primary cloning vehicles. Some of these are plasmid based but a more interesting group is based on phage M13. These M13-based vectors eliminate the need to linearize the template in order to limit transcription to the sequence of interest. This is achieved by using the single-stranded form of the phage as the initial substrate for the transcription reaction as shown in Fig. 15.4 for the mICE series of vectors originated by Eperon (1986). By annealing an appropriate oligonucleotide primer and copying the strand using Klenow polymerase in the presence of dNTPs a partial duplex is formed. This serves as an efficient substrate for T7 RNA polymerase which recognizes the double-stranded form of the promoter and terminates synthesis at the end of the double-stranded region. The use of an oligonucleotide as a transcription terminator allows increased flexibility in choice of run-off point, which is no longer confined to a known restriction enzyme site.

End-labelling of nucleic acids. A wide variety of techniques is available for introducing label at either the 3′ or 5′ ends of linear DNA or RNA. Usually only a single label is introduced so that the specific activities achievable by such techniques are significantly lower than those obtained by the uniform labelling methods discussed above.

Nucleic acids can be 5′-end labelled using T4 polynucleotide kinase (Fig. 15.5). The enzyme catalyses the transfer of the γ-phosphate of a ribonucleoside 5′-triphosphate donor to the 5′-hydroxyl group of a polynucleotide, oligonucleotide or nucleoside-3′-triphosphate. For end-labelling [γ-^{32}P]ATP is most frequently used as the donor to transfer a radiolabelled phosphate group to DNA or RNA containing a 5′-hydroxyl terminus. This is termed the 'forward' reaction. However, most polynucleotides have a 5′-phosphate group which must be removed using alkaline phosphatase before they can be used as substrates for the reaction. This step may be avoided by carrying out an 'exchange' reaction (Berkner & Folk 1977). In this case the reaction is driven by excess ADP which causes the enzyme to transfer the terminal 5′-phosphate from DNA to ADP. The DNA then is rephosphorylated by transfer of the labelled γ-phosphate from the [γ-^{32}P]ATP. Although this reaction is more convenient than dephosphorylation followed by the forward reaction, it usually occurs at a lower efficiency.

326

*Section 4
Applications of
Recombinant
DNA
Technology*

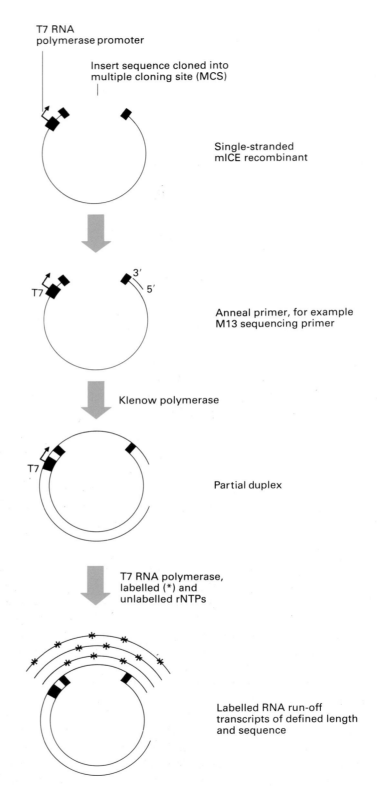

Fig. 15.4 Preparation of probes using a single-stranded mICE vector.

327
Chapter 15
Nucleic Acid
Probes and
their
Applications

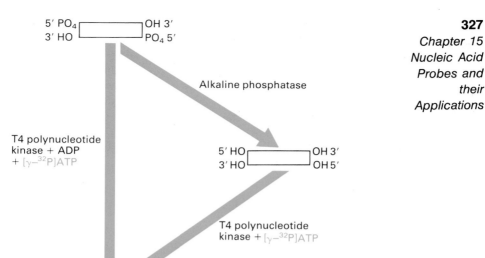

Fig. 15.5 The 5′-end labelling reaction using T4 polynucleotide kinase.

The major advantages of 5′-end labelling are that:
1 both DNA and RNA can be labelled,
2 oligonucleotide can be conveniently labelled, e.g. for detection of mutants following site-directed mutagenesis,
3 the location of the labelled group is known,
4 very small fragments of DNA can be labelled,
5 restriction digest fragments may be labelled, e.g. for Maxam & Gilbert sequencing (see p. 99).

DNA fragments also can be labelled at their 3′ ends by using terminal deoxynucleotidyl transferase (p. 32), but this method is not used widely.

Choice of label

In many probe applications a certain degree of resolution is necessary since information about the relative position of a nucleic acid fragment is required. Since very small amounts of material are normally available and the sequence of interest may be present at low abundance, there usually is a requirement for high sensitivity. Thus there are two important parameters which largely determine the choice of label—resolution and sensitivity. Different applications require a balance between these parameters. For example, ultimate sensitivity may be compromised for the sake of high resolution in dideoxy DNA sequencing (p. 100) while the converse is true when the detection of a single-copy mammalian gene on a Southern blot is the aim. The sensitivity required in the latter case is demonstrated by the fact that a single-copy 'gene' 1 kb long accounts for only 3 pg of a 10-μg sample of total human genomic DNA. In a Southern blot only a percentage of that DNA would be transferred to the membrane. Broadly speaking, the choice of label is a trade-off between sensitivity and resol-

328

Section 4
Applications of
Recombinant
DNA
Technology

ution, in combination with other factors such as probe stability, safety, and ease of use.

Radioactive labels. The most widely used labels for probes are radioactive. Autoradiography usually is used to detect probes which have hybridized to the target DNA and this can provide positional information as well. The important features which determine sensitivity in the detection of radio-nuclides by film methods are shown in Table 15.3. The data indicate that radioactive methods using ^{32}P are easily able to detect single-copy genes in 10 μg of human DNA, and in practice detection in 0.5 μg of total DNA is possible.

Non-radioactive labels. A number of non-radioactive labels for probes are available but the only one in widespread use is biotin. Biotin can be incorporated into polynucleotides enzymatically using biotinylated nucleo-tides as the substrate (Langer *et al*. 1981). An alternative labelling method is to use a photoactivatable analogue of biotin. Upon brief irradiation with visible light, stable linkages are formed with both single- and double-stranded nucleic acids (Forster *et al*. 1985).

Biotin-labelled probes are detected via avidin or streptavidin through a variety of signal-generating systems. A typical example is shown in Fig. 15.6. Avidin is a glycoprotein with an extremely high affinity for biotin. Streptavidin is an avidin-like protein widely used in place of avidin. Its main advantage is a near neutral isoelectric point which leads to a reduction in non-specific binding to DNA and thereby mini-mizes background signal. An alternative method is to label DNA with digoxigenin-labelled deoxyuridine triphosphate. The dUTP is linked via a spacer arm to the steroid hapten digoxigenin (dig-dUTP). After hybrid-ization to the target DNA the hybrids are detected by enzyme-linked immunoassay using an antibody conjugate, e.g. anti-digoxigenin alkaline phosphatase. Enzyme reporter molecules can be directly cross-linked to DNA, thereby obviating the need for biotin labelling (Jablonski *et al*. 1986; Renz & Kurz 1984). For example, horseradish peroxidase or alka-line phosphatase can be cross-linked to polyethyleneimine with *p*-benzoquinone and the resulting conjugates covalently linked to DNA using glutaraldehyde. There are a number of advantages to using covalently-bound horseradish peroxidase as a signal-generating system. First, direct labelling of probes is much faster than methods involving incorporation of modified nucleotides. Second, on addition of substrate (luminol plus hydrogen peroxide), light is emitted and this light can be detected using photographic film in a similar fashion to the detection of radioactive emissions by autoradiography. Finally, so much light is emitted that detection of bound probe is possible in 5–20 minutes compared with 5–20 hours when ^{32}P-labelled probes are used.

The major disadvantage of biotin, and of the other non-radioactive labels, is low sensitivity in comparison to radioactive labels. Values quoted in the literature generally refer to ultimate sensitivity obtainable by ex-perienced users. Using a suitable chromogenic substrate and an enzyme such as alkaline phosphatase, an experienced worker should be able to

329

Chapter 15
Nucleic Acid
Probes and
their
Applications

Table 15.3 Characteristics of radionuclides used for nucleic acid labelling.

Radio-nuclide	$T\frac{1}{2}$	Type/max energy of emission (MeV)	Specific activity range of nucleotides (Ci/mmol)	Labelling methods	Typical spec. act. of probe (dpm/μg)	Detection limit (dpm/cm^2)	Equivalent amount of probe (pg/cm^2)
^{32}P	14.3d	β/1.71	400–6000	nick translation	5×10^8	50[a]	0.1
				random primer	5×10^9		0.1
				phage RNA pol.	1.3×10^9		0.04
^{35}S	87.4d	β/0.167	400–1500	end-labelling	5×10^6		10
				nick translation	1×10^8	400[b]	4
				random primer	7×10^8		0.57
				phage RNA pol.	1.3×10^9		0.3
^{125}I	60d	γ/0.035 β/0.035	1000–2000	nick translation	1×10^8	100[a]	1
				random primer	1.5×10^9		0.067
				direct iodination.	2×10^8		0.5
^{3}H	12.35y	β/0.018	25–100	nick translation	5×10^7	8000[b]	160
				random primer	1.5×10^8		53.3
				phage RNA pol.	5×10^7		160

[a] Detection using an intensifying screen.
[b] Detection by fluorography.

330

Section 4
Applications of
Recombinant
DNA
Technology

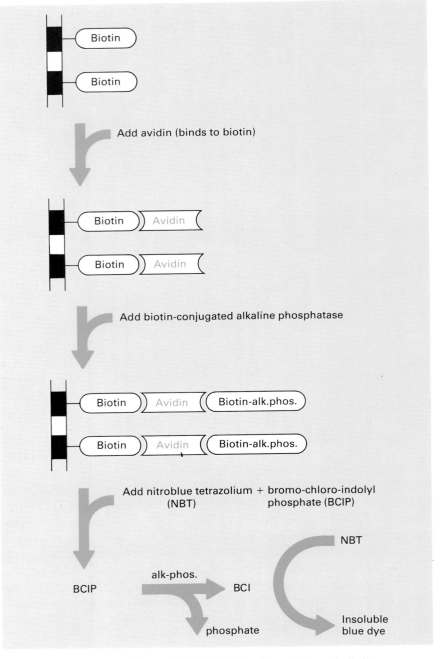

Fig. 15.6 The detection of biotin-labelled DNA using avidin and alkaline phosphatase. The insoluble blue dye produced by the reduction of nitroblue tetrazolium 'stains' the nitrocellulose in the vicinity of biotin-labelled DNA.

detect single-copy gene sequences in filter hybridization (1–10 pg), but for a less experienced user a figure of 50 pg is more realistic.

331

Chapter 15
Nucleic Acid
Probes and
their
Applications

Specific DNA amplification: the polymerase chain reaction

The low sensitivity of the currently available non-radioactive labelling methods means that radioactive probes are required for most commercial applications of hybridization technology. This has hampered the realization of the potential of probe technology for diagnostic applications. The disadvantages of radiolabelled probes are twofold. First, the radioisotopes used have short half-lives (see Table 15.3) and hence the probes have to be freshly prepared at regular intervals. Second, unlike molecular biologists, workers in diagnostic laboratories have an innate dislike of radioactivity despite the fact they often use radioimmunoassay methods based on [125]I.

The recent development of a method for amplifying DNA *in vitro* eliminates the requirement for very sensitive detection systems (Sakai *et al.* 1985). The method, called the polymerase chain reaction (PCR), is shown in Fig. 15.7. PCR amplification involves two oligonucleotide primers that flank the DNA segment to be amplified. These primers hybridize to opposite strands of the target sequence and are oriented so that DNA synthesis by the polymerase proceeds across the region between the primers, effectively doubling the amount of that DNA segment. Since the extension products are also complementary to and capable of binding primers, the cycle can be repeated after a denaturation step. By repeated cycles of denaturation, priming and extension there is a rapid exponential accumulation of the specific target fragment. Thus after only 10 cycles the target sequence will have been amplified 1000-fold, which brings the sensitivity of detection within the range of most non-radioactive labels. This technique has been used in a whole variety of applications (see Sakai *et al.* 1988) and specific examples are given later in this chapter.

In the original description of the PCR method, Klenow polymerase was used to catalyse the extension of the primers. Because of the heat denaturation step fresh enzyme had to be added during each cycle. Sakai *et al.* (1988) have shown that the DNA polymerase from the thermophile *Thermus aquaticus (Taq)* is not inactivated by the heat denaturation step and so does not need to be replenished at each cycle. Furthermore, by enabling the amplification reaction to be performed at higher temperatures, the specificity, sensitivity, and yield of the product was significantly improved.

THE APPLICATIONS OF PROBES

There are three basic applications of probes. The first of these is the detection of specific nucleic acid sequences as in the detection of microorganisms in clinical specimens. The sequences detected are usually large (i.e. an entire gene or cluster of genes) and considerable mismatching between the target and probe sequences is expected. In the second application, most commonly found in clinical genetics, the objective is to detect

332

Section 4
Applications of
Recombinant
DNA
Technology

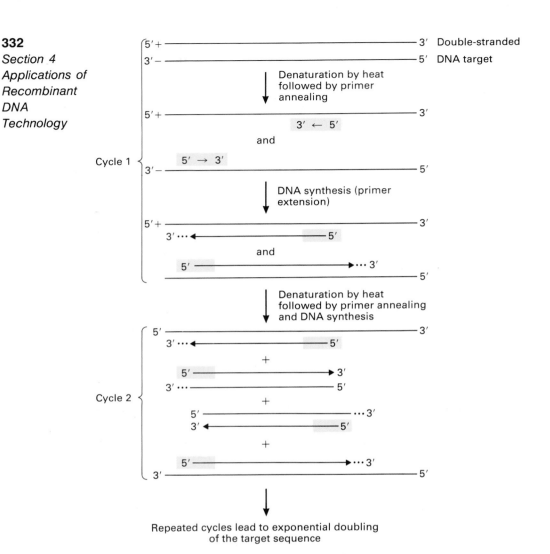

Fig. 15.7 The polymerase chain reaction. Note that although the amplified DNA is not a full copy of the starting DNA it can still bind fresh primer and so participate in further amplification steps.

changes to specific sequences. These changes can include deletions or rearrangements in the test DNA, but the methods used must be precise enough to routinely detect single base changes. The third type of application is a forensic one. Here the need is to detect the size distribution of DNA fragments bearing repetitive DNA sequences. This technique is referred to as *DNA profiling or fingerprinting*. In this technique a considerable degree of sequence mismatch is expected.

The concept of *stringency* is very important to understanding the specificity of nucleic acid probe reactions. Nucleic acid molecules can tolerate a certain number of mismatched base pairs (e.g. A with C and G with T) and still form stable duplexes, provided a significant number of base pairs do match and form bonds. The greater the degree of mismatch the more likely the two molecules are to come apart. The degree of mismatch which can be tolerated in a hybridization reaction and still maintain a double-stranded molecule depends on the environmental conditions. Conditions favouring considerable mismatch (low stringency) are low temperature, low concentrations of formamide, and high salt concentrations. Under conditions of high stringency (high temperatures, low salt concentration) only exact matches of DNA will anneal, particularly when the length of the probe is short.

333
Chapter 15
Nucleic Acid
Probes and
their
Applications

Detection of sequences at the gross level

There are many applications of hybridization technology where all that is required is to determine if a particular nucleic acid sequence is present. In almost all instances the requirement is to detect a specific microorganism. The greatest potential is seen in clinical microbiology (Tenover 1988) where a single technology (hybridization) could replace a whole variety of disparate test procedures. For example, a patient with a disorder of the gastrointestinal tract could be infected with any one of the infectious agents shown in Table 15.4. In conventional diagnostic practice, identification of the causative organism would require cultivation of stool samples on a variety of different media in a variety of different ways, microscopy, animal cell culture and immunoassay. Clearly, hybridization of the test sample with a battery of probes should be much simpler.

There are other advantages in the use of hybridization for the diagnosis of infectious disease. First, if sufficient of the infectious agent is present then no cultivation of the microbe is necessary. This eliminates hazards and should shorten the time required for identification of the pathogen. The polymerase chain reaction is particularly powerful in this respect. For example, Ou *et al.* (1988) were able to detect HIV-1 DNA in peripheral blood mononuclear cells in less than 1 day whereas virus isolation took

Table 15.4 Pathogens causing infection of the gastrointestinal tract.

Bacteria
  Salmonella spp.
  Shigella spp.
  Yersinia enterocolitica
  Enterotoxigenic *E. coli*
  Clostridium difficile
  Clostridium welchii
  Campylobacter spp.
  Aeromonas spp.
  Vibrio spp.

Viruses
  Rotaviruses
  Enteroviruses
  Enteric adenoviruses

Protozoa
  Entamoeba histolytica
  Giardia lamblia

334

Section 4
Applications of
Recombinant
DNA
Technology

3–4 weeks. A second advantage of hybridization technology is that it can be used even when the target organism cannot be cultured. This is often the case with viral infections, and the current diagnostic method is to look for rising titres of antibody in the blood. The method will also work with latent infections when no antibody may be present. A third advantage is that a single probe can identify all of the serotypes of an infectious agent whereas a single antibody may not. Finally, it is easy to clone a gene known to be involved in pathogenicity and to develop a suitable probe. Generating an antibody to the product of that gene can be much more difficult.

The probes currently approved for use by the US Food and Drug Administration are shown in Table 15.5. The regulatory issues surrounding the use of these probes is discussed by Young (1987).

An important diagnostic tool is *in-situ* hybridization. In most instances *in-situ* hybridization is carried out on formalin-fixed, paraffin-embedded tissues. This technique has proven particularly useful for detection of viral pathogens (Brigatti *et al.* 1983). It allows one to examine the tissue first by traditional staining methods, such as haematoxylin and eosin staining, and then to correlate the cytopathology seen with the presence of infectious agents as shown by a non-radioactive probe.

Hybridization as a diagnostic tool is not restricted to clinical micro-biology. There are many applications in plant pathology (Hull & Al-Hakim 1988). For example, the identification of viruses is important in the prediction of plant diseases in annual crops, the prevention of infection in planting stock, in monitoring disease control methods and in diagnosing disease in plants held in quarantine. Hybridization is also being used as a tool in the study of microbial ecology (Holben & Tiedje 1988, Stahl *et al.* 1988).

The detection of genetic disorders

There are several hundred recognized genetic diseases in man which result from single recessive mutations. In some of these a protein product that is defective or absent has been identified but in many others the nature of the mutation is unknown. For many of these genetic diseases

Table 15.5 Probes approved for diagnosis of microorganisms.

Bacteria
  Legionella
  Mycobacterium tuberculosis complex
  Mycobacterium spp.
  Mycobacterium avium-sutracellulare
  Mycoplasma pneumoniae
  Chlamydia
  Campylobacter spp.

Viruses
  herpes virus type 2
  herpes virus type 1 and 2

335

*Chapter 15
Nucleic Acid
Probes and
their
Applications*

there is no definitive treatment although in some cases human gene therapy may become possible (see p. 363). Their prevention is the current strategy in many countries. Primary prevention by identifying heterozygous carriers and dissuading them from reproducing with other carriers is not practicable, and where it has been attempted it has failed. Usually the discovery that both partners are carriers is made only after the birth of an affected child. When a pregnant woman has been identified as at risk, as indicated by a previous birth or other factors, the approach is to offer antenatal diagnosis and the possibility of an abortion.

An essential prerequisite for prenatal diagnosis is the availability of fetal DNA. Fetal cells can be obtained by amniocentesis (Orkin 1982) but this method is not entirely satisfactory for it cannot be carried out early in pregnancy. An alternative approach is to obtain fetal DNA from biopsies of trophoblastic villi in the first trimester of pregnancy (Williamson *et al.* 1981). In this technique some villi of the trophoblast—this is an external part of the human embryo which functions in implantation and becomes part of the placenta—are biopsied with the aid of an endoscope passed through the cervix of the uterus. Up to 100 μg of pure fetal DNA can be obtained between the 6th and 10th weeks of pregnancy (Old *et al.* 1982a). The efficiency of these techniques has been greatly improved with the advent of the polymerase chain reaction. First, no longer is it necessary to grow fetal cells in culture following amniocentesis in order to isolate sufficient DNA for analysis (Bugawan *et al.* 1988). Second, amplification can be done without purification of the cellular DNA (Kogan *et al.* 1987). This improvement circumvents losses of DNA during purification and leads to more rapid results. In their study Kogan *et al.* (1987) were able to determine the sex of the fetus by using primers surrounding a DNA segment specific to the Y chromosome. Amplified 'male' DNA could be directly visualized on a gel.

Of all genetic diseases, the inherited haemoglobin disorders have been the most extensively studied at the DNA level. In what follows, haemoglobinopathies will be taken as examples. Their antenatal diagnosis by recombinant DNA techniques has served as a prototype for other genetic disorders (Weatherall & Old 1983). Clinically the most important haemoglobinopathies are sickle-cell anaemia and the thalassaemias. The latter are a group of disorders in which there is an imbalance in the synthesis of globin chains due to the low, or totally absent, synthesis of one of them. Many α-thalassaemias are caused by gene deletions, although several non-deletion forms have been identified (Weatherall & Clegg 1982). The β-thalassaemias are complex and more than 20 different molecular lesions have been identified. Most are nonsense or frameshift mutations in exons of the β-globin gene. Others are point mutations affecting transcript processing, or point mutations affecting the promoter. Only one form has been identified as due to a major deletion of the β-globin gene (Weatherall & Clegg 1982).

Fetal DNA analysis. Clearly, if a mutation either removes or produces a restriction enzyme site in genomic DNA, this can be used as a marker for the presence or absence of the defect. The mutation from GAG to GTG in

336

Section 4
Applications of
Recombinant
DNA
Technology

sickle-cell anaemia eliminates a restriction site for the enzyme *Dde* I (CTNAG) or the enzyme *Mst* II (CCTNAGG) (Chang & Kan 1981, Orkin *et al*. 1982). The mutation can therefore be detected by digesting mutant and normal DNA with the restriction enzyme and performing a Southern-blot hybridization with a cloned β-globin DNA probe (Fig. 15.8). Such an approach is applicable only to those disorders where there is an alteration in a restriction site, or where a major deletion or rearrangement alters the restriction pattern. It is not applicable to most β-thalassaemias.

There are many polymorphic restriction sites scattered throughout the β-globin gene cluster (Weatherall & Old 1983). These are revealed as restriction-fragment-length polymorphisms in Southern-blot experiments. They can be used as linkage markers for antenatal diagnosis, i.e. the close physical linkage with a β-globin gene will mean that the polymorphic site will trace the inheritance of that gene. These polymorphisms can be used in two ways. First, some polymorphisms are linked to specific globin mutations; for example in the USA among the black population the sickle

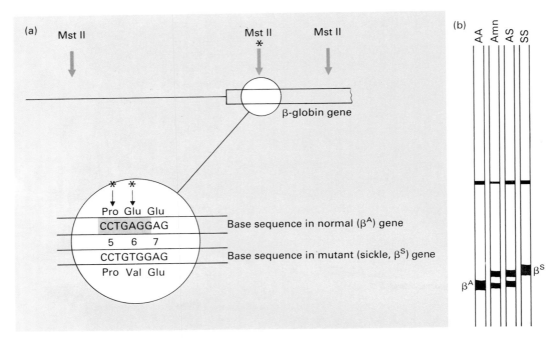

Fig. 15.8 Antenatal detection of sickle cell genes. Normal individuals are homozygous for the β[A] allele, while sufferers from sickle-cell anaemia are homozygous for the β[S] allele. Heterozygous individuals have the genotype β[A]β[S]. In sickle-cell anaemia, the 6th amino acid of β-globin is changed from glutamate to valine. (a) Location of recognition sequences for restriction endonuclease *Mst* II in and around the β-globin gene. The change of A → T in codon 6 of the β-globin gene destroys the recognition site (shaded) for *Mst* II as indicated by the asterisk. (b) Elecrophoretic separation of *Mst* II-generated fragments of human control DNAs (AA, AS, SS) and DNA from amniocytes (Amn). After Southern blotting and probing with a cloned β-globin gene, the normal gene and the sickle gene can be clearly distinguished. Examination of the pattern for the amniocyte DNA indicates that the fetus has the genotype β[A]β[S], i.e. it is heterozygous.

mutation is associated 60% of the time with a polymorphic mutation near the gene that eliminates a Hpa I site (Kan & Dozy 1978). The linkage could form the basis of diagnosis, but is inferior to the direct analysis with Mst I. Examples of such associations (so-called linkage disequilibrium) are rare. A second approach is required in which it is necessary to establish linkages between polymorphic restriction sites and a particular β-globin gene mutation by carrying out a family study before antenatal diagnosis (Fig. 15.9). This is not always possible, and in any case suitable polymorphic markers may not be present. It has been estimated that this approach is feasible in no more than 50% of β-thalassaemia cases in the UK (Weatherall & Old 1983).

Recently a very powerful and direct approach to analysing point mutations has been devised. Conner *et al.* (1983) synthesized two 19-mer oligonucleotides, one of which was complementary to the amino-terminal region of the normal β-globin (β^A) gene, and one of which was complementary to the sickle cell β-globin gene (β^S). These oligonucleotides were radiolabelled and used to probe Southern blots. Under appropriate conditions, the probes could distinguish the normal and mutant alleles. The DNA from normal homozygotes only hybridized with the β^A probe, and DNA from sickle-cell homozygotes only hybridized with the β^S probe. DNA of heterozygotes hybridized with both probes. These experiments, therefore, showed that oligonucleotide hybridization probes can discriminate between a fully complementary DNA and one containing a single mismatched base (for a related example see p. 88). Similar results have subsequently been obtained with a point mutation in the α-antitrypsin gene which is implicated in pulmonary emphysema (Cox *et al.* 1985). The generality of this approach is very impressive. It should be applicable to other genetic disorders provided that the nucleotide sequences around the mutation site, which could be a substitution, insertion or deletion, can be established.

For a more detailed description of the applications of nucleic acid probes in clinical genetics the reader should consult the monograph of Weatherall (1985).

DNA profiling (fingerprinting)

The human genome contains many polymorphic loci known as hypervariable regions (HVRs) or variable number tandem repeats. Polymorphisms at such loci are the result of variations in the number of tandem repeats of a short core sequence. These regions of mini-satellites may occur dispersed throughout the human genome (Jeffreys *et al.* 1985a) or they may be clustered on a single chromosome (Das *et al.* 1987, Buroker *et al.* 1987). DNA probes have been developed that at low stringency hybridize simultaneously to a number of these loci to produce individual-specific fingerprints (Jeffreys *et al.* 1985a). The technique has been shown to have considerable potential in forensic applications (Gill *et al.* 1985, 1987, Hill & Jeffreys 1985). Of particular value, suitable high-mol. wt DNA can be isolated from such stains as blood and semen made on clothing several years previously. Also, sperm nuclei can be separated from the vaginal

337
Chapter 15
Nucleic Acid
Probes and
their
Applications

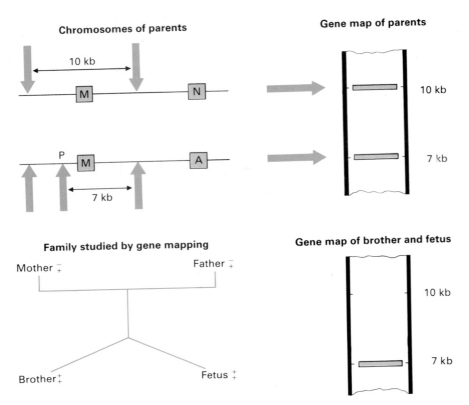

Chromosomes of parents

10 kb

M   N

P

M   A

7 kb

Gene map of parents

10 kb

7 kb

Family studied by gene mapping

Mother ⁻₊   Father ⁻₊

Brother ⁻₊   Fetus ⁻₊

Gene map of brother and fetus

10 kb

7 kb

Fig. 15.9 An example of prenatal diagnosis using restriction fragment length polymorphism (RFLP) linkage analysis. The parents are both carriers for a deleterious gene (A): one of their chromosomes carries this determinant, the other its normal allele (N). One of the parental chromosomes carries a polymorphic restriction enzyme site P which is close enough to A or N so that they will not be separated in successive generations. On the chromosome which does not contain this site (−) a particular restriction enzyme cuts out a piece of DNA 10 kb long which contains another locus (M) for which we have a radioactive probe. On the chromosome containing the polymorphism (+) a single base change produces a new site and hence the DNA fragment containing locus M is now only 7 kb. On gene mapping of the parents' DNA using probe M we see two bands representing either the + or − chromosomes. A previously born child had received the deleterious gene A from both parents and on mapping we find that it has the + + chromosome arrangement, i.e. only a single 7 kb band. Hence the mutation must be on the + chromosome in both parents. To identify the disease in a fetus in subsequent pregnancies we will be looking for an identical pattern, i.e. the 7 kb band only. (Reproduced courtesy of Professor D. Weatherall and Oxford University Press.)

cellular debris present in semen-contaminated vaginal swabs taken from rape victims.

The potential of DNA profiling is best illustrated by the following example. A Ghanaian boy was initially refused entry into Britain because the immigration authorities were not satisfied that the woman claiming him as her son was in fact his mother. Analysis of serum proteins and erythrocyte antigens and enzymes showed that the alleged mother and son were related but could not determine whether the woman was the

339

Chapter 15
Nucleic Acid
Probes and
their
Applications

boy's mother or aunt. To complicate matters, the father was not available for analysis nor was the mother certain of the boy's paternity. DNA profiles from blood samples taken from the mother and three children who were undisputedly hers as well as the alleged son were prepared by Southern blot hybridization to two of the mini-satellite probes shown in Fig. 15.10. Although the father was absent, most of his DNA profile could be reconstructed from paternal-specific DNA fragments present in at least one of the three undisputed siblings but absent from the mother (Fig. 15.11). The DNA profile of the alleged son contained 61 scorable fragments, all of which were present in the mother and/or at least one of the siblings. Analysis of the data showed the following.

1 The probability that either the mother or the father by chance possess all 61 of the alleged son's bands is 7×10^{-22}. Clearly the alleged son is part of the family.

2 There were 25 maternal-specific fragments in the 61 identified in the alleged son and the chance probability of this is 2×10^{-15}. Thus the mother and alleged son are related.

3 If the alleged mother of the boy in question is in fact a maternal aunt, the chance of her sharing the 25 maternal-specific fragments with her sister is 6×10^{-6}.

When presented with the above data (Jeffreys *et al.* 1985b), as well as results from conventional marker analysis, the immigration authorities allowed the boy residence in Britain. In a similar kind of investigation a man originally charged with murder was shown to be innocent (Gill & Werrett 1987).

The probes described by Jeffreys and colleagues (Fig. 15.10) were derived from a tandemly repeated sequence within an intron of the myoglobin region on chromosome and shown subsequently to hybridize to autosomal gene. Other suitable probes include the insulin, inter-ζ and α-globin 3' HVRs (Fowler *et al.* 1988) and DYXS15 (Simmler *et al.* 1987). Under the right conditions it is possible to use bacteriophage M13 DNA as a probe (Vassart *et al.* 1987) the effective sequence being two clusters of 15 base repeats within the protein III gene.

DNA profiling is not restricted to human DNA. It also is applicable to cats and dogs (Jeffreys & Morton 1987) and birds (Burke & Bruford 1987) and has been used to confirm cell line identity in a cell line collection (Devor *et al.* 1988).

Core sequence	GGAGGTGGGCAGGA(A)GG	
Probe 33.6	$[(AGGGCTGGAGG)_3]_{18}$	
Probe 33.15	$(AGAGGTGGGCAGGTGG)_{29}$	
Probe 33.5	$(GGG(C)AGTGGGCAGGAGG)_{14}$	

Fig. 15.10 Probes used for DNA profiling.

340

Section 4
Applications of
Recombinant
DNA
Technology

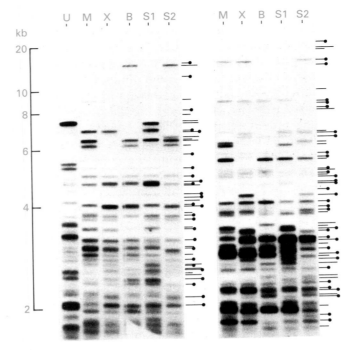

Fig. 15.11 DNA profiles of a Ghanaian family involved in an immigration dispute. Profiles of blood DNA are shown for the mother (M), the boy in dispute (X), his brother (B), sisters (S1, S2) and an unrelated individual (U). Fragments present in the mother's (M) DNA are indicated by a short horizontal line (to the right of each profile); paternal fragments absent from M but present in at least one of the undisputed siblings (B, S1, S2) are marked with a long line. Maternal and paternal fragments transmitted to X are shown with a dot.

The lefthand profile was obtained with probe 33.15 and the righthand profile with probe 33.6 (Photograph courtesy of Dr A. Jeffreys and the editor of *Nature*.)

Probes as therapeutic agents

An unusual application of an oligonucleotide has been described by Verspierin *et al*. (1987). They synthesized oligonucleotides complementary to part of the 35-nucleotide sequence present at the 5′ end of all mRNAs of the blood parasite *Trypanosoma brucei*. An acridine was then linked to this antimessenger oligonucleotide. When tested on cultured trypanosomes the acridine-linked oligonucleotide had a lethal effect which was not seen with the unmodified oligonucleotide or with an acridine-linked oligomer not complementary to mRNA. With luck this observation will lead to the discovery of new trypanocidal drugs of high specificity and efficiency.

Formats for hybridization reactions

There are two basic formats for hybridization reactions. Either one of the two nucleic acid molecules in the hybridization reaction is immobilized to a solid support, as in filter hybridizations, or the hybridization is carried out with both nucleic acids in solution. In solution hybridization both the

341

*Chapter 15
Nucleic Acid
Probes and
their
Applications*

target and the probe nucleic acid are free to move, thus maximizing the chance that complementary sequences will align and bind. Consequently solution hybridizations go to completion five- to 10-fold faster than those on solid supports (Bryan *et al.* 1986). This can be particularly important in many diagnostic microbiology applications where the concentration of the target sequence is very low and speed is essential.

At the end of the hybridization step it is essential to separate duplexes from unbound probe. If one of the two sequences in the hybridization reaction has been immobilized this separation step is achieved by a simple washing procedure. This explains the popularity of filter hybridization, of which there have been numerous examples in previous chapters.

A variation of the filter hybridization reaction is to attach the probe to the bottom of a microtitre plate well or to a tube (Polsky-Cynkin *et al.* 1985). This facilitates the washing step, reduces the total volume of the hybridization, and facilitates the automated reading of results if a colorimetric detection system is used. A clever utilization of this format is the sandwich hybridization reaction (Ranki *et al.* 1983, Palva & Ranki 1985). Here one probe is attached to the solid support and serves to capture homologous nucleic acids. A second DNA probe, which recognizes a contiguous sequence carries the reporter molecule (Fig. 15.12). Although the sandwich format is far less dependent upon the sample composition than direct blotting methods, it is a relatively slow process. Also, the capture is inefficient, since after denaturing the target DNA, the rate of reassociation of the target with itself in solution is considerably greater than its rate of association with the solid-phase probe (Syvanen *et al.* 1986). For the latter reason RNA probes are preferable to DNA probes.

If solution hybridization has been used, removal of unbound probe is not easy. One method is to digest the single-stranded nucleic acid that remains after hybridization with an appropriate nuclease. Although attractive in principle, in practice efficient digestion and separation are difficult to achieve reliably, particularly with crude samples. An alternative method is to separate duplexes from single-stranded by means of differential binding to hydroxyapatite or antibodies specific for double-stranded nucleic acids. A different approach has been adopted by Gingeras *et al.* (1987) who carried out sandwich hybridization in solution. The target DNA is hybridized in solution to a labelled probe and to an unlabelled capture probe which is immobilized by covalent attachment at its 5′ end to a solid support. If the solid support is in the form of beads, subsequent

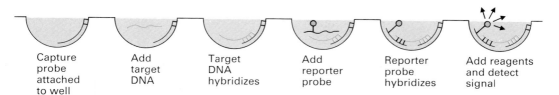

| Capture probe attached to well | Add target DNA | Target DNA hybridizes | Add reporter probe | Reporter probe hybridizes | Add reagents and detect signal |

Fig. 15.12 The sandwich hybridization method as carried out in a microtitre tray. The different wells of the microtitre tray show different stages in the procedure.

342

*Section 4
Applications of
Recombinant
DNA
Technology*

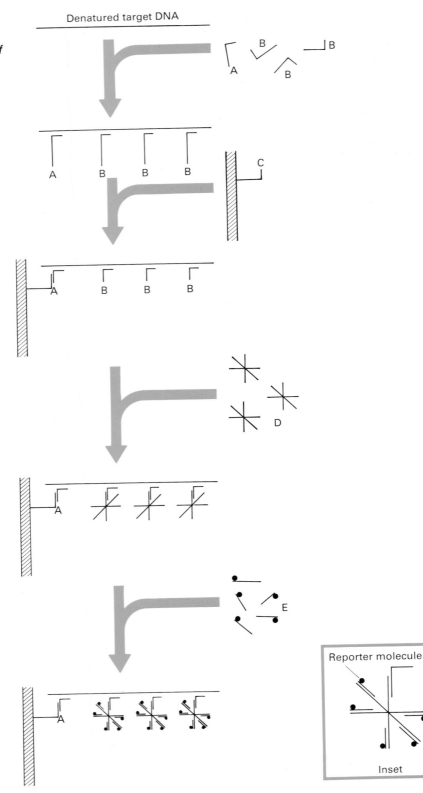

343

*Chapter 15
Nucleic Acid
Probes and
their
Applications*

handling of the complexes is easy. The efficiency of this format is limited by target reassociation and leaching of the immobilized probe.

Urdea *et al.* (1987) have devised a method employing solution phase hybridization of unmodified synthetic oligonucleotides with the target nucleic acid followed by solid-phase capture, labelling and signal amplification steps. The method which is rather complex is shown in Fig. 15.13. Probes A and B carry sequences complementary to different parts of the target genome as well as unique sequences not found in the target DNA. After hybridization of probes A and B, a third probe (C) is added which is linked to a solid support and which is complementary to the unique sequence of A. Thus probe C serves to capture the hybridization complex and to immobilize it. Next probe D is added. Probe D serves as an amplification multimer and is a chemically cross-linked oligonucleotide complex (star structure). Finally probe E is added which hybridizes to probe D and carries the detection label, e.g. horseradish peroxidase.

The ultimate choice of format will depend on the specific application and the speed and sensitivity required. Filter hybridization is very convenient but the rate of hybridization is slow by comparison with reaction in solution. The sensitivity of hybridization is affected by the choice of label, regardless of the format used, but in the case of filter hybridization the efficiency of retention of target DNA by the filter also is important. These problems can be minimized by use of the polymerase chain reaction to increase the target DNA concentration. As the target DNA concentration increases so does the rate of hybridization and the need for a sensitive detection system decreases. Ultimately hybridization assays will be automated to permit the high throughput required in the average diagnostic laboratory. The format ultimately chosen will thus depend not just on speed and sensitivity, but also on suitability for automation.

Fig. 15.13 (*Facing page*) Non-blotting assay configuration and multimer assay method. The inset shows detail for the star complex following hybridization of the reporter probe. Note that probe B is heterogeneous. The target-specific sequences on B are complementary to different parts of the target.

16 The Impact of Recombinant DNA Technology: the Generation of Novelty

Genetics is the fundamental biological science, for without genes there is no life. Thus a full understanding of any biological process can be achieved only when there has been a detailed analysis of gene structure and function. Classically this analysis was undertaken by making mutants, studying their properties, mapping them and generating hypotheses for further testing. The archetypal example of the success of this approach was the development of the operon concept from studies on the regulation of lactose metabolism. What recombinant DNA technology has done is to add new weapons to the armoury of the geneticist. For example, hypothesizing what has happened at the DNA level is no longer necessary: the genes now can be cloned and sequenced and the location and nature of the mutation identified precisely, be it base change, deletion or addition. Given that attempts are under way to sequence the human genome, how long will it be before the basis for the evolution of Darwin's finches in the Galapagos islands is understood at the molecular level?

Mutation still remains an essential tool for the geneticist but instead of seeking mutants with interesting phenotypes, mutations of different types are introduced at will and at predetermined locations (see Chapter 5). The net effect is that the whole analytical process is speeded up by several orders of magnitude.

Because of the speed and precision of the techniques of gene manipulation, biologists now are making major advances in the analysis of fundamental but much more complex biological systems. Examples range from the control of mitosis and division of individual cells to the differentiation and development of whole animals. These studies are being facilitated by the impact of gene manipulation on biochemical methods. In the cell, proteins play a key role because they are intermediaries between gene and phenotype. Traditionally proteins have been purified from cell extracts

Table 16.1 Recombinant proteins approved for clinical use.

Protein	Application
human growth hormone	pituitary dwarfism
human insulin	diabetes
IFN-α_2	hairy cell leukaemia; genital warts
tissue plasminogen activator	coronary thrombosis
hepatitis B surface antigen	vaccination
erythropoietin	anaemia associated with renal failure
interleukin-2	cancer

345

Chapter 16
The Impact of
Recombinant
DNA
Technology

and their properties studied *in vitro*. However, the behaviour of a purified protein in the test tube ('*in-vitro* biochemistry') may be quite different from that of the same protein in the complex milieu of the cell. Now it is possible to do '*in-vivo* biochemistry' by under- or overproducing natural and mutant proteins inside the cell and studying their effects on key cellular processes. An excellent treatise on this topic is the text by Kingsman & Kingsman (1988). The traditional approach also presupposes that enough of the protein is made in the cell for it to be isolated in the first place. However, many key cellular proteins are made transiently and at very low levels, e.g. proteins involved in cell division, lymphokines, etc. In principle, using recombinant DNA technology it is possible to produce any protein in quantity and Table 16.1 lists some of the successes which have been achieved. The impact of this goes far beyond understanding cellular processes: many of them have commercial value as pharmaceuticals and their current status is shown in Table 16.2.

Novel problems associated with the overproduction of proteins

From previous chapters it should be clear that there is a wide range of cloning hosts and, theoretically, any one could be used to overproduce a protein of interest. So, what governs the ultimate choice? If overproduction is all that is required then it will be convenience. More often than not, though, the deciding factor will be the degree of authenticity required. Ideally a recombinant-derived protein would have the same amino acid sequence, the same post-translational modifications, the same 3-dimensional structure and the same range of biological activities as its natural counterpart. In practice this is difficult to achieve and what deviation from the ideal is acceptable depends on the use to which the protein will be put. For an enzyme to be used as a detergent additive the key parameters will be specific activity and stability. For a therapeutic protein which will be administered parenterally the criteria are much more stringent. Whatever use the protein is put to, authenticity is defined operationally. Is the recombinant-derived protein identical to its natural counterpart as determined by a variety of analytical techniques? Does it have the same specific activity in particular assays? Answers to questions such as these presuppose that we have sufficient of the natural material for comparative purposes, that suitable analytical techniques are available, and that we even know what questions to ask in the first place!

None of the cloning systems currently available is ideal. Each has its advantages and disadvantages. These are discussed below and summarized in Table 16.3. The interaction of the different factors discussed is shown in Fig. 16.1.

Expression. Overproduction means that expression needs to be maximized and, for reasons of stability (see p. 166), should be easily regulated by environmental factors. In this context *E. coli* offers the greatest advantage. Many different constructs are available with strong, regulatable promoters controlling both gene expression and copy number. Yeast is the next choice because high levels of expression can be obtained, up to 2 g/l at

Table 16.2 Current status of selected recombinant proteins.

Protein	Size structure	Expression system	Clinical indications	Comments
human insulin	two peptide chains: **A**, 21 amino acids long, and **B**, 30 amino acids long	E. coli	juvenile-onset diabetes	A and B chains made separately as fusion proteins and joined in vitro already on market
human somatotrophin	191 amino acids	E. coli	pituitary dwarfism	if useful in treatment of osteoporosis, then market size will be much larger has additional methionine residue at N-terminus, but technology for removing this now available already on market
IFN-α₂	166 amino acids	E. coli	hairy cell leukaemia prophylaxis of common cold	Over 80% success in treatment of hairy cell leukaemia; success with other cancers lower and more variable market size may be limited unpleasant ('flu-like') side-effects already on market
IFN-γ	143 amino acids, glycosylated	E. coli	treatment of cancers treatment of virus diseases?	in clinical trials
tissue plasminogen activator		E. coli yeast animal cells	thrombosis	animal cell culture most effective way of producing active enzyme on the market

Protein	Structure	Source	Application	Notes
relaxin	53 amino acids; insulin-like (two protein chains)	E. coli	facilitates childbirth	prepares endometrium for parturition and reduces fetal distress; pig relaxin shown to be clinically effective
α₁-antitrypsin	394 amino acids glycosylated	E. coli; yeast	treatment of emphysema	prevents cumulative damage to lung tissue caused by leukocyte elastase in clinical trials on the market
interleukin-2	133 amino acids	E. coli; animal cells	treatment of cancer	
tumour necrosis factor	157 amino acids	E. coli; animal cells	treatment of cancer	
human serum albumin	582 amino acids; 17 disulphide bridges	yeast	plasma replacement therapy	normally obtained from plasma but now concern over potential contamination with AIDS virus
factor VIII	2332 amino acids	mammalian cells	treatment of haemophilia	normally obtained from plasma but now concern over potential contamination with AIDS virus
factor IX	415 amino acids glycosylated; modified residues	mammalian cells	treatment of Christmas disease	must be made in mammalian cells since glycosylation and conversion of first 12 glutamate residues to pyroglutamate essential for activity
erythropoietin	166 amino acids glycosylated	mammalian cells	treatment of anaemia	without glycosylation protein is cleared very quickly from plasma on the market
hepatitis B surface antigen	monomer has 226 amino acids	yeast; mammalian cells	vaccination	monomer self-assembles into structure resembling virus particles now on market

Table 16.3 Comparison of different organisms as cloning hosts.

Organism	Advantages	Disadvantages
Escherichia coli	ease of manipulation promoters and gene regulation well understood many high-expression vectors available easy to culture on large scale already used in manufacture of insulin, interferon and human somatotrophin	do not usually get export of proteins into growth medium overexpressed foreign proteins often form aggregates ('inclusions') of denatured protein many foreign proteins rapidly degraded many post-translational modifications do not occur
Bacillus subtilis	many proteins naturally exported into growth medium non-pathogenic easy to culture some *Bacillus* enzymes excreted at high level (> 5 g l^{-1})	still not much known about gene regulation good, high-level expression vectors lacking high-level export of heterologous proteins not achieved
Saccharomyces cerevisiae	widely used industrial organism which is easy to culture glycosylates proteins can get export into growth medium of heterologous proteins High-level expression systems developed	much still to be learned about control of gene expression post-translational modifications of proteins not necessarily the same as those in the animal cell heterologous proteins can form inclusions
filamentous fungi	large surface area to volume ratio should favour protein export have been used in industrial microbiology for over 40 years	promoters/gene regulation poorly understood but may be similar to yeast good expression systems lacking rheology of fermentations important
actinomycetes	large surface area to volume ratio should favour protein export widely used in industrial microbiology good expression systems being developed	promoters/gene regulation still poorly understood rheology of fermentations important
mammalian cells	get export of proteins get desired post-translational modification and products not likely to be immunogenic to humans good expression systems available	large-scale growth of animal cells costly great care needed to avoid contamination of cultures

349

*Chapter 16
The Impact of
Recombinant
DNA
Technology*

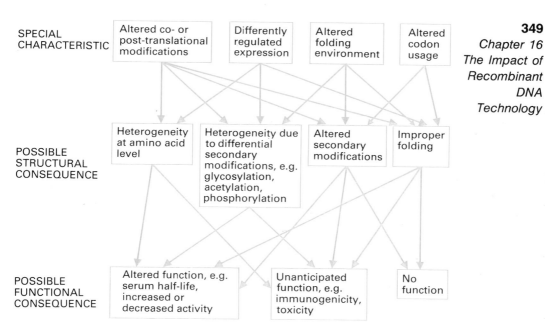

Fig. 16.1 The structural and functional consequences of heterologous gene expression. (Reproduced from *Biotechnology*, 1987, **5**, 883–890, by permission of the editor.)

high cell density in *Pichia pastoris* (Tschopp *et al*. 1987). However, unlike the situation with *E. coli*, many heterologous proteins cannot be expressed at high levels in *S. cerevisiae* for reasons which are not yet understood (see p. 211). In all other systems not enough is known about gene regulation, and good expression vectors are lacking.

Generation of the correct primary sequence. In theory, the primary sequence of a protein should match the DNA sequence from which it was derived, but this is not always the case for high-level expression of heterologous proteins. In *E. coli* the normal error rate for translation is about 1 codon in 3000 read (Edelmann & Gallant 1977). Under conditions of amino acid or charged-tRNA limitation the error frequency increases significantly. An artificial tRNA imbalance may be set up if a highly expressed recombinant gene contains a high proportion of infrequently used codons. Minor tRNA species which now are required in much larger quantities may be sequestered on ribosomes to such an extent that they cause a potentially serious starvation of the cognate tRNA. High-frequency mistranslation of this type has been reported by McPherson (1988). In many instances mistranslation may not have been detected because the techniques for protein analysis are not particularly sensitive. One solution is to use artificially synthesized genes with fully optionized codons for the system in question but, as pointed out earlier (p. 162), this is more difficult than it appears at first sight. The alternative is to clone the gene in the cells from which it was derived. This however may not be possible, not least because of poor expression.

A second factor which can influence the primary structure of a protein

350

Section 4
Applications of
Recombinant
DNA
Technology

is proteolysis. All cells contain a variety of proteases which can sequentially remove amino acids from the C- or N-termini or cleave the molecule internally. One method of minimizing proteolytic degradation, and which can be used in any cloning system, is to generate fusion proteins (p. 53). Another is to increase the level of expression to such an extent that the proteases are swamped with protein but, as stated above, not all systems permit high-level expression as yet. In *E. coli* it is possible to use strains that are deficient in certain proteases, e.g. *lon* mutants (p. 151). An alternative to the use of protease-deficient mutants is to secrete the protein. This option is available for *B. subtilis, S. cerevisiae* and mammalian cells, and possibly with *E. coli* (see p. 149).

A different kind of sequence disparity can occur in bacterial expression systems. The initiating codon for most genes is ATG, and in bacteria this encodes N-formyl-methionine. In their native form the proteins of most cells lack the amino-terminal methionine residue. However, many of the proteins overproduced in *E. coli* as a result of recombinant DNA technology retain the fMet residue, e.g. human growth hormone, rabbit β-globin, SV40t-antigen and λ repressor (Backmann & Ptashne 1978, Goeddel *et al.* 1979a, Roberts *et al.* 1979a,b, Guarente *et al.* 1980). One solution to this problem has been presented by the identification of the enzyme which removes the N-terminal methionine from peptides and the isolation of overproducing mutants (Ben-Bassat *et al.* 1987, Miller *et al.* 1987).

An alternative approach to the removal of unwanted N-terminal methionine residues is based on the ubiquitin — ubiquitin hydrolase system of yeast. The hydrolase will cleave ubiquitin-containing protein fusions immediately beyond the ubiquitin carboxy-terminal residues Arg−Gly−Gly. Sabin *et al.* (1989) constructed a yeast expression vector that contains a synthetic gene for ubiquitin. A unique restriction site at the 3′ end of the gene allowed the precise in-frame insertion of a heterologous gene. With this system they were able to produce interferon-gamma (IFN-γ) and a proteinase inhibitor with authentic N-termini. By cloning and expressing a yeast ubiquitin hydrolase gene in *E. coli* Miller *et al.* (1989) achieved a similar result.

Generating the correct 3-dimensional structure. In many instances high expression of recombinant proteins leads to the formation of high mol. wt. aggregates often referred to as inclusions (Fig. 16.2). They were first reported in strains of *E. coli* overproducing insulin chains A and B (Williams *et al.* 1982). At first their formation was thought to be restricted to the overexpression of heterologous proteins in *E. coli* but they can form in the presence of high levels of normal *E. coli* proteins, e.g. the λ subunit of RNA polymerase (Gribskov & Burgess 1983) and the product of the *envZ*

Fig. 16.2 (*Facing page*) Inclusions of Trp polypeptide-proinsulin fusion protein in *E. coli*. (*Upper*) Scanning electron micrograph of cells fixed in the late logarithmic phase of growth; the inset shows normal *E. coli* cells. (*Lower*) Thin section through *E. coli* cells producing Trp polypeptide-insulin A chain fusion protein. (Photographs reproduced from *Science* courtesy of Dr D. C. Williams (Eli Lilly & Co) and the American Association for the Advancement of Science.)

351
*Chapter 16
The Impact of
Recombinant
DNA
Technology*

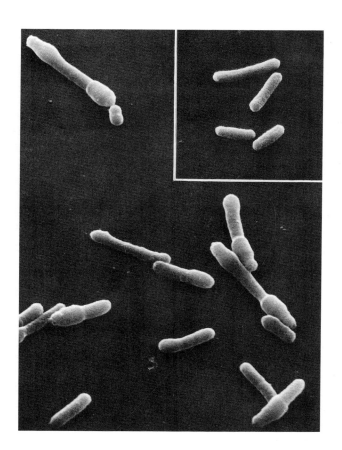

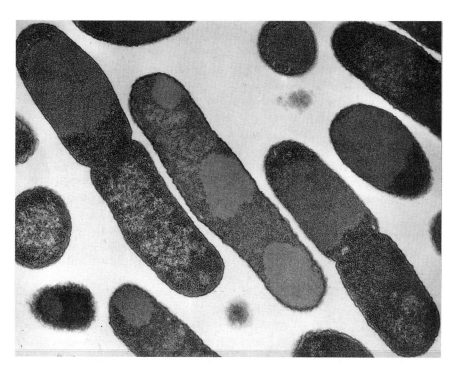

352

*Section 4
Applications of
Recombinant
DNA
Technology*

gene (Masui *et al.* 1984). More recently the formation of inclusions has been observed in yeast (Hayakawa *et al.* 1987).

Inclusions fall into two categories. First paracrystalline arrays in which the protein is presumably in a stable conformation, although not necessarily native. Second, amorphous aggregates which contain partially and completely denatured proteins as well as aberrant proteins synthesized as a result of inaccurate translation. The factors influencing the formation of inclusions in both homologous and heterologous cytoplasms have been reviewed by Mitraki & King (1989). Although inclusions can facilitate purification of the protein (Schoner *et al.* 1985) and may afford protection from proteases, they do present problems of extraction and purification. In most instances denaturants such as SDS, urea or guanidinium HCl have to be used to extract the protein. Subsequently the protein has to be renatured and this may prove difficult, if not impossible. With some overexpression systems it may be possible to keep the protein in a soluble form by altering the fermentation conditions, e.g. lowering the growth temperature (Schein & Noteborn 1988; Takagi *et al.* 1988). The alternative is to use a system where the protein is secreted into the growth medium. This is not without its disadvantages also. With the exception of cultured animal cells, if the producing organism is grown to a high cell density the culture will need vigorous aeration and agitation. This will result in many air—liquid interfaces which can denature the protein.

The codon distribution in a gene may influence the way in which a protein folds. Organisms may use sets of rare codons to influence the rate at which nascent polypeptide chains elongate. Positioning these codons at key junctures in the protein, thus introducing pauses in translation, may allow discrete functional domains to fold more efficiently. Purvis *et al.* (1987a) have suggested that some proteins may have evolved this mechanism as a means of avoiding secondary folding pathways leading to stable but aberrant conformations. If this is a general phenomenon it has significant implications for the accurate folding of proteins in heterologous systems. In such circumstances the codon bias of the original organism is unlikely to be reflected in the tRNA pools of the new host.

Post-translational modifications of a protein also can affect its folding. Thus tissue-derived IFN-β is glycosylated and has a specific activity in excess of 10^8 antiviral units per mg protein. The protein can be denatured and renatured repeatedly and still retain full specific activity. Recombinant-derived IFN-β is unglycosylated and tends to have a specific activity about 100-fold lower unless special renaturation techniques are used.

Post-translational modifications. Many different post-translational modifications have been described for proteins. These include N- or O-linked glycosylation, phosporylation, acetylation, amidation, sulphation, attachment of fatty acids and the formation of unusual amino acids. As far as we know, none of these modifications occurs on proteins from *E. coli* or other bacteria. Many of these modifications do not occur in yeast either, e.g. hirudin from leeches is sulphated but the recombinant protein made in yeast is not (Loison *et al.* 1988). Glycosylation is possible in yeast and heterologous proteins known to be glycosylated following secretion from

353

*Chapter 16
The Impact of
Recombinant
DNA
Technology*

yeast include influenza viral haemagglutinin (Jabbar *et al.* 1985), immuno-globulins (Wood *et al.* 1985), the acetylcholine receptor (Fujita *et al.* 1986), marcophage colony stimulating factor (Miyajima *et al.* 1986) and α_1-antitrypsin (Moir & Dumais 1987). Only in the latter case has it been shown that the same residues are glycosylated. However, the composition of the glycosyl residues in yeast differ from those found in mammalian cells (Dunphy & Rothman 1985). Although it might seem that mammalian cells offer the greatest promise for the production of authentic mammalian proteins this is not necessarily the case. Recombinant erythropoietin produced in two different mammalian cell lines was glycosylated, but the carbohydrate composition varied with the cell line and in both instances differed from that found in erythropoietin from urine (Goto *et al.* 1988). Ultimately it may be possible to achieve the correct post-translational modifications in any microorganism. Considerable progress is being made in cloning the genes involved in glycosylation (Kukuruzinska & Robbins 1987) and these genes could be introduced into a wide range of organisms.

A different kind of post-translational modification is exemplified by hormones such as insulin and relaxin. These molecules are normally synthesized as proproteins with an additional amino acid sequence which dictates the final three-dimensional structure. In the case of proinsulin, proteolytic attack cleaves out a stretch of 35 amino acids in the middle of the molecule to generate insulin (Fig. 16.3). Bacteria and yeasts have no mechanism for specifically cutting out the folding sequences from prohormones. A solution has been found for insulin which might be applicable for other similar proteins.

For the production of insulin in *E. coli*, chemically synthesized genes for the insulin A and B chains were cloned separately and attached in phase to the gene for β-galactosidase, just as with somatostatin. Thus two bacterial strains were constructed: one producing a fused protein carrying the human insulin A chain and the other a similar fused protein carrying the B chain (Goeddel *et al.* 1979b). The insulin chains were clipped from the β-galactosidase precursor by treatment with cyanogen bromide, purified and chemically linked *in vitro*. An alternative approach has been to clone the entire proinsulin gene which is then synthesized as a hybrid protein.

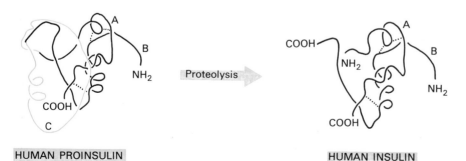

HUMAN PROINSULIN   HUMAN INSULIN

Fig. 16.3 The 3-dimensional structure of human proinsulin and insulin. The three disulphide bonds which stabilize the correctly folded structures are shown as dotted lines.

354

Section 4

Applications of

Recombinant

DNA

Technology

The proinsulin is chemically cleaved from the *E. coli* carrier protein with cyanogen bromide and purified. Since the primary structure of human proinsulin has in it all the information needed to assume the proper configuration, proinsulin synthesized by bacteria can be stabilized *in vitro* by formation of the disulphide bonds. The proinsulin formed in this way is subjected to proteolysis to generate human insulin. To facilitate the conversion of the prohormone to the native hormone Wetzel *et al.* (1981) constructed a gene encoding an analogue of human proinsulin in which the normal 35 amino acid C peptide was replaced with a 'mini-C' peptide of 6 amino acids (Arg−Arg−Gly−Ser−Lys−Arg).

Use of transgenic animals. An alternative approach to producing authentic proteins has been described by Gordon *et al.* (1987). They fused a gene encoding human tissue plasminogen activator (tPA) to the promotor and upstream regulatory sequences of murine whey acid protein. The hybrid gene then was injected into mouse embryos. The resultant transgenic mice were mated and the milk obtained from lactating females shown to contain biologically active tPA. In a similar fashion Clark *et al.* (1989) obtained expression of human anti-haemophilic factor IX in the milk of transgenic sheep.

The generation of novel proteins

One of the most exciting aspects of recombinant DNA technology is that its use need not be restricted to introducing intact genes of eukaryotes into microbes. Genes for completely novel proteins also can be constructed. The simplest example of the generation of novel proteins involves redesigning enzyme structure by site-directed mutagenesis. The investigations of Winter and colleagues demonstrated the power of this approach (Winter *et al.* 1982, Wilkinson *et al.* 1984). From a detailed knowledge of the enzyme tyrosyl-tRNA synthetase from *Bacillus stearothermophilus*, including its crystal structure, they were able to predict point mutations in the gene which should increase the enzyme's affinity for the substrate ATP. These changes were introduced, and in one case a single amino acid change improved the affinity for ATP by a factor of 100. Using a similar approach the stability of an enzyme can be increased. Thus Perry & Wetzel (1984) were able to increase the thermostability of T4 lysozyme by the introduction of a disulphide bond. However, although new cysteine residues can be introduced at will, they will not necessarily lead to increased thermal stability (Wetzel *et al.* 1988). Proof of the power of gene manipulation coupled with the techniques of site-directed mutagenesis is provided by the work on subtilisin (Carter & Wells 1987, Wells & Estell 1988). Almost every property of this serine protease has been altered including its catalysis, substrate specificity, pH/rate profile and stability to oxidative, thermal and alkaline inactivation. For example, a variant was constructed which was more than 1000-fold more resistant to oxidation inactivation than was the wild-type enzyme. This was achieved by replacing a critical methionine residue with alanine. A similar kind of modification, the replacement of methionine with valine generated an oxidation-resistant variant of α₁-antitrypsin (Courtney *et al.* 1985).

355

*Chapter 16
The Impact of
Recombinant
DNA
Technology*

Once the rules governing protein folding and enzyme catalysis are known sufficiently well to permit prediction, it should be possible to design functional protein molecules from scratch. A small but promising start on this major undertaking has been made by Gutte and his co-workers (Gutte *et al.* 1979). Using secondary structure prediction rules and model building, they designed and then synthesized a neutral, artificial 34-residue polypeptide with potential nucleic acid binding activity. The monomeric polypeptide and a covalent dimer of it interacted strongly with single-stranded DNA. Interestingly, the dimer displayed considerable ribonuclease activity. More recently, a polypeptide which binds the insecticide DDT was successfully designed and synthesized (Moser *et al.* 1983).

A standard technique in conventional medicinal chemistry is to identify a chemical entity with a desirable pharmacological effect and then to construct analogues and determine if any of them have improved therapeutic properties. A similar technique can be applied to proteins by means of recombinant DNA technology. For example, in the neutral solutions used for therapy, insulin is mostly assembled as zinc-containing hexamers. This self-association may limit absorption. By making single amino acid substitutions Brange *et al.* (1988) were able to generate insulins which are essentially monomeric at pharmaceutical concentrations. Not only have these insulins preserved their biological activity, they are absorbed two to three times faster. Other modifications are possible. Thus, by deliberately introducing sequence variations into a protein with multiple biological activities, novel proteins which retain only one or a few of these activities may be obtained. A slightly different approach has been used by Streuli *et al.* (1982) and Weck *et al.* (1982). By combining fragments from different interferon IFN-α genes they were able to construct genes which encode novel hybrid interferons, e.g. a hybrid combining the amino-terminal half of IFN-α₂ with the carboxy-terminal half of IFN-α₁. These hybrid interferons had antiviral properties distinct from the two parental interferon molecules. The commercial success of this approach remains to be seen, for such novel proteins may be immunogenic.

Novel routes to vaccines

An effective vaccine generates humoral and/or cell-mediated immunity which prevents the development of disease upon exposure to the corresponding pathogen. This is accomplished by presenting pertinent antigenic determinants to the immune system in a fashion which mimics that in natural infections. Conventional viral vaccines consist of inactivated, virulent strains or live-attenuated strains, but they are not without their problems. For example, many viruses have not been adapted to grow to high titre in tissue culture, e.g. hepatitis B virus. There is a danger of vaccine-related disease when using inactivated virus since replication-competent virus may remain in the inoculum. Outbreaks of foot-and-mouth disease in Europe have been attributed to this cause. Finally, attenuated virus strains have the potential to revert to a virulent phenotype upon replication in the vaccinee. This occurs about once or twice in every

356

Section 4
Applications of
Recombinant
DNA
Technology

million people who receive live polio vaccine. Recombinant DNA technology offers some interesting solutions.

Given the ease with which heterologous genes can be expressed in various prokaryotic and eukaryotic systems it is not difficult to produce large quantities of purified immunogenic material for use as a subunit vaccine. A whole series of immunologically pertinent genes have been cloned and expressed but in general the results have been disappointing. For example, of all the polypeptides of foot-and-mouth disease virus, only VP1 has been shown to have immunizing activity. However, polypeptide VP1 produced by recombinant means was an extremely poor immunogen (Kleid *et al.* 1981). Perhaps it is not too surprising that subunit vaccines produced in this way do not generate the desired immune response, for they lack authenticity (see p. 344). The hepatitis B vaccine, which is commercially available (Valuenzuela *et al.* 1982), differs in this respect, for expression of the surface antigen in yeast results in the formation of virus-like particles. A similar phenomenon is seen with a yeast Ty vector carrying a gene for HIV coat protein (Adams *et al.* 1987). These subunit vaccines also have another disadvantage. Being inert they do not multiply in the vaccinee and so they do not generate the effective cellular immune response essential for the recovery from infectious disease.

A better approach to vaccine preparation is to make use of an animal virus as a vector to express immunologically pertinent proteins. Vaccinia is the strongest candidate because of its successful history as an immunizing agent. Vaccinia virus was the immunogen used to accomplish the global eradication of smallpox. Among the properties which contributed to its success as a live vaccine were its stability in a freeze-dried preparation, its low production cost, and the ability to administer it by simple dermal abrasion. A number of viral proteins have been expressed by vaccinia virus recombinants, basically using the method described on page 287. These recombinants show great promise for they induce the correct immunological responses in experimental animals (Table 16.4).

Table 16.4 Immune response to heterologous antigens expressed by vaccinia virus recombinants.

Antigen	Neutralizing antibodies	Cellular immunity	Animal protection
rabies virus glycoprotein	+	+	+
vesicular stomatitis glycoprotein	+	+	+
herpes simplex virus glycoprotein D	+	+	+
hepatitis B surface antigen	+	+	+
influenza virus haemagglutinin	+	+	+
human immunodeficiency virus envelope	+	+	not determined

Reproduced with permission from Tartaglia & Paoletti (1988).

357

Chapter 16
The Impact of
Recombinant
DNA
Technology

A major disadvantage associated with the use of live virus vaccines is that persons with congenital or acquired immunodeficiency risk severe infections. In particular, children in developing countries often are immunodeficient because of malnutrition and/or infection with parasites or viruses. Ramshaw *et al.* (1987) and Flexner *et al.* (1987) have shown that one solution to this problem is to incorporate a gene for interleukin-2 expression into the vaccinia vector. There may be an interesting spin-off from these experiments. Normally interleukins and IFN-γ have very short half-lives when administered *in vivo*. However, Flexner *et al.* (1987) noted that interleukin-2 was produced for several days following infection with vaccinia virus carrying the interleukin-2 gene.

A different procedure for attenuating a vaccine organism, in this case a bacterium rather than a virus, has been proposed by Kaper *et al.* (1984). They attenuated a pathogenic strain of *Vibrio cholerae* by deletion of DNA sequences encoding the A_1 subunit of the cholera enterotoxin. A restriction endonuclease fragment encoding the A_1, but not the A_2 or B sequences was deleted *in vitro* from cloned cholera toxin genes. The mutation then was recombined into the chromosome of a pathogenic strain. The resulting strain produces the immunogenic but non-toxic B subunit of cholera toxin but is incapable of producing the A subunit.

An alternative approach to generating bacterial vaccines has been suggested by d'Enfert & Pugsley (1987). They fused a gene for alkaline phosphatase to the *Klebsiella pneumoniae* gene for the secreted lipoprotein pullulanase (see p. 150). When this construct was expressed in *E. coli* the hybrid protein was located on the outer face of the outer membrane.

New routes to small molecules

Recombinant DNA technology does not just offer novel methods for the generation of proteins, it also provides new ways of making low-molecular-weight compounds. A good example of this is the microbial synthesis of the blue dye indigo (Ensley *et al.* 1983). This compound is not normally produced by microbes. However, the cloning of a single gene from *Pseudomonas putida*, that encoding naphthalene dioxygenase, resulted in the generation of an *E. coli* strain able to synthesize indigo in medium containing tryptophan (Fig. 16.4). A similar approach has yielded a new route to vitamin C. The conventional process starts with glucose and comprises one microbiological and four chemical steps (Fig. 16.5). By cloning in *Erwinia* a single gene, that from *Corynebacterium* encoding 2, 5-diketogluconic acid reductase, the process can be simplified to a single microbiological and a single chemical step (Anderson *et al.* 1985).

Bioconversions (or biotransformations) are processes in which microorganisms convert a compound to a structurally related product. They comprise only one or a small number of enzymatic reactions as opposed to the multireaction sequences of fermentations. Although hundreds of different bioconversions have been described they are only used commercially when conventional chemical approaches are too costly or too difficult, e.g. when stereoselective transformation is required. A key feature of bioconversions is that they are easy to scale-up, since the only parameter of

358

*Section 4
Applications of
Recombinant
DNA
Technology*

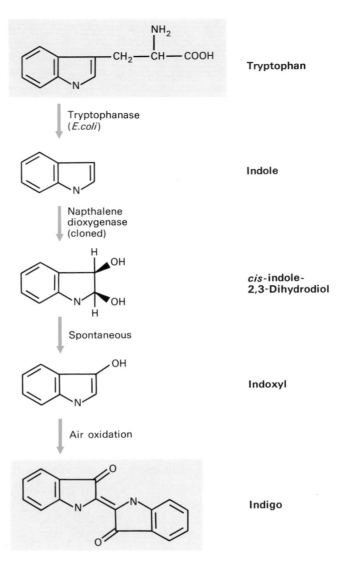

Fig. 16.4 Proposed pathway for indigo biosynthesis by a strain of *E. coli* carrying a cloned gene for naphthalene dioxygenase.

interest is the level of the enzyme mediating the transformation. This makes bioconversions obvious candidates for the application of recombinant DNA technology. Not only can the enzyme levels be increased in existing processes, but processes which previously were uneconomic can be reconsidered. Thus the levels of aspartate aminotransferase in wild-type *E. coli* are too low to make this organism of much use for L-amino acid synthesis from L-keto acids. By cloning the relevant gene enzyme levels can be raised to over 20% of total soluble protein and bioconversion becomes feasible. Once the gene is cloned it is possible to use site-directed mutagenesis to produce enzyme variants with enhanced thermo-stability or altered substrate profile.

359

*Chapter 16
The Impact of
Recombinant
DNA
Technology*

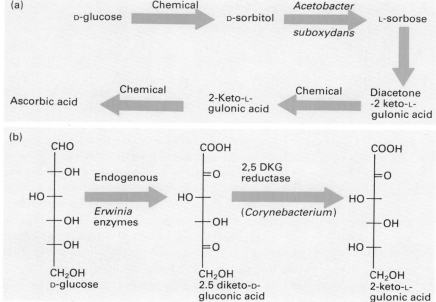

Fig. 16.5 Simplified route to vitamin C (ascorbic acid) developed by cloning in *Erwinia* the *Corynebacterium* gene for 2,5-diketogluconic acid reductase. (a) Classical route to vitamin C. (b) The simplified route to 2-ketogulonic acid, the immediate precursor of vitamin C.

Just as novel proteins can be produced by recombinant DNA techniques so too can novel small molecules. The *Streptomyces coelicolor* gene cluster encoding the biosynthesis of the isochromanequinone antibiotic actino-rhodin has been cloned. When the cloned genes were introduced into a variety of other *Streptomyces* spp. producing different isochromanequi-nones, at least three new antibiotics were detected (Hopwood *et al.* 1985b). Clearly actinorhodin, or one of its precursors, is a novel metabolite in these other *Streptomyces* spp. and is subject to further or different enzymatic modifications.

The generation of new routes to low mol. wt molecules is not restricted to bacteria. Plant cells also can be manipulated. Plants remain a major source of structurally complex, high-value small molecules, e.g. atropine, hyoscyamine. There has been great interest in the exploitation of plant cell cultures to produce such fine chemicals, but progress has been slow. The main reason for this is that the productivity of many undifferentiated plant cell cultures is low and/or unstable. However, when plant cells are infected with *Agrobacterium rhizogenes*, a close relative of *A. tumefaciens* (see p. 232), they become transformed. These transformed cells form a callus from which hairy roots appear. Hairy root cultures can be established and they are both rapidly growing and productive (Hamill *et al.* 1987). Since plasmid transfer is involved in the formation of hairy roots the system is amenable to further genetic manipulation.

360

Section 4

Applications of
Recombinant
DNA
Technology

The generation of novel plants

Although the techniques for expressing heterologous genes in transgenic plants were described for the first time in only 1983 we are already witnessing the practical use of these methods for agriculturally relevant plant breeding (Gasser & Fraley 1989). Not surprisingly, these examples have to do with the transfer and expression of single genes. The engineering of herbicide resistance in plants is a good example of what has been done.

Herbicides generally affect processes that are unique to plants, e.g. photosynthesis or amino acid biosynthesis (see Table 16.5). These processes are shared by both crops and weeds, and developing herbicides which are selective for weeds is very difficult. An alternative approach is to modify crop plants so that they become resistant to broad-spectrum herbicides. Two approaches to engineering herbicide resistance have been adopted. In the first of these the target molecule in the cell either is rendered insensitive or is overproduced. In the second approach a pathway that degrades or detoxifies the herbicide is introduced into the plant.

Glyphosate inhibits EPSP synthase, a key enzyme in the biosynthesis of aromatic amino acids in plants and bacteria. A glyphosate-tolerant *Petunia* cell line obtained after selection for glyphosate resistance was found to overproduce the EPSP synthase as a result of gene amplification. A gene encoding the enzyme was isolated and introduced into petunia plants under the control of a CaMV promoter. Transgenic plants expressed increased levels of EPSP synthase in their chloroplasts and were significantly more tolerant to glyphosate (Shah *et al*. 1986). An alternative approach to glyphosate resistance has been to introduce a gene encoding a mutant EPSP synthase. This mutant enzyme retains its specific activity but has decreased affinity for the herbicide. Transgenic tomato plants expressing this gene under the control of an opine promoter were also glyphosate tolerant (Comai *et al*. 1985).

Phosphinothricin (PPT) is an irreversible inhibitor of glutamine synthetase in plants and bacteria. Bialaphos, produced by *Streptomyces hygroscopicus*, consists of PPT and two alanine residues. When these residues are removed by peptidases the herbicidal component PPT is released. To prevent self-inhibition of growth, bialaphos-producing strains of *S. hygroscopicus* produce an acetyltransferase that inactivates PPT by acetylation. The *bar* gene that encodes the acetylase has been introduced into potato, tobacco and tomato cells using *Agrobacterium*-mediated transformation. The resultant plants were resistant to commercial formulations of PPT and bialaphos in the laboratory (De Block *et al*. 1987) and in the field (De Greef *et al*. 1989) (Fig. 16.6). Other examples of these approaches to engineering herbicide resistance have been reviewed by Botterman & Leemans (1988).

Plants can be made resistant to insects as well as to herbicides. Insect-resistant plants are particularly desirable for a number of reasons. First, innate resistance eliminates the need for insecticides which are both costly and time consuming to administer. Second, insecticides are not selective and can kill off desirable as well as undesirable insects. Finally, many insecticides accumulate in the environment and can lead to long-

Tables 16.5 Mode of action of herbicides and method of engineering herbicide-resistant plants.

Herbicide	Pathway inhibited	Target enzyme	Basis of engineered resistance to herbicide
glyphosate	aromatic amino acid biosynthesis	5-enol-pyruvyl shikimate-3-phosphate (EPSP) synthase	overexpression of plant EPSP gene or introduction of bacterial glyphosate-resistant aroA gene
sulphonylurea	branched-chain amino acid biosynthesis	acetolactate synthase (ALS)	introduction of resistant ALS gene
imidazolinones	branched-chain amino acid biosynthesis	acetolactate synthase (ALS)	introduction of mutant ALS gene
phosphinothricin	glutamine biosynthesis	glutamine synthetase	overexpression of glutamine synthetase or introduction of the bar gene, which detoxifies the herbicide
atrazine	photosystem II	Q_B	introduction of mutant gene for Q_B protein or introduction of gene for glutathione-S-transferase, which can detoxify atrazines
bromoxynil	photosynthesis		introduction of nitrilase gene which detoxifies bromoxynil

(a) (b)

Fig. 16.6 Evaluation of phosphin othricin resistance in transgenic tobacco plants under field conditions. (a) Untransformed control plants (b) Transgenic plants. (Photographs courtesy of Dr J. Botterman and the editor of *Biotechnology*.)

term changes in fauna. Fischhoff *et al.* (1987) and Vaeck *et al.* (1987) have generated transgenic tomato and tobacco plants which expressed the insect control protein from *Bacillus thuringiensis*. When plants expressing the insect toxin were infected with caterpillars, damage to the leaves was minimal and all the larvae died within 3 days. Control plants not expressing the toxin gene were stripped bare. One problem with the use of *B. thuringiensis* toxins is that they are very specific. Different strains produce different toxins, each with a different but very restricted spectrum of activity. The approach used by Hilder *et al.* (1987) has been to exploit the resistance to a wide range of pests which is conferred on cowpeas by the presence of trypsin inhibitors. A gene for cowpea trypsin inhibitor was cloned and expressed in tobacco plants which then exhibited enhanced resistance to caterpillars.

Major crop losses occur every year as a result of viral infections, e.g. tobacco mosaic virus (TMV) causes losses of tomato plants of over $50 million per annum. Currently there is little the farmer can do to control such infections and he is dependent on the plant breeder for the production of novel, virus-resistant cultivars. Unfortunately little is known about the genetics of virus resistance. However, there is a useful phenomenon known as cross-protection in which infection of a plant with one strain of virus protects against superinfection with a second, related strain. The mechanism of cross-protection is not understood but it is believed that the viral coat protein is important. Transgenic plants have been developed which express either the TMV protein (Powell-Abel *et al.* 1986) or the alfalfa mosaic virus coat protein (Tumer *et al.* 1987). In both instances disease symptoms were greatly ameliorated following infection with the respective viruses. A different method of minimising the effects of plant virus infection was developed by Gehrlach *et al.* (1987). They developed plants which expressed the satellite RNA of tobacco ringspot virus and such plants were phenotypically resistant to infection with tobacco ringspot virus itself.

Many plants contain storage proteins and in many third world countries a single crop can be the staple diet. Since most plant storage proteins are deficient in one or more essential amino acids, malnutrition is common.

Given the current status of gene manipulation in plants it should not be long before it is possible to alter the amino acid composition of the major storage proteins and hence increase their nutritional value.

Transgenesis: the generation of novel animals

The use of gene manipulation to permanently modify the germ cells of animals ('transgenesis') was described in Chapter 14. The archetypal example of transgenesis in animals is the production of 'supermice' which are extra-large as a result of the overproduction of human growth hormone (see Fig. 14.5). Although of great academic interest, such supermice have no commercial value. Other transgenic mice have been produced which synthesize human tissue plasminogen activator (tPA) and secrete it in their milk (Gordon *et al*. 1987, Pittius *et al*. 1988). Although this tPA is produced at a high concentration (50 000 ng/ml) and is biologically active, mice are unlikely to be acceptable as a pharmaceutical production system. However, other transgenic mice may be of particular value in medical research (Friedmann 1987). Currently there are many human diseases for which animal models are lacking, thereby hindering the development of suitable therapeutic agents. By means of transgenesis a number of mice have been produced which carry genetic lesions identical to those existing in certain human inherited diseases (Table 16.6).

363

*Chapter 16
The Impact of
Recombinant
DNA
Technology*

Table 16.6 Human disease equivalents derived from genetic alterations in the mouse.

Genetic alteration	Method of genetic alteration	Human disease equivalent	Reference
introduction of mutant collagen gene into wild-type mice	nuclear microinjection of inducible minigene	osteogenesis imperfecta	Sinn *et al*. 1987
inactivation of mouse gene encoding hypoxanthine-guanine phosphoribosyl transferase (HPRT)	insertion of retrovirus into HPRT locus in embryonic stem cells	HPRT deficiency	Kuehn *et al*. 1987.
mutation at locus for X-linked muscular dystrophy	male mutagenesis followed by identification of female carriers	X-linked muscular dystrophy	Chamberlain *et al*. 1988
introduction of activated human *ras* and c-*myc* oncogenes	Nuclear microinjection of inducible minigene	induction of malignancy	Sinn *et al*. 1987
introduction of mutant (Z) allele of human α_1-antitrypsin gene	microinjection of DNA fragment bearing mutant allele	neonatal hepatitis	Dycaico *et al*. 1988
introduction of HIV *tat* gene	microinjection of DNA fragment	Kaposis sarcoma	Vogel *et al*. 1988

364

Section 4
Applications of
Recombinant
DNA
Technology

Transgenesis in large mammals. While transgenic mice are of interest as an experimental tool, the technique needs to be applied to farm animals if any commercial benefit is to ensue. Working with large domestic animals is much more difficult than with mice. First, they do not produce as many eggs, e.g. a ewe on average only gives three or four. Second, reimplantation of manipulated embryos is more difficult because sheep and cattle will not give birth to more than two offspring. Finally, the eggs of many domestic animals have such an opaque cytoplasm that it is impossible to see their pronuclei or nuclei without resorting to special techniques. One technique which has been used to visualize the pronucleus in pig embryos is centrifugation to separate the cytoplasmic contents. This centrifugation step does not appear to affect their subsequent development. The difficulties cited above are evident from the data in Table 16.7, which show the results obtained when attempts were made to isolate transgenic rabbits, sheep and pigs.

So far, the enhanced growth of mice after the transfer of the human growth hormone gene is an effect that has not been as easy to mimic in other animals. For example, two successive generations of pigs expressing the bovine growth hormone gene showed significant improvements in both daily weight gain and feed efficiency. They also exhibited changes in carcass composition that included a marked reduction in subcutaneous fat. However, long-term elevation of growth hormone levels was detrimental to health (Pursel *et al.* 1989). The pigs had a high incidence of gastric ulcers, arthitis, cardiomegaly, dermatitis and renal disease.

However, the real goal is not to produce cows as big as elephants but to introduce economically significant traits into livestock. Desirable features of transgenic farm animals include increased efficiency of feed utilization, leaner meat, more rapid growth to marketable size and increased disease resistance. In sheep, a good choice would be the Merino booroolla gene which confers increased ovulation rates and large litter sizes—up to seven lambs per litter—without detriment to body size or quality and quantity of wool. That success is possible is shown by the generation of transgenic sheep which secrete human clotting factor IX and human α_1-antitrypsin (Simons *et al.* 1988).

Novel therapies for human diseases

The characterization of a growing number of heritable disorders at the gene and protein level has led to the conceptualization of therapies directly

Table 16.7 Efficiency of production of transgenic animals by microinjection of a growth hormone gene. (Adapted from Hammer *et al.* 1985.)

Animal species	No. of ova injected	No. of offspring	No. of transgenic offspring
Rabbit	1907	218	28
Sheep	1032	73	1
Pig	2035	192	20

365

Chapter 16
The Impact of
Recombinant
DNA
Technology

aimed at correcting the molecular defects. However, translating vision into reality has proved trickier than most people imagined. The major genetic diseases like cystic fibrosis, sickle-cell anaemia and thalassaemia remain unassailable—the genetics are too complex, or too poorly understood, to even contemplate therapeutic strategies at this time. Instead, investigators have focused on a handful of diseases that are simple enough to tackle now (Table 16.8). By and large these are very rare disorders that affect only a few people worldwide.

Getting cells in and out of the body so that a new gene can be inserted has proved formidable. For that reason the first experimental forays were made on bone marrow cells (Williams *et al*. 1984), which can be removed from the body and reinserted with relative ease.

Bone marrow consists of a heterogeneous population of cells, most of which are committed to differentiation into erythrocytes, lymphocytes, etc. Less than 1% of nucleated bone marrow cells are the undifferentiated progenitor cells known as stem cells. In gene therapy it is the stem cells that would be the primary target for exogenous DNA. Since they are low in number and not easily recognizable, an efficient gene delivery system is needed. With bone marrow cells calcium phosphate-mediated DNA uptake is very ineffective: the transfection frequency is as low as one cell in 10^6-10^7. Approximately 10^{10} cells can be obtained from the marrow of a patient undergoing transplantation and only 10^7-10^8 of them will be stem cells. Thus, at best, only 100 stem cells would be transfected by the calcium phosphate method and these would have little effect when transferred back into the marrow of the patient unless they had some extraordinary growth advantage.

Retrovirus vectors (see p. 291) offer a number of advantages over calcium phosphate-mediated DNA uptake. First, up to 100% of recipient cells can be infected and express the cloned gene. Second, as many cells as desired can be infected. Third, under appropriate conditions the DNA can integrate as a single gene copy, albeit randomly. Finally, the infection with the retroviral vector does not appear to harm cells.

A model system for human gene therapy. One candidate for human gene therapy is adenosine deaminase (ADA) deficiency, and as a model system

Tables 16.8 Target diseases for human gene therapy.

Protein	Disease state in absence of protein
factor VIII	haemophilia
hypoxanthine−guanine phosphoribosyl transferase	Lesch−Nyhan disease
pyrimidine nucleoside phosphorylase	immunodeficiency
adenosine deaminase	combined immunodeficiency disease

366

Section 4
Applications of
Recombinant
DNA
Technology

the human adenosine deaminase gene has been cloned in bone marrow cells of mice. The protocol used is outlined in Fig. 16.7. Proviral DNA from a retrovirus was cloned in a plasmid and some of the proviral DNA replaced with the ADA gene and a gene encoding resistance to the antibiotic G418. The recombinant DNA was then introduced into fibroblasts in tissue culture using the calcium phosphate-mediated uptake method. Since the retroviral genome is now defective the fibroblast line selected was one in which the missing viral genes were incorporated into the cell genome. Once viral particles began to appear in the growth medium, mouse bone marrow cells were added. The marrow cells were subsequently removed and those carrying the antibiotic resistance gene selected by growth in the presence of G418. These marrow cells were then injected back into the mouse and synthesis of human ADA detected.

Retroviruses are already at the point of being applied in human gene therapy, but improved vectors are being reported constantly (for review see Eglitis & Anderson 1988). A major disadvantage of retroviruses is that they are prone to delete gene sequences. Another problem is that one retrovirus can exchange gene sequences with another retrovirus. Thus a retroviral vector might recombine with an endogenous viral sequence to produce infectious recombinant virus. The risk is small, particularly if murine retroviruses are used in humans for there is little homology between murine and primate retroviruses. Nevertheless, murine retroviruses are considered a hazard in pharmaceuticals produced using monoclonal antibodies! Recently a clinical study has begun of the fate of tumour-infiltrating lymphocytes on re-implantation into donor cancer patients after *in-vitro* infection with a retrovirus expressing neomycin resistance.

Like bone marrow cells, fibroblasts are relatively easy to manipulate. Garner *et al.* (1987) have shown how a major genetic disease, hereditary emphysema associated with α_1-antitrypsin (AAT) deficiency, might be cured by using genetically modified fibroblasts. A retroviral vector was used to insert human AAT cDNA into the genome of mouse fibroblasts to create a clonal population of mouse fibroblasts secreting human AAT. After demonstrating that this clone of fibroblasts produced AAT after more than 100 population doublings in the absence of selection pressure, the clone was transplanted into the peritonial cavities of nude mice. When the animals were evaluated 4 weeks later, human AAT was detected in both sera and the epithelial surface of the lungs. The location of the AAT in the lungs is particularly important since this is its normal body location.

If gene therapy is to become widespread then cells other than bone marrow cells and fibroblasts must be readily manipulated. Recently progress has been made in treating a liver defect, familial hypercholesterolaemia. Victims of this disease lack a receptor on the surface of their liver cells that is necessary for cholesterol metabolism. This leads to severely elevated cholesterol levels and premature heart disease. The disease also occurs in a particular type of rabbit, the Watanabe rabbit. A retrovirus has been used to insert a normal copy of the receptor gene into Watanabe rabbit hepatocytes in culture (Wilson *et al.* 1988a,b). Following insertion of the gene, hepatocytes synthesized the low-density lipoprotein (LDL) receptor which took up LDL cholesterol. Hepatocytes can be re-introduced into animals

367
*Chapter 16
The Impact of
Recombinant
DNA
Technology*

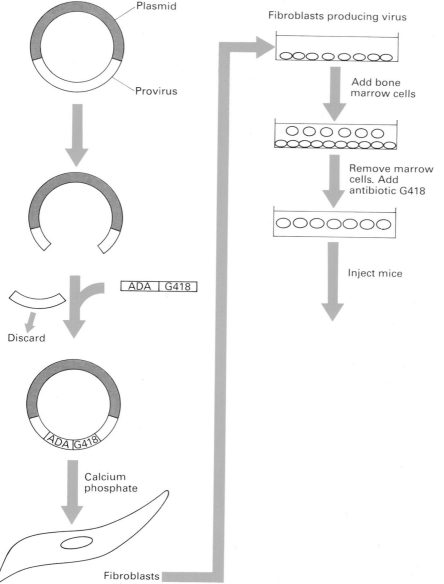

Fig. 16.7 Expression of human adenosine deaminase (ADA) gene in mouse bone marrow cells. (See text for details.)

by attaching them to collagen-coated dextran microcarriers and injecting them intraperitoneally (Demetriou *et al.* 1986).

Transgenesis in man. At present only somatic cell gene therapy is being considered. Ethics apart, are transgenic humans a possibility in the near future? The answer must be in the negative for three basic reasons. First, microinjection has a high failure rate even in experienced hands, and this presupposes that sufficient human fertilized eggs could be obtained.

368

Section 4
Applications of
Recombinant
DNA
Technology

Second, microinjection of eggs can produce deleterious effects because there is no control over where the injected DNA will integrate in the genome. A high level of spontaneous abortion may be tolerated in farm animals but it would not be acceptable in humans. Finally, there is the question of limited usefulness. Most of the serious genetic disorders result in death before puberty, or infertility in homozygous patients. Therefore the worst case that can be imagined is that both parents are heterozygous and, on average, only one-quarter of the offspring will suffer the consequences of homozygosity. The question is, which ova are affected? Currently there is no way to tell and manipulating all ova poses far greater hazards!

Section 5
Appendices

Appendix 1
Restriction Endonuclease Target Sites

Aat II   5′...GACGT▾C...3′
3′...C▴TGCAG...5′

Acc I   5′...GT▾AT/CG AC...3′
3′...CA GC/TA TG...5′

Afl II   5′...C▾TTAAG...3′
3′...GAATT▴C...5′

Aha I   5′...GPu▾CGPyC...3′
3′...CPyGC▴PuG...5′

Alu I   5′...AG▾CT...3′
3′...TC▴GA...5′

Alw I   5′...GGATC(N)₄▾...3′
3′...CCTAG(N)₅▴...5′

*Alw*N I   5′...CAGNNN▾CTG...3′
3′...GTC▴NNNGAC...5′

Apa I   5′...GGGCC▾C...3′
3′...C▴CCGGG...5′

*Apa*L I   5′...G▾TGCAC...3′
3′...CACGT▴G...5′

Ase I   5′...AT▾TAAT...3′
3′...TAAT▴TA...5′

Ava I   5′...C▾PyCGPuG...3′
3′...GPuGCPy▴C...5′

Ava II   5′...G▾G A/T CC...3′
3′...CC T/A G▴G...5′

Avr II   5′...C▾CTAGG...3′
3′...GGATC▴C...5′

Bal I   5′...TGG▾CCA...3′
3′...ACC▴GGT...5′

*Bam*H I   5′...G▾GATCC...3′
3′...CCTAG▴G...5′

Ban I   5′...G▾GPyPuCC...3′
3′...CCPuPyG▴G...5′

Ban II   5′...GPuGCPy▾C...3′
3′...C▴PyCGPuG...5′

Bbv I   5′...GCAGC(N)₈▾...3′
3′...CGTCG(N)₁₂▴...5′

Bcl I   5′...T▾GATCA...3′
3′...ACTAG▴T...5′

Bgl I   5′...GCCNNNN▾NGGC...3′
3′...CGGN▴NNNNCCG...5′

Bgl II   5′...A▾GATCT...3′
3′...TCTAG▴A...5′

Bsm I   5′...GAATGCN▾...3′
3′...CTTAC▴GN...5′

*Bsm*A I   5′...GTCTC(N₁)▾...3′
3′...CAGAG(N₅)▴...5′

Bsp 1286 I   5′...GAGCA▾C...3′ (with G/C, T/T above; C/G, A/A below)
3′...C▴TCGTG...5′

*Bsp*H I   5′...T▾CATGA...3′
3′...AGTAC▴T...5′

*Bsp*M I   5′...ACCTGC(N)₄▾...3′
3′...TGGACG(N)₈▴...5′

*Bsp*M II   5′...T▾CCGGA...3′
3′...AGGCC▴T...5′

Bsr I   5′...ACTGGN▾...3′
3′...TGAC▴CN...5′

*Bss*H II   5′...G▾CGCGC...3′
3′...CGCGC▴G...5′

Bst I   5′...TT▾CGAA...3′
3′...AAGC▴TT...5′

*Bst*E II   5′...G▾GTNACC...3′
3′...CCANTG▴G...5′

*Bst*N I   5′...CC▾A/T GG...3′
3′...GG T/A ▴CC...5′

*Bst*U I   5′...CG▾CG...3′
3′...GC▴GC...5′

*Bst*X I   5′...CCANNNNN▾NTGG...3′
3′...GGTN▴NNNNNACC...5′

*Bst*Y I   5′...Pu▾GATCPy...3′
3′...PyCTAG▴Pu...5′

371

Bsu36 I
5'...CC▼TNAGG...3'
3'...GGANT▲CC...5'

Cla I
5'...AT▼CGAT...3'
3'...TAGC▲TA...5'

Dde I
5'...C▼TNAG...3'
3'...GANT▲C...5'

Dpn I
CH$_3$
5'...GA▼TC...3'
3'...CT▲AG...5'
CH$_3$

Dra I
5'...TTT▼AAA...3'
3'...AAA▲TTT...5'

Dra III
5'...CACNNN▼GTG...3'
3'...GTG▲NNNCAC...5'

Eae I
5'...Py▼GGCCPu...3'
3'...PuCCGG▲Py...5'

Eag I
5'...C▼GGCCG...3'
3'...GCCGG▲C...5'

Ear I
5'...CTCTTCN▼NNN...3'
3'...GAGAAGNNNN▲...5'

EcoN I
5'...CCTNN▼NNNAGG...3'
3'...GGANNN▲NNTCC...5'

EcoO 109 I
5'...PuG▼GNCCPy...3'
3'...PyCCNG▲GPu...5'

EcoR I
5'...G▼AATTC...3'
3'...CTTAA▲G...5'

EcoR V
5'...GAT▼ATC...3'
3'...CTA▲TAG...5'

Fnu4H I
5'...GC▼NGC...3'
3'...CGN▲CG...5'

Fok I
5'...GGATG(N)$_9$▼...3'
3'...CCTAC(N)$_{13}$▲...5'

Fsp I
5'...TGC▼GCA...3'
3'...ACG▲CGT...5'

Hae II
5'...PuGCGC▼Py...3'
3'...Py▲CGCGPu...5'

Hae III
5'...GG▼CC...3'
3'...CC▲GG...5'

Hga I
5'...GACGC(N)$_5$▼...3'
3'...CTGCG(N)$_{10}$▲...5'

HgiA I
5'...G$_A^T$GC$_A^T$▼...3'
3'...C▲$_A^T$ACG$_A^T$G...5'

Hha I
5'...GCG▼C...3'
3'...C▲GCG...5'

Hinc II
5'...GTPy▼PuAC...3'
3'...CAPu▲PyTG...5'

Hind III
5'...A▼AGCTT...3'
3'...TTCGA▲A...5'

Hinf I
5'...G▼ANTC...3'
3'...CTNA▲G...5'

HinP I
5'...G▼CGC...3'
3'...CGC▲G...5'

Hpa I
5'...GTT▼AAC...3'
3'...CAA▲TTG...5'

Hpa II
5'...C▼CGG...3'
3'...GGC▲C...5'

Hph I
5'...GGTGA(N)$_8$▼...3'
3'...CCACT(N)$_7$▲...5'

Kpn I
5'...GGTAC▼C...3'
3'...C▲CATGG...5'

Mbo I
5'...▼GATC...3'
3'...CTAG▲...5'

Mbo II
5'...GAAGA(N)$_8$▼...3'
3'...CTTCT(N)$_7$▲...5'

Mlv I
5'...A▼CGCGT...3'
3'...TGCGC▲A...5'

Mnl I
5'...CCTC(N)$_7$▼...3'
3'...GGAG(N)$_7$▲...5'

Mse I
5'...T▼TAA...3'
3'...AAT▲T...5'

Msp I
5'...C▼CGG...3'
3'...GGC▲C...5'

Nae I
5'...GCC▼GGC...3'
3'...CGG▲CCG...5'

Nar I
5'...GG▼CGCC...3'
3'...CCGC▲GG...5'

Nci I
5'...CC▼$_G^C$GG...3'
3'...GG$_G^C$▲CC...5'

Nco I
5'...C▼CATGG...3'
3'...GGTAC▲C...5'

Nde I
5'...CA▼TATG...3'
3'...GTAT▲AC...5'

Nhe I
5'...G▼CTAGC...3'
3'...CGATC▲G...5'

Nla III
5'...CATG▼...3'
3'...▲GTAC...5'

Nla IV
5'...GGN▼NCC...3'
3'...CCN▲NGG...5'

373

*Appendix I
Restriction
Endonuclease
Target
Sites*

Not I	5´...GC▼GGCCGC...3´ 3´...CGCCGG▲CG...5´		*Sca* I	5´...AGT▼ACT...3´ 3´...TCA▲TGA...5´
Nru I	5´...TCG▼CGA...3´ 3´...AGC▲GCT...5´		*Scr*F I	5´...CC▼NGG...3´ 3´...GGN▲CC...5´
Nsi I	5´...ATGCA▼T...3´ 3´...T▲ACGTA...5´		*Sfa*N I	5´...GCATC(N)₅▼...3´ 3´...CGTAG(N)₉▲...5´
*Pae*R7 I	5´...C▼TCGAG...3´ 3´...GAGCT▲C...5´		*Sfi* I	5´...GGCCNNNN▼NGGCC...3´ 3´...CCGGN▲NNNNCCGG...5´
*Pfl*M I	5´...CCANNNN▼NTGG...3´ 3´...GGTN▲NNNNACC...5´		*Sma* I	5´...CCC▼GGG...3´ 3´...GGG▲CCC...5´
Ple I	5´...GAGTC(N₄)▼...3´ 3´...CTCAG(N₅)▲...5´		*Sna*B I	5´...TAC▼GTA...3´ 3´...ATG▲CAT...5´
*Ppu*M I	5´...PuG▼G^A_TCCPy...3´ 3´...PyCC^A_TG▲Pu...5´		*Spe* I	5´...A▼CTAGT...3´ 3´...TGATC▲A...5´
Pst I	5´...CTGCA▼G...3´ 3´...G▲ACGTC...5´		*Sph* I	5´...GCATG▼C...3´ 3´...C▲GTACG...5´
Pvu I	5´...CGAT▼CG...3´ 3´...GC▲TAGC...5´		*Ssp* I	5´...AAT▼ATT...3´ 3´...TTA▲TAA...5´
Pvu II	5´...CAG▼CTG...3´ 3´...GTC▲GAC...5´		*Stu* I	5´...AGG▼CCT...3´ 3´...TCC▲GGA...5´
Rsa I	5´...GT▼AC...3´ 3´...CA▲TG...5´		*Sty* I	5´...C▼C$^{AA}_{TT}$GG...3´ 3´...GG$^{AA}_{TT}$C▲C...5´
Rsr II	5´...CG▼G^A_TCCG...3´ 3´...GCC^A_TG▲GC...5´		*Taq*ᵅ I	5´...T▼CGA...3´ 3´...AGC▲T...5´
Sac I	5´...GAGCT▼C...3´ 3´...C▲TCGAG...5´		*Tth*111 I	5´...GACN▼NNGTC...3´ 3´...CTGNN▲NCAG...5´
Sac II	5´...CCGC▼GG...3´ 3´...GG▲CGCC...5´		*Xba* I	5´...T▼CTAGA...3´ 3´...AGATC▲T...5´
Sal I	5´...G▼TCGAC...3´ 3´...CAGCT▲G...5´		*Xho* I	5´...C▼TCGAG...3´ 3´...GAGCT▲C...5´
*Sau*3A I	5´...▼GATC...3´ 3´...CTAG▲...5´		*Xma* I	5´...C▼CCGGG...3´ 3´...GGGCC▲C...5´
*Sau*96 I	5´...G▼GNCC...3´ 3´...CCNG▲G...5´		*Xmn* I	5´...GAANN▼NNTTC...3´ 3´...CTTNN▲NNAAG...5´

Appendix 2
The Genetic Code and Single-Letter Amino Acid Designations

5'-OH terminal base	Middle base				3'-OH terminal base
	U	C	A	G	
U	Phe	Ser	Tyr	Cys	U
	Phe	Ser	Tyr	Cys	C
	Leu	Ser	STOP	STOP	A
	Leu	Ser	STOP	Trp	G
C	Leu	Pro	His	Arg	U
	Leu	Pro	His	Arg	C
	Leu	Pro	Gln	Arg	A
	Leu	Pro	Gln	Arg	G
A	Ile	Thr	Asn	Ser	U
	Ile	Thr	Asn	Ser	C
	Ile	Thr	Lys	Arg	A
	Met	Thr	Lys	Arg	G
G	Val	Ala	Asp	Gly	U
	Val	Ala	Asp	Gly	C
	Val	Ala	Glu	Gly	A
	Val[a]	Ala	Glu	Gly	G

[a] Codes for Met if in the initiator position.

Alanine	A	Leucine	L
Arginine	R	Lysine	K
Asparagine	N	Methionine	M
Aspartic Acid	D	Phenylalanine	F
Cysteine	C	Proline	P
Cystine	C	Serine	S
Glycine	G	Threonine	T
Glutamic Acid	E	Tryptophan	W
Glutamine	Q	Tyrosine	Y
Histidine	H	Valine	V
Isoleucine	I		

References

Aaij C. & Borst P. (1972) The gel electrophoresis of DNA. *Biochim. biophys. Acta* **269**, 192–200.

Abrahmsen L., Moks T., Nilsson B., Hellman U. & Uhlen M. (1985) Analysis of signals for secretion in the staphylococcal protein A gene. *EMBO J.* **4**, 3901–6.

Abrahmsen L., Moks T., Nilsson B. & Uhlen M. (1986) Secretion of heterologous gene products to the culture medium of *Escherichia coli*. *Nucleic Acids Res.* **14**, 7487–500.

Adams J. M., Harris A. W., Pinkert C. A., Corcoran L. M., Alexander W. S., Cory S., Palmiter R. D. & Brinster R. L. (1985) The c-*myc* oncogene driven by immunoglobulin ehancers induces lymphoid malignancy in transgenic mice. *Nature* **318**, 533–8.

Adams S. E., Dawson K. M., Gull K., Kingsman S. M. & Kingsman A. J. (1987) The expression of hybrid HIV: Ty virus-like particles in yeast. *Nature* **329**, 68–70.

Ahlquist P., French R. & Bujarski J. J. (1987) Molecular studies of brome mosaic virus using infectious transcripts from cloned cDNA. *Adv. Virus Res.* **32**, 215–42.

Akiyoshi D. E., Klee H., Amasino R. M., Nester E. W. & Gordon M. P. (1984) T-DNA of *Agrobacterium tumefaciens* encodes an enzyme of cytokinin biosynthesis. *Proc. Natl. Acad. Sci. USA* **81**, 5994–8.

Altenbuchner J. & Cullum J. (1987) Amplification of cloned genes in *Streptomyces*. *Biotechnology* **5**, 1328–9.

Alwine J. C., Kemp D. J., Parker B. A., Reiser J., Renart J., Stark G. R., & Wahl G. M. (1979) Detection of specific RNAs or specific fragments of DNA by fractionation in gels and transfer to diazobenzyloxymethyl paper. *Methods Enzymol.* **68**, 220–42.

Amann E., Brosius J. & Ptashne M. (1983) Vectors bearing a hybrid *trp–lac* promoter useful for regulated expression of cloned genes in *Escherichia coli*. *Gene* **25**, 167–78.

Anderson S., Gait M. J., Mayol L. & Young I. G. (1980) A short primer for sequencing DNA cloned in the single-stranded phage vector M13mp2. *Nucleic Acids Res.* **8**, 1731–43.

Anderson S., Marks C. B., Lazarus R., Miller J., Stafford K., Seymour J., Light D., Rastetter W. & Estell D. (1985) Production of 2-keto-L-gulonate, an intermediate in L-ascorbate synthesis by a genetically modified *Erwinia herbicola*. *Science* **230**, 144–9.

Apostol B. & Greer C. L. (1988) Copy number and stability of yeast 2μ-based plasmids carrying a transcription-conditioned centromere. *Gene* **67**, 59–68.

Axel R., Fiegelson P. & Schutz G. (1976) Analysis of the complexity and diversity of mRNA from chicken oviduct and liver. *Cell* **11**, 247–54.

Bachmair A., Finley D. & Varshavsky A. (1986) *In vivo* half-life of a protein is a function of its amino-terminal residue. *Science* **234**, 179−86.

Backman K. & Boyer H. W. (1983) Tetracycline resistance determined by pBR322 is mediated by one polypeptide. *Gene* **26**, 197−203.

Backman K. & Ptashne M. (1978) Maximizing gene expression on a plasmid using recombination *in vitro*. *Cell* **13**, 65−71.

Backman K., Ptashne M. & Gilbert W. (1976) Construction of plasmids carrying the cI gene of bacteriophage λ. *Proc. Natl. Acad. Sci. USA* **73**, 4174−8.

Bagdasarian M. M., Amann E., Lurz R., Ruckert B. & Bagdasarian M. (1983) Activity of the hybrid *trp−lac* (*tac*) promoter of *Escherichia coli* in *Pseudomonas putida*. Construction of broad-host-range, controlled-expression vectors. *Gene* **26**, 273−82.

Bagdasarian M., Bagdasarian M. M., Coleman S. & Timmis K. N. (197⁻ vector plasmids for gene cloning in *Pseudomonas*. In *Plasmids of Me. Environmental and Commercial Importance*, eds Timmis K. N. & Pühler A., pp. 411−22. Elsevier/North Holland Biomedical Press, Amsterdam.

Bagdasarian M., Lurz R., Rückert B., Franklin F. C. H., Bagdasarian M. M., Frey J. & Timmis K. N. (1981) Specific-purpose plasmid cloning vectors. II. Broad host range, high copy number, RSF1010-derived vectors, and a host-vector system for gene cloning in *Pseudomonas*. *Gene* **16**, 237−47.

Baigori M., Sesma R., de Ruiz A. P. & de Mendoza D. (1988) Transfer of plasmids between *Bacillus subtilis* and *Streptococcus lactis*. *Appl. Environ. Microbiol.* **54**, 1309−11.

Balazs E., Bouzoubaa S., Guilley H., Jonasd G., Paszkowski J. & Richards K. (1985) Chimeric vector construction for higher plant transformation. *Gene* **40**, 343−8.

Balazs E., Guilley H., Jonard G. & Richards K. (1982) Nucleotide sequence of DNA from an altered-virulence isolate D/H of the cauliflower mosaic virus. *Gene* **19**, 239−49.

Balbás P., Soberon X., Merino E., Zurita M., Lomeli H., Valle F., Flores N. & Bolivar F. (1986) Plasmid vector pBR322 and its special-purpose derivatives — a review. *Gene* **50**, 3−40.

Band L. & Henner D. J. (1984) *Bacillus subtilis* requires a 'stringent' Shine−Dalgarno region for gene expression. *DNA* **3**, 17−21.

Bandyopadhyay P. K., Studier W. F., Hamilton D. L. & Yuan R. (1985) Inhibition of the type I restriction-modification enzymes Eco B and Eco K by the gene 0.3 protein of bacteriophage T7. *J. Mol. Biol.* **182**, 567−78.

Banerji J., Rusconi S. & Schaffner W. (1981) Expression of a β-globin gene is enhanced by remote SV40 DNA sequences. *Cell* **27**, 299−308.

Banner C. D. B., Moran C. P. & Losick R. (1983) Deletion analysis of a complex promoter for a developmentally regulated gene from *Bacillus subtilis*. *J. Mol. Biol.* **168**, 351−65.

Barbour A. G. & Garon C. F. (1987) Linear plasmids of the bacterium *Borrelia burgdorferi* have covalently closed ends. *Science* **237**, 409−11.

Barcack G. J. & Wolf R. E. (1986) A method for unidirectional deletion mutagenesis with application to nucleotide sequencing and preparation of gene fusions. *Gene* **49**, 119−28.

Barry G. F. (1988) A broad-host-range shuttle system for gene insertion into the chromosomes of Gram-negative bacteria. *Gene* **71**, 75−84.

Barnes W. M. (1980) DNA cloning with single-stranded phage vectors. In *Genetic Engineering*, vol. 2, ed. Setlow J. K. & Hollaender A., pp. 185−200. Plenum Press, New York.

Bates P. F. & Swift R. A. (1983) Double *cos* site vectors: simplified cosmid cloning. *Gene* **26**, 137−46.

Bebehani A. M. (1983) The smallpox story: life and death of an old disease. *Microbiol. Rev.* **47**, 455—509.

Beck E. & Bremer E. (1980) Nucleotide sequence of the gene *ompA* encoding the outer membrane protein II of *Escherichia coli* K-12, *Nucleic Acids Res.* **8**, 3011—24.

Beggs J. D. (1978) Transformation of yeast by a replicating hybrid plasmid *Nature* **275**, 104—9.

Beggs J. D., van den Berg J., van Ooyen A. & Weissmann C. (1980) Abnormal expression of chromosomal rabbit β-globin gene in *Saccharomyces cerevisiae*. *Nature* **283**, 835—40.

Belasco J. G. & Higgins C. F. (1988) Mechanisms of mRNA decay in bacteria: a perspective. *Gene* **72**, 15—23.

Ben-Bassat A., Bauer K., Chang S-Y., Myambo K., Boosman A. & Chang S. (1987) Processing of the initiation methionine from proteins: properties of the *Escherichia coli* methionine aminopeptidase and its gene structure. *J. Bacteriol.* **169**, 751—7.

Bender W., Spierer P. & Hogness D. S. (1983) Chromosomal walking and jumping to isolate DNA from the *Ace* and *rosy* loci and the bithorax complex in *Drosophila melanogaster*. *J. Mol. Biol.* **168**, 17—33.

Bendig M. M. & Williams J. G. (1983) Replication and expression of *Xenopus laevis* globin genes injected into fertilized *Xenopus* eggs. *Proc. Natl. Acad. Sci. USA* **80**, 6197—201.

Bendig M. M. & Williams J. G. (1984) Fidelity of transcription of *Xenopus laevis* globin genes injected into *Xenopus laevis* oocytes and unfertilized eggs. *Mol. Cell. Biol.* **4**, 2109—19.

Bennetzen J. L. & Hall B. D. (1982) Codon selection in yeast. *J. Biol. Chem.* **257**, 3026—31.

Benton W. D. & Davis R. W. (1977) Screening λgt recombinant clones by hybridization to single plaques *in situ*. *Science* **196**, 180—2.

Berg P. (1981) Dissections and reconstructions of genes and chromosomes. *Science* **213**, 296—303.

Berkner K. L. & Folk W. R. (1977) Polynucleotide kinase exchange reaction. Quantitative assay for restriction endonuclease-generated 5′-phosphoryl termini in DNAs. *J. Biol. Chem.* **252**, 3176—84.

Bernhard K., Schrempf H. & Goebel W. (1978) Bacteriocin and antibiotic resistance plasmids in *Bacillus cereus* and *Bacillus subtilis*. *J. Bacteriol.* **133**, 897—903.

Bevan M. (1984) Binary *Agrobacterium* vectors for plant transformation. *Nucleic Acids Res.* **12**, 8711—21.

Bevan M., Barnes W. & Chilton M. D. (1983a) Structure and transcription on the nopaline synthase gene region of T-DNA. *Nucleic Acids Res.* **11**, 369—85.

Bevan M. W. & Chilton M. D. (1982) T-DNA of the Agrobacterium Ti and Ri plasmids. *Ann. Rev. Genet.* **16**, 357—84.

Bevan M. W., Flavell R. B. & Chilton M. D. (1983b) A chimaeric antibiotic resistance gene as a selectable marker for plant cell transformation. *Nature* **304**, 184—7.

Bickle T. A., Brack C. & Yuan R. (1978) ATP-induced conformational changes in restriction endonuclease from *Escherichia coli* K12. *Proc. Natl. Acad. Sci. USA* **75**, 3099—103.

Bienz M. & Pelham H. R. B. (1982) Expression of a *Drosophila* heat-shock protein in *Xenopus* oocytes: conserved and divergent regulatory signals. *EMBO J.* **1**, 1583—8.

Bingham A. H. A., Bruton C. J. & Atkinson T. (1980) Characterization of *Bacillus stearothermophilus* plasmid pAB124 and construction of deletion variants. *J. Gen. Microbiol.* **119**, 109–115.

Bingham A. H. A. & Busby S. J. W. (1987) Translation of *gal*E and coordination of galactose operon expression in *Escherichia coli*: effects of insertions and deletions in the non-translated leader sequence. *Molec. Microbiol.* **1**, 117–24.

Bingham P. M., Kidwell M. G. & Rubin G. M. (1982) The molecular basis of P–M hybrid dysgenesis: the role of the P element, a P-strain-specific transposon family. *Cell* **29**, 995–1004.

Bird A. P. & Southern E. M. (1978) Use of restriction enzymes to study eukaryotic DNA methylation: I. The methylation pattern in ribosomal DNA from *Xenopus laevis*. *J. Mol. Biol.* **118**, 27–47.

Birnboim H. C. & Doly J. (1979) A rapid alkaline extraction procedure for screening recombinant plasmid DNA. *Nucleic Acids Res.* **7**, 1513–23.

Birnstiel M. L. & Chipchase M. (1977) Current work on the histone operon. *Trends Biochem. Sci.* **2**, 149–52.

Blancou J., Kieny M. P., Lathe R., Lecocq J. P., Pastovet P. P., Soulebot J. P. & Desmettre P. (1986) Oral vaccination of the fox against rabies using a live recombinant vaccinia virus. *Nature* **322**, 373–5.

Blattner F. R., Williams B. G., Blechl A. E., Deniston-Thompson K., Faber H. E., Furlong L. A., Grunwald D. J., Kiefer D. O., Moore D. D., Schumm J. W., Sheldon E. L. & Smithies O. (1977) Charon phages: safer derivatives of bacteriophage lambda for DNA cloning. *Science* **196**, 161–9.

Blobel G. & Dobberstein B. (1975) Transfer of proteins across membranes. I. Presence of proteolytically processed and unprocessed nascent immunoglobulin light chains on membrane-bound ribosomes of murine myeloma. *J. Cell Biol.* **67**, 835–51.

Bloom K. S. & Carbon J. (1982) Yeast centromere DNA is a unique and highly ordered structure in chromosomes and small circular minichromosomes. *Cell* **29**, 305–17.

Boe L., Gerdes K. & Molin S. (1987) Effects of genes exerting growth inhibition and plasmid stability on plasmid maintenance. *J. Bacteriol.* **169**, 4646–50.

Boeke J. D., Vovis G. F. & Zinder N. D. (1979) Insertion mutant of bacteriophage f1 sensitive to *Eco* RI. *Proc. Natl. Acad. Sci. USA* **76**, 2699–702.

Boeke J. D., Xu H. & Fink G. R. (1988) A general method for the chromosomal amplification of genes in yeast. *Science* **239**, 280–2.

Bolivar F. (1978) Construction and characterization of new cloning vehicles. III. Derivatives of plasmid pBR322 carrying unique *Eco* RI sites for selection of *Eco* RI generated recombinant DNA molecules. *Gene* **4**, 121–36.

Bolivar F., Rodriguez R. L., Betlach M. C. & Boyer H. W. (1977a) Construction and characterization of new cloning vehicles. I. Ampicillin-resistant derivatives of the plasmid pMB9. *Gene* **2**, 75–93.

Bolivar F., Rodriguez R. L., Greene P. J., Betlach M. V., Heynecker H. L., Boyer H. W., Crosa J. H. & Falkow S. (1977b) Construction and characterization of new cloning vehicles. II. A multipurpose cloning system. *Gene* **2**, 95–113.

Bomhoff G. H., Klapwijk F. M., Kester M. C. M., Schilperoort R. A., Hernalsteens J. P. & Schell J. (1976) Octopine and nopaline synthesis and breakdown genetically controlled by a plasmid of *Agrobacterium tumefaciens*. *Mol. Gen. Genet.* **145**, 177–81.

Bossi L. & Smith D. M. (1984) Conformational change in the DNA associated with an unusual promoter mutation in a tRNA operon of *Salmonella*. *Cell* **39**, 643–52.

Bostian K. A., Ellio O., Bussey H., Burn V., Smith A. & Tipper D. J. (1984) Sequence of the prepro-toxin ds RNA gene of Type 1 killer yeast: multiple processing events produce a two-component toxin. *Cell* **36**, 741–51.

Botstein D. & Davis R. W. (1982) Principles and practice of recombinant DNA research with yeast. In *The Molecular Biology of the Yeast Saccharomyces*, eds Strathern J. N., Jones E. W. & Broach J. R. Cold Spring Harbor Press, Cold Spring Harbor, New York.

Botterman J. & Leemans J. (1988) Engineering herbicide resistance in plants. *Trends Genet.* **4**, 219–22.

Boyko W. L. & Ganschow R. E. (1982) Rapid identification of *Escherichia coli* transformed by pBR322 carrying inserts at the *Pst* I site. *Anal. Biochem.* **122**, 85–8.

Brange J., Ribel U., Hansen J. F., Dodson G., Hansen M. T., Haveland S., Melberg S. G., Norris F., Norris K., Snel L., Sørenson A. R. & Voist H. O. (1988) Monomeric insulins obtained by protein engineering and their medical implications. *Nature* **333**, 679–82.

Braun A. C. & Wood H. N. (1976) Suppression of the neoplastic state with the acquisition of specialised functions in cells, tissues and organs of crown gall teratoma of tobacco. *Proc. Natl. Acad. Sci. USA* **73**, 496–500.

Brenner S., Cesareni G. & Karn J. (1982) Phasmids: hybrids between Col E1 plasmids and *E. coli* bacteriophage lambda. *Gene* **17**, 27–44.

Breunig K. D., Mackedonski V. & Hollenberg C. P. (1982) Transcription of the bacterial β-lactamase gene in *Saccharomyces cerevisiae*. *Gene* **20**, 1–10.

Brigatti D. J., Myerson D., Leary J. J., Spalholz B., Travis S. Z., Fong C. K. Y., Hsuing G. D. & Ward D. C. (1983) Detection of viral genomes in cultured cells and paraffin-embedded tissue sections using biotin-labelled hybridization probes. *Virology* **126**, 32–50.

Brinster R. L., Chen H. Y., Trumbauer M., Senear A. W., Warren R. & Palmiter R. D. (1981) Somatic expression of Herpes thymidine kinase in mice following injection of a fusion gene into eggs. *Cell* **27**, 223–31.

Brinster R. L., Chen H. Y., Warren R., Sarthy A. & Palmiter R. D. (1982) Regulation of metallothionein-thymidine kinase fusion plasmids injected into mouse eggs. *Nature* **296**, 39–42.

Brisson N., Paszkowski J., Penswick J. R., Gronenborn B., Potrykus I. & Hohn T. (1984) Expression of a bacterial gene in plants by using a viral vector. *Nature* **310**, 511–14.

Broda P. (1979) *Plasmids*. W. H. Freeman & Co., San Francisco.

Broglie R., Coruzzi G., Fraley R. T., Rogers S. G., Horsch R. B., Niedermeyer J. G., Fink C. L., Flick J. S. & Chua N. H. (1984) Light-regulated expression of a pea ribulose-1,5-bisphosphate carboxylase small subunit gene in transformed plant cells. *Science* **224**, 838–43.

Bron S., Bosma P., Van Belkum M. & Luxen E. (1987) Stability function in the *Bacillus subtilis* plasmid pTA 1060. *Plasmid* **18**, 8–16.

Bron S. & Luxen E. (1985) Segregational instability of pUB110-derived recombinant plasmids in *Bacillus subtilis*. *Plasmid* **14**, 235–44.

Broome S. & Gilbert W. (1978) Immunological screening method to detect specific translation products. *Proc. Natl. Acad. Sci. USA* **75**, 2746–9.

Bryan R. N., Ruth J. L., Smith R. D. & Le Bon J. M. (1986) Diagnosis of clinical samples with synthetic oligonucleotide hybridization probes. In *Microbiology 1986* ed. Leive L., pp. 113–16. American Society for Microbiology, Washington.

Buchanan-Wollaston V., Passiatore J. E. & Channon F. (1987) The *mob* and *ori* T functions of a bacterial plasmid promote its transfer to plants. *Nature* **328**, 172–5.

Bugawan T. L., Saiki R. K., Levenson C. H., Watson R. M. & Ehrlich H. A. (1988). The use of non-radioactive oligonucleotide probes to analyze enzymatically amplified DNA for prenatal diagnosis and forensic HLA typing. *Biotechnology* **6**, 943–7.

Burke D. T., Carle G. F. & Olson M. V. (1987) Cloning of large segments of exogenous DNA into yeast by means of artificial chromosome vectors. *Science* **236**, 806–13.

Burke T. & Bruford M. W. (1987) DNA fingerprinting in birds. *Nature* **327**, 149–52.

Burnette W. N. (1981) Western blotting: electrophoretic transfer of proteins from sodium dodecyl sulphate–polyacrylamide gels to unmodified nitrocellulose and radiographic detection with antibody and radioiodinated protein A. *Anal. Biochem.* **112**, 195–203.

Buroker N., Bestwick R., Haight G., Magenis R. E. & Litt M. (1987) A hypervariable repeated sequence on human chromosome 1p36. *Hum. Genet.* **77**, 175–81.

Butler E. T. & Chamberlain M. J. (1982) Bacteriophage SP6-specific RNA polymerase I Isolation and characterization of the enzyme. *J. Biol. Chem.* **257**, 5772–8.

Cabezon T., De Wilde M., Herion P., Lorian R. & Bollen A. (1984) Expression of human alpha$_1$-antitrypsin cDNA in the yeast *Saccharomyces cerevisiae*. *Proc. Natl. Acad. Sci. USA* **81**, 6594–8.

Cameron J. R., Panasenko S. M., Lehman I. R. & Davis R. W. (1975) *In vitro* construction of bacteriophage λ carrying segments of the *Escherichia coli* chromosome: selection of hybrids containing the gene for DNA ligase. *Proc. Natl. Acad. Sci. USA* **72**, 3416–20.

Campo M. S. (1985) Bovine papillomavirus DNA: a eukaryotic cloning vector. In *DNA Cloning: A Practical Approach*, vol. II, ed. Glover D., pp. 213–39. IRL Press, Oxford.

Canosi U., Iglesias A. & Trautner T. A. (1981) Plasmid transformation in *Bacillus subtilis*: effects of insertion of *Bacillus subtilis* DNA into plasmid pC194. *Mol. Gen. Genet.* **181**, 434–40.

Canosi U., Morelli G. & Trautner T. A. (1978) The relationship between molecular structure and transformation efficiency of some *S. aureus* plasmids isolated from *B. subtilis*. *Mol. Gen. Genet.* **166**, 259–67.

Carle G. R., Frank M. & Olson M. V. (1986) Electrophoretic separations of large DNA molecules by periodic inversion of the electric field. *Science* **232**, 65–8.

Carle G. R. & Olson M. V. (1984) Separation of chromosomal DNA molecules from yeast by orthogonal-field-alternation gel electrophoresis. *Nucleic Acids Res.* **12**, 5647–64.

Carlton B. C. & Helinski D. R. (1969) Heterogeneous circular DNA elements in vegetative cultures of *Bacillus megaterium*. *Proc. Natl. Acad. Sci. USA* **64**, 592–9.

Carter P., Bedouelle H. & Winter G. (1985) Improved oligonucleotide site-directed mutagenesis using M13 vectors. *Nucleic Acids Res.* **13**, 4431–43.

Carter P. & Wells J. A. (1987) Engineering enzyme specificity by "substrate-assisted catalysis". *Science* **237**, 394–9.

Cesarini G., Muesing M. A. & Polisky B. (1982) Control of Col E1 DNA replication: the *rop* gene product negatively affects transcription from the replication primer promoter. *Proc. Natl. Acad. Sci. USA* **79**, 6313–7.

Chada K., Magram J. & Constantini F. (1986) An embryonic pattern of expression of a human fetal globin gene in transgenic mice. *Nature* **319**, 685–9.

Chakrabarti S., Robert-Guroff M., Wong-Staal F., Gallo R. C. & Moss B. (1986) Expression of the HTLV-III envelope gene by a recombinant vaccinia virus. *Nature* **320**, 535–7.

Chamberlain J. S., Pearlman J. A., Muzny D. M., Gibbs R. A., Ranier J. E., Reeves A. A. & Caskey C. T. (1988) Expression of murine muscular dystrophy gene in muscle and brain. *Science* **239**, 1416–18.

Chambers A. & Old R. (1988) RNA 3' cleavage and polyadenylation in oocytes and unfertilized eggs of *Xenopus laevis*. *Dev. Biol.* **125**, 237–45.

Chang A. C. Y. & Cohen S. N. (1974) Genome construction between bacterial species *in vitro*: replication and expression of *Staphylococcus* plasmid genes in *Escherichia coli*. *Proc. Natl. Acad. Sci. USA* **71**, 1030–4.

Chang A. C. Y., Ehrlich H. A., Gunsalus R. P., Nunberg J. H., Kaufman R. J., Schimke R. T. & Cohen S. N. (1980) Initiation of protein synthesis in bacteria at a translational start codon of mammalian cDNA: effects of the preceding nucleotide sequence. *Proc. Natl. Acad. Sci. USA* **77**, 1442–6.

Chang A. C. Y., Nunberg J. H., Kaufman R. K., Ehrlich H. A., Schimke R. T. & Cohen S. N. (1978) Phenotypic expression in *E. coli* of a DNA sequence coding for mouse dihydrofolate reductase. *Nature* **275**, 617–24.

Chang J. C. & Kan Y. W. (1981) Antenatal diagnosis of sickle-cell anaemia by direct analysis of the sickle mutation. *Lancet* **2**, 1127–9.

Chang L. M. S. & Bollum F. J. (1971) Enzymatic synthesis of oligo-deoxynucleotides. *Biochemistry* **10**, 536–42.

Chang S. & Cohen S. N. (1979) High-frequency transformation of *Bacillus subtilis* protoplasts by plasmid DNA. *Mol. Gen. Genet.* **168**, 111–15.

Charnay P., Perricaudet M., Galibert F. & Tiollais P. (1978) Bacteriophage lambda and plasmid vectors allowing fusion of cloned genes in each of the three translational phases. *Nucleic Acids Res.* **5**, 4479–94.

Chen C. Y. & Hitzeman R. A. (1987) Human, yeast and hybrid 3-phosphoglycerate kinase gene expression in yeast. *Nucleic Acids Res.* **15**, 643–60.

Chen C. Y., Oppermann H. & Hitzeman R. A. (1984) Homologous versus heterologous gene expression in the yeast *Saccharomyces cerevisiae*. *Nucleic Acids Res.* **12**, 8951–70.

Chen E., Howley P., Levenson A. & Seeburg P. (1982) The primary structure and genetic organisation of the bovine papilloma virus type I genome. *Nature* **299**, 529–34.

Chen J.-D. & Morrisson D. A. (1987) Cloning of *Streptococcus pneumoniae* DNA fragments in *Escherichia coli* requires vectors protected by strong transcriptional terminators. *Gene* **55**, 179–87.

Chilton M.-D., Drummond M. H., Merlo D. J., Sciaky D., Montoya A. L., Gordon M. P. & Nester E. W. (1977) Stable incorporation of plasmid DNA into higher plant cells: the molecular basis of crown gall tumorigenesis. *Cell* **11**, 263–71.

Chilton M.-D., Tepfer D. A., Petit A., David C., Casse-Delbart F. & Tempe J. (1982) *A. rhizogenes* inserts T-DNA into the genomes of host plant root cells. *Nature* **295**, 432–4.

Chlebowicz-Sledziewska E. & Sledziewski A. Z. (1985) Construction of multicopy yeast plasmids with regulated centromere function. *Gene* **39**, 25–31.

Chollet A. & Kawashima E. H. (1985) Biotin-labeled synthetic oligodeoxyribonucleotides: chemical synthesis and uses as hybridization probes. *Nucleic Acids Res.* **13**, 1529–41.

Christman J. K., Gerber M., Price P. M., Flordellis C., Edelman J. & Acs G. (1982) Amplification of expression of hepatitis B surface antigen in 3T3 cells cotransfected with a dominant-acting gene and cloned viral DNA. *Proc. Natl. Acad. Sci. USA* **79**, 1815—19.

Chu G., Hayakawa H. & Berg P. (1987) Electroporation for the efficient transfection of mammalian cells with DNA. *Nucleic Acids Res.* **15**, 1311—26.

Chu, G. & Sharp P. A. (1981) SV40 DNA transfection of cells in suspension: analysis of the efficiency of transcription and translation of T-antigen. *Gene* **13**, 197—202.

Cigan A. M. & Donahue T. F. (1987) Sequence and structural features associated with translational initiator regions in yeast—a review. *Gene* **59**, 1—18.

Clark A. J., Bessos H., Bishop J. O., Brown P., Harris S., Lathe R., McClenaghan M., Prowse C., Simons J. P., Whitelaw C. B. A. & Wilmut I. (1989) Expression of human anti-hemophilic factor IX in the milk of transgenic sheep. *Biotechnology* **7**, 487—92.

Clarke L. & Carbon J. (1976) A colony bank containing synthetic Col E1 hybrid plasmids representative of the entire *E. coli* genome. *Cell* **9**, 91—9.

Clarke L. & Carbon J. (1980) Isolation of a yeast centromere and construction of functional small circular chromosomes. *Nature* **287**, 504—9.

Cohen J. D., Eccleshall T. R., Needleman R. B., Federoff H., Buchferer B. A. & Marmur J. (1980) Functional expression in yeast of the *Escherichia coli* plasmid gene coding for chloramphenicol acetyl-transferase. *Proc. Natl. Acad. Sci. USA* **77**, 1078—82.

Cohen, S. N., Chang A. C. Y. & Hsu L. (1972) Nonchromosomal antibiotic resistance in bacteria: genetic transformation of *Escherichia coli* by R-factor DNA. *Proc. Natl. Acad. Sci. USA* **69**, 2110—4.

Colbère-Garapin F., Horodniceanu F., Kourilsky P. & Garapin A. C. (1981) A new dominant hybrid selective marker for higher eukaryotic cells. *J. Mol. Biol.* **150**, 1—14.

Collins F. S., Drumm M. L., Cole J. L., Lockwood W. K., Vande Woude G. F. & Iannuzzi M. C. (1987) Construction of a general human chromosome jumping library, with application to cystic fibrosis. *Science* **235**, 1046—9.

Collins F. S. & Weissman S. M. (1984) Directional cloning of DNA fragments at a large distance from an initial probe: a circularization method. *Proc. Natl. Acad. Sci. USA* **81**, 6812—16.

Collins J. & Brüning H. J. (1978) Plasmids usable as gene-cloning vectors in an *in vitro* packaging by coliphage λ: 'cosmids'. *Gene* **4**, 85—107.

Collins J. & Hohn B. (1979) Cosmids: a type of plasmid gene-cloning vector that is packageable *in vitro* in bacteriophage λ heads. *Proc. Natl. Acad. Sci. USA* **75**, 4242—6.

Comai L., Facciotti D., Hiatt W. R., Thompson G., Rose R. T. & Stalker D. M. (1985) Expression in plants of a mutant *aro*A gene from *Salmonella typhimurium* confers tolerance to glyphosate. *Nature* **317**, 741—4.

Cone R. D. & Mulligan R. C. (1984) High-efficiency gene transfer into mammalian cells: generation of helper-free recombinant retrovirus with broad mammalian host range. *Proc. Natl. Acad. Sci. USA* **81**, 6349—53.

Conner B. J., Reyes, A. A., Morin C., Itakura K., Teplitz R. & Wallace R. B. (1983) Detection of sickle cell β^S-globin allele by hybridization with synthetic oligonucleotides. *Proc. Natl. Acad. Sci. USA* **80**, 278—82.

Contente S. & Dubnau D. (1979) Characterization of plasmid transformation in *Bacillus subtilis*: kinetic properties and the effect of DNA conformation. *Mol. Gen. Genet.* **167**, 251—8.

Cooley L., Berg C. & Spradling A. (1988) Controlling P element insertional mutagenesis. *Trends Genet.* **4**, 254−8.

Coppella S. J., Acheson C. M. & Dhurjati P. (1987) Measurement of copy number using HPLC. *Biotechnol. Bioeng.* **XXIX**, 646−7.

Cosloy S. D. & Oishi M. (1973) Genetic transformation in *Escherichia coli* K12. *Proc. Natl. Acad. Sci. USA* **70**, 84−7.

Courtney M., Jallat S., Terrier L.-H., Benavente A., Crystal R. G. & Lecocq J.-P. (1985) Synthesis in *E. coli* of alpha₁-antitrypsin variants of therapeutic potential for emphysema and thrombosis. *Nature* **313**, 149−51.

Cousens L. S., Shuster J. R., Gallegos C., Ku L., Stempien M. M., Urden M. S., Sanchez-Pescador R., Taylor A. & Tekamp-Olson P. (1987) High level of expression of proinsulin in the yeast, *Saccharomyces cerevisiae*. *Gene* **61**, 265−75.

Covey S. & Hull R. (1981) Transcription of cauliflower mosaic virus DNA. Detection of transcripts, properties and location of the gene encoding the virus inclusion body. *Virology* **111**, 463−74.

Cox K. H., DeLeon D. V., Angerer L. M. & Angerer R. C. (1984) Detection of mRNAs in sea urchin embryos by *in situ* hybridization using assymetric RNA probes. *Dev. Biol.* **101**, 485−502.

Cox D. W., Woo S. L. C. & Mansfield T. (1985) DNA restriction fragments associated with alpha-antitrypsin indicate a single origin for deficiency allele PIZ. *Nature* **316**, 79−81.

Crabeel M., Huygen R., Cunin R. & Glansdorff N. (1983) The promoter region of the *arg3* gene in *Saccharomyces cerevisiae*. Nucleotide sequence and regulation in an *arg3−lac* Z gene fusion. *EMBO J.* **2**, 205−12.

Cross G. S., Wilson C., Erba H. P. & Woodland H. R. (1988) Cytoskeletal actin gene families of *Xenopus borealis* and *Xenopus laevis*. *J. Mol. Evol.* **27**, 17−28.

Crouse G. F., Simonsen C. C., McEwan R. N. & Schrinke R. T. (1982) Structure of amplified normal and variant dihydrofolate reductase genes in mouse sarcoma S180 cells. *J. Biol. Chem.* **257**, 7887−97.

Currier T. C. & Nester E. W. (1976) Evidence for diverse types of large plasmids in tumor-inducing strains of *Agrobacterium*. *J. Bacteriol.* **126**, 157−65.

Dalbadie-McFarland G., Cohen L. W., Riggs A. D., Morin C., Itakura K. & Richards J. H. (1982) Oligonucleotide-directed mutagenesis as a general and powerful method for studies of protein functions. *Proc. Natl. Acad. Sci. USA* **79**, 6409−13.

Dalbøge H., Dahl H.-H. M., Pedersen J., Hansen J. W. & Christensen T. (1987) A novel enzymatic method for production of authentic bGH from an *Escherichia coli* produced hGH-precursor. *Biotechnol.* **5**, 161−4.

Dani G. M. & Zakian U. A. (1983) Mitotic and meiotic stability of linear plasmids in yeast. *Proc. Natl. Acad. Sci. USA* **80**, 3406−10.

Das H. K., Jackson C. L., Miller D. A., Leff T. & Breslow J. L. (1987) The human apolipoprotein C-11 gene sequence contains a novel chromosome 19-specific minisatellite in its third intron. *J. Biol. Chem.* **262**, 4787−93.

Davanloo P., Rosenberg A. H., Dunn J. J. & Studier F. W. (1984) Cloning and expression of the gene for bacteriophage T7 RNA polymerase. *Proc. Natl. Acad. Sci. USA* **81**, 2035−9.

Davidson E. H. (1986) *Gene Activity in Early Development*, 3rd edn. Academic Press, New York.

Davis A. R., Kostek B., Mason B. B., Hsiao C. L., Morin J., Dheer S. K. & Hung P. P. (1985) Expression of hepatitis B surface antigen with a recombinant adenovirus. *Proc. Natl. Acad. Sci. USA* **82**, 7560−4.

Davis B. & MacDonald R. J. (1988) Limited transcription of rat elastase I transgene repeats in transgenic mice. *Genes Dev.* **2**, 13−22.

Davison J., Brunel F. & Merchez M. (1979) A new host-vector system allowing selection for foreign DNA inserts in bacteriophage λ gt *WES. Gene* **8**, 69−80.

De Block M., Botterman J., Vandewiele M., Dockx J., Thoen C., Gossele V., Rao Movra N., Thompson C., Van Montagu M. & Leemans J. (1987) Engineering herbicide resistance in plants by expression of a detoxifying enzyme. *EMBO J.* **6**, 2513−18.

De Block M., Herrera-Estrella L., Van Montagu M., Schell J. & Zambryski P. (1984) Expression of foreign genes in regenerated plants and their progeny. *EMBO J.* **3**, 1681−9.

De Block M., Schell J. & Van Montagu M. (1985) Chloroplast transformation by *Agrobacterium tumefaciens. EMBO J.* **4**, 1367−72.

de Boer H. A., Comstock L. J. & Vasser M. (1983a) The *tac* promoter: a functional hybrid derived from the *trp* and *lac* promoters. *Proc. Natl. Acad. Sci. USA* **80**, 21−5.

de Boer H. A., Hui A., Comstock L. J., Wong E. & Vasser M. (1983b) Portable Shine−Dalgarno regions: a system for a systematic study of defined alterations of nucleotide sequences within *E. coli* ribosome binding sites. *DNA* **2**, 231−41.

d'Enfert C., Chapon C. & Pugsley A. P. (1987) Export and secretion of the lipoprotein pullulanase by *Klebsiella pneumoniae. Mol. Microbiol.* **1**, 107−16.

d'Enfert C. & Pugsley A. P. (1987) A gene fusion approach to the study of pullulanase export and secretion in *Escherichia coli. Mol. Microbiol.* **1**, 159−68.

de Framond A. J., Barton K. A. & Chilton M.-D. (1983) Mini-Ti: a new vector strategy for plant genetic engineering. *Biotechnology* **1**, 262−9.

De Greef W., Delon R., De Block M., Leemans J. & Botterman J. (1989) Evaluation of herbicide resistance in transgenic crops under field conditions. *Biotechnology* **7**, 61−4.

De Greve H., Decraemer H., Senrinck J., Van Montagu M. & Schell J. (1981) The functional organization of the octopine *Agrobacterium tumefaciens* plasmid pTi B653. *Plasmid* **6**, 235−48.

De Greve H., Leemans J., Hernalsteens J. P., Thia-Toong L., De Benckeleer M., Willmitzer L., Olten L., Van Montagu M. & Schell J. (1982a) Regeneration of normal fertile plants that express octopine synthase from tobacco crown galls after deletion of tumor-controlling functions. *Nature* **300**, 752−5.

De Greve H., Phaese P., Seurwick J., Lemmers M., Van Montagu M. & Schell J. (1982b) Nucleotide sequence and transcript map of the *Agrobacterium tumefaciens* Ti plasmid-encoded octopine synthase gene. *J. Mol. Appl. Genet.* **1**, 499−511.

de la Peña A., Lorz H. & Schell J. (1987) Transgenic rye plants obtained by injecting DNA into young floral tillers. *Nature* **325**, 274−6.

De Maeyer E., Skup D., Prasad K. S. N., De Maeyer-Guignard J., Williams B., Meacock P., Sharpe G., Pioli D., Hennam J., Schuch W. & Atherton K. (1982) Expression of a chemically synthesized human α1 interferon gene. *Proc. Natl. Acad. Sci. USA* **79**, 4256−9.

de Saint Vincent B. R., Delbruck S., Eckhart W., Meinkoth J., Vitto L. & Wahl G. (1981) The cloning and reintroduction into animal cells of a functional CAD gene, a dominant amplifiable genetic marker. *Cell* **27**, 267−77.

de Vos W. M., Venema G., Canosi U. & Trautner T. A. (1981) Plasmid transformation in *Bacillus subtilis*: fate of plasmid DNA. *Mol. Gen. Genet.* **181**, 424−33.

Dean D. (1981) A plasmid cloning vector for the direct selection of strains carrying recombinant plasmids. *Gene* **15**, 99−102.

Deans R. J., Denis K. A., Taylor A. & Wall R. (1984) Expression of an immunoglobulin heavy chain gene transfected into lymphocytes. *Proc. Natl. Acad. Sci. USA* **81**, 1292−6.

Delseny M. & Hull R. (1983) Isolation and characterization of faithful and altered clones of the genomes of cauliflower mosaic virus isolates Cabb B−J1, CM4−184 and Bari 1. *Plasmid* **9**, 31−41.

Demetriou A. A., Whiting J. F., Feldman D., Levenson S. M., Chowdhury N. R., Moscioni A. D., Kram M. & Chowdhury J. R. (1986) Replacement of liver function in rats by transplantation of microcarrier-attached hepatocytes. *Science* **233**, 1190−2.

Deng G. & Wu R. (1981) An improved procedure for utilizing terminal transferase to add homopolymers to the 3′ termini of DNA. *Nucleic Acids Res.* **9**, 4173−88.

Denhardt D. T. (1966) A membrane-filter technique for the detection of complementary DNA. *Biochem. Biophys. Res. Commun.* **23**, 641−6.

Dente L., Cesareni G. & Cortese R. (1983) pEMBL: a new family of single stranded plasmids. *Nucleic Acids Res.* **11**, 1645−55.

Depicker A., Stachel S., Dhaese P., Zambryski P. & Goodman H. M. (1982) Nopaline synthase: transcript mapping and DNA sequence. *J. Mol. Appl. Genet.* **1**, 561−74.

Deretic V., Chandrasekharappa S., Gill J. F., Chaterjee D. K. & Chakrabarty A. M. (1987) A set of cassettes and improved vectors for genetic and biochemical characterization of *Pseudomonas* genes. *Gene* **57**, 61−72.

Derynck R., Singh A. & Goeddel D. (1983) Expression of the human interferon-γ cDNA in yeast. *Nucleic Acids Res.* **11**, 1819−37.

Deshayes A., Herrera-Estrella L. & Caboche M. (1985) Liposome-mediated transformation of tobacco mesophyll protoplasts by an *Escherichia coli* plasmid. *EMBO J.* **4**, 2731−7.

Devor E. J., Ivanovich A. K., Hickok J. M. & Todd R. D. (1988) A rapid method for confirming cell line identity: DNA fingerprinting with a minisatellite probe from M13 bacteriophage. *Biotechniques* **6**, 200−1.

Di Maio D., Triesman R. & Maniatis T. (1982) A bovine papillomavirus vector which propagates as an episome in both mouse and bacterial cells. *Proc. Natl. Acad. Sci. USA* **79**, 4030−4.

Ditta G., Stanfield S., Corbin D. & Helinski D. R. (1980) Broad host range DNA cloning system for Gram-negative bacteria. Construction of a gene bank of *Rhizobium meliloti*. *Proc. Natl. Acad. Sci. USA* **77**, 7347−51.

Dobson M. J., Tuite M. F., Roberts N. A., Kingsman A. J., Kingsman S. M., Perkins R. E., Conroy S. C., Dunbar B. & Fothergill L. A. (1982) Conservation of high efficiency promoter sequences in *Saccharomyces cerevisiae*. *Nucleic Acids Res.* **10**, 2625−37.

Dobson M. J., Tuite M. F., Mellor J., Roberts N. A., King R. M., Burke D. C., Kingsman A. J. & Kingsman S. M. (1983) Expression in *Saccharomyces cerevisiae* of human interferon-alpha directed by the *TRP* 5′ region. *Nucleic Acids Res.* **11**, 2287−302.

Doel M. T., Eaton M., Cook E. A., Lewis H., Patel T. & Carey N. H. (1980) The expression in *E. coli* of synthetic repeating polymeric genes coding for poly-(L-aspartyl-L-phenylalanine). *Nucleic Acids Res.* **8**, 4575−92.

Donis-Keller H. (1979) Site specific enzymatic cleavage of RNA. *Nucleic Acids Res.* **7**, 179−92.

Doolittle R. F. (1986) *Of Urfs and Orfs: a primer on how to analyze derived amino acid sequences*. University Science Books, Mill Valley, California.

Dotto G. P., Enea V. & Zinder N. D. (1981) Functional analysis of bacteriophage f1 intergenic region. *Virology* **114**, 463−73.

Dotto G. P. & Horiuchi K. (1981) Replication of a plasmid containing two origins of bacteriophage f1. *J. Mol. Biol.* **153**, 169−76.

Drummond M. H. & Chilton M.-D. (1978) Tumor-inducing (Ti) plasmids of *Agrobacterium* share extensive regions of DNA homology. *J. Bacteriol.* **136**, 1178−83.

Dueschle U., Kammerer W., Genz R. & Bujard H. (1986) Promoters of *Escherichia coli*: a hierarchy of *in vivo* strength indicates alternate structure. *EMBO J.* **5** 2987−94.

Dugaiczyk A., Boyer H. W. & Goodman H. M. (1975) Ligation of *Eco* RI endonuclease-generated DNA fragments into linear and circular structures. *J. Mol. Biol.* **96**, 171−84.

Duncan C. H., Wilson G. A. & Young F. E. (1978) Mechanism of integrating foreign DNA during transformation of *Bacillus subtilis*. *Proc. Natl. Acad. Sci. USA* **75**, 3664−8.

Dunphy W. G. & Rothman J. E. (1985) Compartmental organisation of the Golgi stack. *Cell* **42**, 13−21.

Durnam D. M. & Palmiter R. D. (1981) Transcriptional regulation of the mouse metallothionein-1 gene by heavy metals. *J. Biol. Chem.* **256**, 5712−16.

Dussoix D. & Arber W. (1962) Host specificity of DNA produced by *Escherichia coli*. II. Control over acceptance of DNA from infecting phage λ. *J. Mol. Biol.* **5**, 37−49.

Duvoisin R. M., Belin D. & Krisch H. M. (1986) A plasmid expression vector that permits the stabilization of both mRNAs and proteins encoded by the cloned genes. *Gene* **45**, 193−201.

Dworkin M. B. & Dawid I. B. (1980) Use of a cloned library for the study of abundant poly(A)$^+$ RNA during *Xenopus laevis* development. *Dev. Biol.* **76**, 449−64.

Dycaico M. J., Grant S. G. O., Felts K., Nichols W. S., Geller S. A., Hager J. M., Pollard A. J., Kohler S. W., Short H. P., Jirik F. R., Hanahan D. & Sorge J. A. (1988) Neonatal hepatitis induced by α_1-antitrypsin: a transgenic mouse model. *Science* **242**, 1409−12.

Eckhardt T. (1978) A rapid method for the identification of plasmid deoxyribonucleic acid in bacteria. *Plasmid* **1**, 584−8.

Edelmann P. & Gallant J. (1977) Mistranslation in *E. coli*. *Cell* **10**, 131−7.

Edens L., Heslinga L., Klok R., Ledeboer A. M., Maat J., Toonen M. Y., Visser C. & Verrips C. T. (1982) Cloning of cDNA encoding the sweet-tasting plant protein thaumatin and its expression in *Escherichia coli*. *Gene* **18**, 1−12.

Efstratiadis A., Kafatos F. C., Maxam A. M. & Maniatis T. (1976) Enzymatic *in vitro* synthesis of globin genes. *Cell* **7**, 279−88.

Eglitis M. A. & Anderson W. F. (1988) Retroviral vectors for introduction of genes into mammalian cells. *Biotechniques* **6**, 608−14.

Ehrlich H. A., Cohen S. N. & McDevitt H. O. (1978a) A sensitive radioimmunoassay for detecting products translated from cloned DNA fragments. *Cell* **13**, 681−9.

Ehrlich S. D. (1977) Replication and expression of plasmids from *Staphylococcus aureus* in *Bacillus subtilis*. *Proc. Natl. Acad. Sci. USA* **74**, 1680−2.

Ehrlich S. D. (1978) DNA cloning in *Bacillus subtilis*. *Proc. Natl. Acad. Sci. USA* **75**, 1433−6.

Ehrlich S. D., Jupp S., Niaudet B. & Goze A. (1978b) *Bacillus subtilis* as a host for DNA cloning. In *Genetic Engineering*, eds Boyer H. W. & Nicosia S., pp. 25–32. Elsevier/North Holland Biomedical Press, Amsterdam.

Ehrlich S. D., Niaudet B. & Michel B. (1981) Use of plasmids from *Staphylococcus aureus* for cloning of DNA in *Bacillus subtilis*. *Curr. Top Microbiol. Immunol.* **96**, 19–29.

Emerman M. & Temin H. M. (1986) Quantitative analysis of gene suppression in integrated retrovirus vectors. *Mol. Cell. Biol.* **6**, 792–800.

Emtage J. S., Tacon W. C. A., Catlin G. H., Jenkins B., Porter A. G. & Carey N. H. (1980) Influenza antigenic determinants are expressed from haem-agglutinin genes cloned in *Eschericia coli*. *Nature* **283**, 171–4.

England T. E., Gumport R. I. & Uhlenbeck O. C. (1977) Dinucleoside pyro-phosphates are substrates for T4-induced RNA ligase. *Proc. Natl. Acad. Sci. USA* **74**, 4839–42.

Engler G., Depicker A., Maenhaut R., Villarroel R., Van Montagu M. & Schell J. (1981) Physical mapping of DNA base sequence homologies between an octopine and a nopaline Ti plasmid of *Agrobacterium tumefaciens J. Mol. Biol.* **152**, 183–208.

Ensley B. D., Ratzkin B. J., Osslund T. D., Simon M. J., Wackett L. P. & Gibson D. T. (1983) Expression of naphthalene oxidation genes in *Escherichia coli* results in biosynthesis of indigo. *Science* **222**, 167–9.

Eperon I. C. (1986) M13 vectors with T7 polymerase promoters: transcription limited by oligonucleotides. *Nucleic Acids Res.* **14**, 2830.

Ernst J. F. (1988) Codon usage and gene expression. *Trends Biotechnol.* **6**, 196–9.

Errington J. (1984) Efficient *Bacillus subtilis* cloning system using bacteriophage vector φ105 J9. *J. Gen. Microbiol.* **130**, 2615–28.

Errington J. & Pughe N. (1987) Upper limit for DNA packaging by *Bacillus subtilis* bacteriophage φ105: isolation of phage deletion mutants by induction of oversized prophages. *Mol. Gen. Genet.* **210**, 347–51.

Etkin L. D., Pearman B. & Ansah-Yiadom R. (1987) Replication of injected DNA templates in *Xenopus* embryos. *Exp. Cell Res.* **169**, 468–77.

Falkow S. (1975) *Infectious multiple drug resistance*. London, Pion.

Feinberg A. P. & Vogelstein B. (1983) a technique for radiolabelling DNA restriction endonuclease fragments to high specific activity. *Anal. Biochem.* **132**, 6–13.

Feinberg A. P. & Vogelstein B. (1984) Addendum: a technique for radiolabelling DNA restriction endonuclease fragments to high specific activity. *Anal. Biochem.* **137**, 266–267.

Felgner P. L., Gadek T. R., Holm M., Roman R., Chan H. W., Wenz M., Northrop J. P., Ringold G. M. & Danielsen Ml (1987) Lipofection: a highly efficient lipid-mediated DNA-transfection procedure. *Proc. Natl. Acad. Sci. USA* **84**, 7413–17.

Ferrari E., Henner D., Pereso M. & Hoch J. (1987) Transcription of *B. subtilis* Subtilisin and expression of subtilisin in sporulation mutants. *J. Bacteriol.* **170**, 162–7.

Ferretti L. & Sgaramella V. (1981) Temperature dependence of the joining by T4 DNA ligase of termini produced by type II restriction endonucleases. *Nucleic Acids Res.* **9**, 85–93.

Fischhoff D. A., Bowdish K. S., Perlak F. J., Marrone P. G., McCormick S. M., Niedermeyer J. G., Dean D. A., Kusano-Kretzmer K., Mayer E. J., Rochester D. E., Rogers S. G. & Fraley R. T. (1987) Insect tolerant transgenic tomato plants. *Biotechnology* **5**, 807–13.

Fitzgerald-Hayes M., Buhler J. M., Cooper T. G. & Carbon J. (1982) Isolation and subcloning analysis of functional centromere DNA (cen11) from yeast chromosome XI. *Mol. Cell. Biol.* **2**, 82−7.

Flexner C., Hugin A. & Moss B. (1987) Prevention of vaccinia virus infection in immunodeficient mice by vector-directed IL-2 expression. *Nature* **330**, 259−62.

Flexner C., Murphy B. R., Rooney J. F., Wohlenberg C., Yuferov V., Notkins A. L., Moss B. (1988) Successful vaccination with a polyvalent live vector despite existing immunity to an expressed antigen. *Nature* **335**, 259−62.

Flintoff W. F., Davidson S. V. & Siminovitch L. (1976) Isolation and partial characterization of three methotrexate-resistant phenotypes from Chinese hamster ovary cells. *Somat. Cell Genet.* **2**, 245−61.

Folger K. R., Wong E. A., Wahl G. & Capecchi M. (1982) Patterns of integration of DNA microinjected into cultured mammalian cells: evidence for homologous recombination between injected plasmid DNA molecules. *Mol. Cell. Biol.* **2**, 1372−87.

Forbes D. J., Kirschner M. W. & Newport J. W. (1983) Spontaneous formation of nucleus-like structures around bacteriophage DNA microinjected into *Xenopus* eggs. *Cell* **34**, 13−23.

Forster A. C., McInnes J. L., Skingle D. C. & Symons R. H. (1985) Non-radioactive hybridization probes prepared by the chemical labelling of DNA and RNA with a novel reagent, photobiotin. *Nucleic Acids Res.* **13**, 745−61.

Fowler S. J., Gill P., Werrett D. J. & Higgs D. R. (1988) Individual specific DNA fingerprints from a hypervariable region probe: alpha-globin 3'HVR.

Fraley R., Rogers S. G., Horsch R. B., Sanders P. R., Flick J. S., Adams S. P., Bittner M. L., Brand L. A., Fink C. L., Fry J. S., Galuppi G. R., Goldberg S. B., Hoffmann N. L. & Woo S. C. (1983) Expression of bacterial genes in plant cells. *Proc. Natl. Acad. Sci. USA* **80**, 4803−7.

Franck A., Guilley H., Jonard G., Richards K. & Hirth L. (1980) Nucleotide sequence of cauliflower mosaic virus DNA. *Cell* **21**, 285−94.

Franklin F. C. H., Bagdasarian M., Bagdasarian M. M. & Timmis K. N. (1981) Molecular and functional analysis of the TOL plasmid pWWO from *Pseudomonas putida* and cloning of genes for the entire regulated aromatic ring *meta* cleavage pathway. *Proc. Natl. Acad. Sci. USA* **78**, 7458−62.

French R., Janda M. & Ahlquist P. (1986) Bacterial gene inserted in an engineered RNA virus: efficient expression in monocotyledonous plant cells. *Science* **231**, 1294−7.

Frey J., Bagdasarian M., Feiss D., Franklin C. H. & Deshusses J. (1983) Stable cosmid vectors that enable the introduction of cloned fragments into a wide range of Gram-negative bacteria. *Gene* **24**, 299−308.

Friedmann T. (1987) HPRT-deficient mice: a useful new animal model for human disease. *Trends Biotechnol.* **5**, 157−8.

Frischauf A.-M., Lehrach H., Poustka A. & Murray N. (1983) Lambda replacement vectors carrying polylinker sequences. *J. Mol. Biol.* **170**, 827−42.

Fujita N., Nelson N., Fox T. D., Claudio T., Lindstrom J., Riezman H. & Hess G. P. (1986) Biosynthesis of the *Torpedo californica* acetylcholine receptor L subunit in yeast. *Science* **231**, 1284−7.

Fukunaga R., Sokawa Y. & Nagata S. (1984) Constitutive production of human interferons by mouse cells with bovine papillomavirus as vector. *Proc. Natl. Acad. Sci. USA* **81**, 5086−90.

Fulton A. M., Adams S. E., Mellor J., Kingsman S. M. & Kingsman A. J. (1987) The organisation and expression of the yeast transposon, Ty. *Microbiol. Sci.* **4**, 180−185.

Futcher A. B. & Carbon J. (1986) Toxic effects of excess cloned centromeres.

Mol. Cell. Biol. **6**, 2213—22.

Galli G., Hofstetter H., Stunnenberg H. G. & Birnstiel M. L. (1983) Biochemical complementation with RNA in the *Xenopus* oocyte: a small RNA is required for the generation of 3'histone mRNA termini. *Cell* **34**, 823—8.

Gallie D. R., Gay P. & Kado C. I. (1988) Specialized vectors for members of *Rhizobiaceae* and other Gram-negative bacteria. In *Vectors. A survey of molecular cloning vectors and their uses*, eds Rodriguez R. L. & Denhardt D. T., pp. 333—42. Butterworths, London.

Gallwitz D. & Sures I. (1980) Structure of a split yeast gene. Complete nucleotide sequence of the actin gene in *Saccharomyces cerevisiae*. *Proc. Natl. Acad. Sci. USA* **77**, 2546—50.

Ganoza M. C., Kofoid E. C., Marliere P. & Louis B. G. (1987) Potential secondary structure at translation initiation sites. *Nucleic Acids Res.* **15**, 345—59.

Gardner R. C., Howarth A. J., Hahn P., Brown-Luedi M., Shepherd R. J. & Messing J. (1981) The complete nucleotide sequence of an infectious clone of cauliflower mosaic virus by M13 mp7 shotgun sequencing. *Nucleic Acids Res.* **9**, 2871—87.

Garfinkel D. J., Boeke J. D. & Fink G. R. (1985) Ty element transposition: reverse transcription and virus-like particles. *Cell* **42**, 507—17.

Garfinkel D. J. & Nester E. W. (1980) *Agrobacterium tumefaciens* mutants affected in crown gall tumorigenesis and octopine catabolism. *J. Bacteriol.* **144**, 732—43.

Gargiulo G. & Worcel A. (1983) Analysis of the chromatin assembled in germinal vesicles of *Xenopus* oocytes. *J. Mol. Biol.* **170**, 699—722.

Garver R. I., Chytil A., Courtney M. & Crystal R. G. (1987) Clonal gene therapy: transplanted mouse fibroblast clones express human α_1-antitrypsin gene *in vivo*. *Science* **237**, 762—4.

Gasser & Fraley (1989) Genetically engineering plants for crop improvement. *Science* **244**, 1293—9.

Gasson M. J. & Davies F. L. (1985) The genetics of dairy lactic-acid bacteria. In *Advances in the Microbiology and Biochemistry of Cheese and Fermented Milk*, eds Davies F. L. & Law B. A., pp. 99—126. Elsevier, London.

Gehrlach W., Llewellyn D. & Haseloff J. (1987) Construction of a plant disease resistance gene from the satellite RNA of tobacco ringspot virus. *Nature* **328**, 802—5.

Gerard R. D. & Gluzman Y. (1985) A new host cell system for regulated simian virus 40 DNA replication. *Mol. Cell. Biol.* **5**, 3231—40.

Gerbaud C., Fournier P., Blanc H., Aigle M., Heslot H. & Geurineau M. (1979) High frequency of yeast transformation by plasmids carrying part or entire 2 μm yeast plasmid. *Gene* **5**, 233—53.

Gershoni J. M. & Palade G. E. (1982) Electrophoretic transfer of proteins from sodium dodecyl sulfate-polyacrylamide gels to a positively charged membrane filter. *Anal. Biochem.* **124**, 396—405.

Gething M.-J. & Sambrook J. (1981) Cell-surface expression of influenza haemogglutinin from a cloned DNA copy of the RNA gene. *Nature* **293**, 620—5.

Gheysen D., Iserentant D., Derom C. & Fiers W. (1982) Systematic alteration of the nucleotide sequence preceding the translation initiation codon and the effects on bacterial expression of the cloned SV-40 small-t antigen gene. *Gene* **17**, 55—63.

Giebelhaus D. H., Eib D. W. & Moon R. T. (1988). Antisense RNA inhibits expression of membrane skeleton protein 4.1 during embryonic development of *Xenopus*. *Cell* **53**, 601—15.

Gill P., Jeffreys A. J. & Werrett D. J. (1985) Forensic application of DNA "fingerprints". *Nature* **318**, 577−9.

Gill P., Lygo J. E., Fowler S. J. & Werrett D. J. (1987) An evaluation of DNA fingerprinting for forensic purposes. *Electrophoresis* **8**, 38−44.

Gill P. & Werrett D. J. (1987) Exclusion of a man charged with murder by DNA fingerprinting. *Forensic Sci. Int.* **35**, 145−8.

Gillam S., Astell C. R. & Smith M. (1980) Site-specific mutagenesis using oligodeoxyribonucleotides: isolation of a phenotypically silent φX174 mutant, with a specific nucleotide deletion, at very high efficiency. *Gene* **12**, 129−37.

Gingeras T. R., Kwoh D. Y. & Davis G. R. (1987) Hybridization properties of immobilized nucleic acids. *Nucleic Acids Res.* **15**, 5373−90.

Giri I. & Danos O. (1986) Papillomavirus genomes: from sequence data to biological properties. *Trends Genet.* **2**, 227−32.

Gluzman Y. (1981) SV40-transformed simian cells support the replication of early SV40 mutants. *Cell* **23**, 175−82.

Goeddel D. V., Heyneker H. L., Hozumi T., Arentzen R., Itakura K., Yansura D. G., Ross M. J., Miozzari G., Crea R. & Seeburg P. H. (1979a) Direct expression in *Escherichia coli* of a DNA sequence coding for human growth hormone. *Nature* **281**, 544−8.

Goeddel D. V., Kleid D. G., Bolivar F., Heyneker H. L., Yansura D. G., Crea R., Hirose T., Kraszewski A., Itakura K. & Riggs A. D. (1979b) Expression in *Escherichia coli* of chemically synthesized genes for human insulin. *Proc. Natl. Acad. Sci. USA* **76**, 106−10.

Goff S. P. & Berg P. (1976) Construction of hybrid viruses containing SV40 and λ phage DNA segments and their propagation in cultured monkey cells. *Cell* **9**, 695−705.

Goldberg D. A., Posakony J. W. & Maniatis T. (1983) Correct developmental expression of a cloned alcohol dehydrogenase gene transduced into the *Drosophila* germ line. *Cell* **34**, 59−73.

Goldfarb D. S., Doi R. H. & Rodriguez R. L. (1981) Expression of Tn9-derived chloramphenicol resistance in *Bacillus subtilis*. *Nature* **293**, 309−11.

Goldfarb M., Shimizu K., Perucho M. & Wigler M. (1982) Isolation and preliminary characterization of a human transforming gene from T24 bladder carcinoma cells. *Nature* **296**, 404−9.

Gordon J. W. & Ruddle F. H. (1981) Integration and stable germ line transmission of genes injected into mouse pronuclei. *Science* **214**, 1244−6.

Gordon K., Lee E., Vitale J. A., Smith A. E., Westphal H. & Henninghausen L. (1987) Production of human tissue plasminogen activator in transgenic mouse milk. *Biotechnology* **5**, 1183−7.

Gordon M. P. (1980) In *Proteins and Nucleic Acids. The Biochemistry of Plants*, ed. Marcus, A., vol. 6, pp. 531−70. Academic Press, New York.

Gordon M. P., Farrand S. K., Sciaky D., Montoya A. L., Chilton M-D., Merlo D. J. & Nester E. W. (1979) In *Molecular Biology of Plants, Symposium*, University of Minnesota, ed. Rubenstein I. Academic Press, London.

Gorman C. M., Merlino G. T., Willingham M. C., Pastan I. & Howard B. (1982a) The Rous sarcoma virus long terminal repeat is a strong promoter when introduced into a variety of eukaryotic cells by DNA-mediated transfection. *Proc. Natl. Acad. Sci. USA* **79**, 6777−81.

Gorman C. M., Moffat L. F. & Howard B. H. (1982b) Recombinant genomes which express chloramphenicol acetyl transferase in mammalian cells. *Mol. Cell. Biol.* **2**, 1044−51.

Gorman C., Padmanabhan R. & Howard B. (1983) High efficiency DNA-mediated transformation of private cells. *Science* **221**, 551−3.

Gorski K., Roch J.-M., Prentki P. & Krisch H. M. (1985) The stability of bacteriophage T4 gene 32 mRNA: a 5′ leader sequence than can stabilize mRNA transcripts. *Cell* **43**, 461−9.

Goto M., Akai K., Murakami A., Hashimoto C., Tsuda E., Ueda M., Kawanishi G., Takahashi N., Ishimoto A., Chiba H. & Sasaki R. (1988) Production of recombinant human erythropoietin in mammalian cells: host cell dependency of the biological activity of the cloned glycoprotein. *Biotechnology* **6**, 67−71.

Gottesmann M. E. & Yarmolinsky M. D. (1968) The integration and excision of the bacteriophage lambda genome. *Cold Spring Harbor Symp. Quant. Biol.* **33**, 735−47.

Gourse R. L., De Boer H. A. & Nomura M. (1986) DNA determinants of RNA synthesis in *E. coli*: growth rate dependent regulation, feedback inhibition, upstream activation, anti-termination. *Cell* **44**, 197−205.

Gouy M. & Gautier C. (1982) Codon usage in bacteria. Correlation with gene expressivity. *Nucleic Acids Res.* **10**, 7055−74.

Gowland P. C. & Hardmann D. J. (1986) Methods for isolating large bacterial plasmids. *Microbiol. Sci.* **3**, 252−4.

Graham F. L. & van der Eb A. J. (1973) A new technique for the assay of infectivity of human adenovirus 5 DNA. *Virology* **52**, 456−67.

Green M. R., Maniatis T. & Melton D. A. (1983) Human β-globin pre-mRNA synthesized *in vitro* is accurately spliced in *Xenopus* oocyte nuclei. *Cell* **32**, 681−94.

Gribskov M. & Burgess R. R. (1983) Overexpression and purification of the sigma subunit of *Escherichia coli* RNA polymerase. *Gene* **26**, 109−18.

Grimsley N., Hohn T., Davies J. W. & Hohn B. (1987) *Agrobacterium* − mediated delivery of infectious maize streak virus into maize plants. *Nature* **325**, 177−9.

Grinter N. J. (1983) A broad-host-range cloning vector transposable to various replicons. *Gene* **21**, 133−43.

Gronenborn B., Gardner R. C., Schaefer S. & Shepherd R. J. (1981) Propagation of foreign DNA in plants using cauliflower mosaic virus as vector. *Nature* **294**, 773−6.

Gronenborn B. & Messing J. (1978) Methylation of single-stranded DNA *in vitro* introduces new restriction endonuclease cleavage sites. *Nature* **272**, 375−7.

Grosjean J. & Fiers W. (1982) Preferential codon usage in prokaryotic genes. The optimal codon−anticodon interaction energy and the selective codon usage in efficiently expressed genes. *Gene* **18**, 199−209.

Grosschedl R. & Birnstiel M. L. (1980a) Identification of regulatory sequences in the prelude sequences of an H2A histone gene by the study of specific deletion mutants *in vivo*. *Proc. Natl. Acad. Sci. USA* **77**, 1432−36.

Grosschedl R. & Birnstiel M. L. (1980b) Spacer DNA sequences upstream of the TATAAATA sequence are essential for promotion of H2A histone gene transcription *in vivo*. *Proc. Natl. Acad. Sci. USA* **77**, 7102−6.

Grosschedl R., Machler M., Rohrer U. & Birnstiel M. L. (1983) A functional component of the sea urchin H2A gene modulator contains an extended sequence homology to a viral enhancer. *Nucleic Acids Res.* **11**, 8123−36.

Grosveld F. G., Lund T., Murray E. J., Mellor A. L., Dahl H. H. M. & Flavell R. A. (1982) The construction of cosmid libraries which can be used to transform eukaryotic cells. *Nucleic Acids Res.* **10**, 6715−32.

Grosveld F., van Assendelft G. B., Greaves D. R. & Kollias G. (1987). Position

References

independent, high level expression of the human β-globin gene in transgenic mice. *Cell* **51**, 975–85.

Grunstein M. & Hogness D. S. (1975) Colony hybridization: a method for the isolation of cloned DNAs that contain a specific gene. *Proc. Natl. Acad. Sci. USA* **72**, 3961–5.

Gruss A. & Ehrlich S. D. (1988) Insertion of foreign DNA into plasmids from Gram-positive bacteria induces formation of high-molecular-weight plasmid multimers. *J. Bacteriol.* **170**, 1183–90.

Gruss P., Efstratiadis A., Karathanasis S., Konig M. & Khoury G. (1981a) Synthesis of stable unspliced mRNA from an intronless simian virus 40-rat preproinsulin gene recombinant. *Proc. Natl. Acad. Sci. USA* **78**, 6091–5.

Gruss P., Ellis R. W., Shih T. Y., Konig M., Scolnick E. M. & Khoury G. (1981b) SV40 recombinant molecules express the gene encoding p21 transforming protein of Harvey murine sarcoma virus. *Nature* **293**, 486–8.

Gruss P. & Khoury G. (1981) Expression of simian virus 40-rat preproinsulin recombinants in monkey kidney cells: use in preproinsulin RNA processing signals. *Proc. Natl. Acad. Sci. USA* **78**, 133–7.

Gryczan T. J., Contente S. & Dubnau D. (1980a) Molecular cloning of heterologous chromosomal DNA by recombination between a plasmid vector and a homologous resident plasmid in *Bacillus subtilis*. *Mol. Gen. Genet.* **177**, 459–67.

Gryzcan T. J. & Dubnau D. (1978) Construction and properties of chimaeric plasmids in *Bacillus subtilis*. *Proc. Natl. Acad. Sci. USA* **75**, 1428–32.

Guarente L. (1987) Regulatory proteins in yeast. *Ann. Rev. Genet.* **21**, 425–52.

Guarente L., Lauer G., Roberts T. M. & Ptashne M. (1980) Improved methods for maximizing expression of a cloned gene: a bacterium that synthesizes rabbit β-globin. *Cell* **20**, 543–53.

Guarente L. & Ptashne M. (1981) Fusion of *Escherichia coli lac* Z to the cytochrome *c* gene of *Saccharomyces cerevisiae*. *Proc. Natl. Acad. Sci. USA* **78**, 2199–203.

Gubler U. & Hoffman B. J. (1983) A simple and very efficient method for generating cDNA libraries. *Gene* **26**, 263–9.

Gumport R. I. & Lehman I. R. (1971) Structure of the DNA ligase adenylate intermediate: lysine (ε-amino) linked AMP. *Proc. Natl. Acad. Sci. USA* **68**, 2559–63.

Gusella J. F., Keys C., Varsanyi-Breiner A., Kao F.-T., Jones C., Puck T. T. & Housman D. (1980) Isolation and localization of DNA segments from specific human chromosomes. *Proc. Natl. Acad. Sci. USA* **77**, 2829–33.

Gutte B., Daumigen M. & Wittschieber E. (1979) Design, synthesis and characterization of a 34-residue polypeptide that interacts with nucleic acids. *Nature* **281**, 650–5.

Hadfield C., Cashmore A. M. & Meacock P. A. (1987) Sequence and expression characteristics of a shuttle chloramphenicol-resistance marker for *Saccharomyces cerevisiae* and *Escherichia coli*. *Gene* **52**, 59–70.

Hadi S. M., Bachi B., Iida S. & Bickle T. A. (1982) DNA restriction-modification enzymes of phage P1 and plasmid p15B. Subunit functions and structural homologies. *J. Mol. Biol.* **165**, 19–34.

Haenlin M., Steller H., Pirotta V., Momer E. (1985) A 43 kb cosmid P transposon rescues the fs(1)K10 morphogenetic locus and three adjacent *Drosophila* developmental mutants. *Cell* **40**, 827–37.

Hagan C. E. & Warren G. J. (1983) Viability of palindromic DNA is restored by deletions occurring at low but variable frequency in plasmids of *Escherichia coli*. *Gene* **24**, 317–26.

Haima P., Bron S. & Venema G. (1987) The effect of restriction on shotgun

cloning and plasmid stability in *Bacillus subtilis* Marburg. *Mol. Gen. Genet.*
209, 335–42.

Hain R., Stabel P., Czernilofsky A. P., Steinbiss H. H., Herrera-Estrella L. &
Schell J. (1985) Uptake, integration, expression and genetic transmission of
a selectable chimaeric gene by plant protoplasts. *Mol. Gen. Genet.* **199**,
161–8.

Hall M. N., Hereford L. & Herskowitz I. (1984) Targeting of *E. coli* β-galactosidase
to the nucleus in yeast. *Cell* **36**, 1057–65.

Hamer D. H. (1977) SV40 carrying an *Escherichia coli* suppressor gene. In
Recombinant Molecules: Impact on Science and Society, eds Beers R. G. &
Bassett E. G., pp. 317–35. Raven, New York.

Hamer D. H. & Leder P. (1979a) Expression of the chromosomal mouse β maj-
globin gene cloned in SV40. *Nature* **281**, 35–40.

Hamer D. H. & Leder P. (1979b) SV40 recombinants carrying a functional RNA
splice junction and polyadenylation site from the chromosomal mouse
β maj-globin gene. *Cell* **17**, 737–47.

Hamer J. E. & Timberlake W. E. (1987) Functional organisation of the *Aspergillus
nidulans trp*C promoter. *Mol. Cell. Biol.* **7**, 2352–9.

Hamill J. D., Parr A. J., Rhodes M. J. C., Robins R. J. & Walton N. J. (1987) New
routes to plant secondary products. *Biotechnology* **5**, 800–4.

Hammer R. E., Brinster R. L., Palmiter R. D. (1985) Use of gene transfer to
increase animal growth. *Cold Spring Harbor Symp. Quant. Biol.* **50**, 379–87.

Hammer R. E., Krumlauf R., Camper S. A., Brinster R. L. & Tilghman S. M.
(1987) Diversity of alpha-fetoprotein gene expression in mice is generated
by a combination of separate enhancer elements. *Science* **235**, 53–8.

Hammer R. E., Palmiter R. D. & Brinster R. L. (1984) Partial correction of
murine hereditary growth disorder by germ line incorporation of a new
gene. *Nature* **311**, 65–7.

Hanahan D. (1983) Studies on transformation of *Escherichia coli* with plasmids.
J. Mol. Biol. **166**, 557–80.

Hanahan D., Lane D., Lipsich L., Wigler M. & Botchan M. (1980) Characteristics of an SV40-plasmid recombinant and its movement into and out of
the genome of a murine cell. *Cell* **21**, 127–39.

Hanahan D. & Meselson M. (1980) Plasmid screening at high colony density.
Gene **10**, 63–7.

Harbers K., Jahner D. & Jaenisch R. (1981) Microinjection of cloned retroviral
genomes into mouse zygotes: integration and expression in the animal.
Nature **293**, 540–43.

Hardies S. C. & Wells R. D. (1976) Preparative fractionation of DNA by
reversed phase column chromatography. *Proc. Natl. Acad. Sci. USA* **73**,
3117–21.

Hardy K., Stahl S. & Küpper H. (1981) Production in *B. subtilis* of hepatitis B
core antigen and of major antigen of foot and mouth disease virus. *Nature*
293, 481–3.

Harland R. M. & Laskey R. A. (1980) Regulated replication of DNA microinjected
into eggs of *Xenopus laevis. Cell* **21**, 761–71.

Harland R. M., Weintraub H. & McKnight S. L. (1983) Transcription of DNA
injected into *Xenopus* oocytes is influenced by template topology. *Nature*
302, 38–42.

Harley C. B. & Reynolds R. P. (1987) Analysis of *E. coli* promoter sequences.
Nucleic Acids Res. **15**, 2343–61.

Hashimoto-Gotoh T., Franklin F. C. H., Nordheim A. & Timmis K. N. (1981)
Low copy number, temperature-sensitive, mobilization-defective pSC101-

derived containment vectors. *Gene* **16**, 227—35.

Hashimoto-Gotoh T., Kume A., Masahashi W., Takeshita Takeshita S. & Fukuda A. (1986) Improved vector, pHSG664, for direct streptomycin-resistance selection: cDNA cloning with G : C-tailing procedure and subcloning of double digest DNA fragments. *Gene* **41**, 125—8.

Hawley D. K. & McClure W. R. (1983) Compilation and analysis of *Escherichia coli* promoter DNA sequences. *Nucleic Acids Res.* **11**, 2237—55.

Hayakawa T., Toibana A., Marumoto R., Nakahama K., Kikuchi M., Fujimoto K. & Ikehara M. (1987) Expression of human lysozyme in an insoluble form in yeast. *Gene* **56**, 53—9.

Heidecker G. & Messing J. (1983) Sequence analysis of Zein cDNAs by an efficient mRNA cloning method. *Nucleic Acids Res.* **11**, 4891—906.

Heinemann J. A. & Sprague G. F. (1989) Bacterial conjugative plasmids mobilize DNA transfer between bacteria and yeast. *Nature* **340**, 205—9.

Helfman D. M., Feramisco J. R., Fiddes J. C., Thomas G. P. & Hughes S. H. (1983) Identification of clones that encode chicken tropomyosin by direct immunological screening of a cDNA expression library. *Proc. Natl. Acad. Sci. USA* **80**, 31—5.

Henikoff S., Tatchell K., Hall B. D. & Nosmyth K. A. (1981) Isolation of a gene from *Drosophila* by complementation in yeast. *Nature* **289**, 33—7.

Hennam J. F., Cunningham A. E., Sharpe G. S. & Atherton K. T. (1982) Expression of eukaryotic coding sequences in *Methylophilus methylotrophus*. *Nature* **297**, 80—2.

Hennecke H., Günther I. & Binder F. (1982) A novel cloning vector for the direct selection of recombinant DNA in *E. coli*. *Gene* **19**, 231—4.

Hernalsteens J.-P., Thia-Toong L., Schell J. & Van Montagu (1984) An *Agrobacterium*—transformed cell culture from the monocot *Asparagus officinalis* *EMBO J.* **3**, 3039—41.

Herrera-Estrella L., Depicker A., Van Montagu M. & Schell J. (1983a) Expression of chimaeric genes transferred into plant cells using a Ti-plasmid-derived vector. *Nature* **303**, 209—13.

Herrera-Estrella L., DeBlock M., Messens E., Hernalsteens J. P., Van Montagu M. & Schell J. (1983b) Chimeric genes as dominant selectable markers in plant cells. *EMBO J.* **2**, 987—95.

Hershfield V., Boyer H. W., Yanofsky C., Lovett M. A. & Helinski D. R. (1974) Plasmid Col El as a molecular vehicle for cloning and amplification of DNA. *Proc. Natl. Acad. Sci. USA* **71**, 3455—9.

Herskowitz I. (1974) Control of gene expression in bacteriophage lambda. *Annu. Rev. Genet.* **7**, 289—324.

Hicks J. B., Strathern J. N., Klar A. J. S. & Dellaporta S. L. (1982) Cloning by complementation in yeast. The mating type genes. In *Genetic Engineering*, eds Setlow J. K. & Hollaender A., pp. 219—48. Plenum Press, New York.

Hieter P., Mann C., Snyder M. & Davis R. W. (1985) Mitotic stability of yeast chromosomes: a colony color assay that measures nondisjunction and chromosome loss. *Cell* **40**, 381—92.

Higuchi R., Paddock G. V., Wall R. & Salser W. (1976) A general method for cloning eukaryotic structural gene sequences. *Proc. Natl. Acad. Sci. USA* **73**, 3146—50.

Hilder V. A., Gatehouse A. M. R., Sheerman S. E., Barker R. F. & Boulter D. (1987) A novel mechanism of insect resistance engineered into tobacco. *Nature* **300**, 160—3.

Hill A. & Bloom K. (1987) Genetic manipulation of centromere function. *Mol. Cell. Biol.* **7**, 2397—405.

Hills A. V. S. & Jeffreys A. J. (1985) Use of minisatellite DNA probes for determination of twin zygosity at birth. *Lancet* **2**, 1394–5.

Hinnen A., Hicks, J. B. & Fink G. R. (1978) Transformation of yeast. *Proc. Natl. Acad. Sci. USA* **75**, 1929–33.

Hirst T. R. & Welch R. A. (1988) Mechanisms for secretion of extracellular proteins by Gram-negative bacteria. *Trends Biochem. Sci.* **13**, 265–9.

Hitzeman R. A., Hagie F. E., Levine H. L., Goeddel D. V., Ammerer G. & Hall B. D. (1981) Expression of a human gene for interferon in yeast. *Nature* **293**, 717–22.

Hitzeman R. A., Leung D. W., Perry L. J., Kohr W. J., Levine H. L. & Goeddel D. V. (1983) Secretion of human interferons by yeast. *Science* **219**, 620–5.

Hoekma A., Hirsch P. R., Hooykass P. J. J. & Schilperoort R. A. (1983) A binary plant vector strategy based on separation of *vir-* and T-regions of the *Agrobacterium tumefaciens* Ti-plasmid. *Nature* **303**, 179–83.

Hoekma A., Kastelein R. A., Vasser M. & de Boer H. A. (1987) Codon replacement in the PGK1 gene of *Saccharomyces cerevisiae*: experimental approach to study the role of biased codon usage in gene expression. *Mol. Cell. Biol.* **7**, 2914–24.

Hoekstra W. P. M., Bergmans J. E. N. & Zuidweg E. M. (1980) Role of *rec*BC nuclease in *Escherichia coli* transformation. *J. Bacteriol.* **143**, 1031–2.

Hoffman E. P., Monaco A. P., Feener C. C. & Kunkel L. M. (1987) Conservation of the Duchenne Muscular Dystrophy gene in mice and humans. *Science* **238**, 347–50.

Hogan B. & Lyons K. (1988) Gene targeting: getting nearer the mark. *Nature* **336**, 304–5.

Hohn B. (1975) DNA as substrate for packaging into bacteriophage lambda, *in vitro. J. Mol. Biol.* **98**, 93–106.

Hohn B. (1979) *In vitro* packaging of lambda and cosmid DNA. *Methods enzymol.* **68**, 299–309.

Hohn B. & Murray K. (1977) Packaging recombinant DNA molecules into bacteriophage particles *in vitro. Proc. Natl. Acad. Sci. USA* **74**, 3259–63.

Hohn T., Hohn B. & Pfeiffer P. (1985) Reverse transcription in CaMV. *TIBS*, 205–9.

Holben W. E. & Tiedje (1988) Applications of nucleic acid hybridization in microbial ecology. *Ecology* **69**, 561–8.

Holsters M., Silva B., Van Vliet F., Genetello C., De Block M., Dhaese P., Depicker A., Inze D., Engler G., Villarroel R., Van Montagu M. & Schell J. (1980) The functional organization of the nopaline *A. tumefaciens* plasmid pTi C58. *Plasmid* **3**, 212–30.

Hooykaas-Van Slogteren G. M. S., Hooykaas P. J. J. & Schilperoort R. A. (1984) Expression of Ti plasmid genes in monocotyledonous plants infected with *Agrobacterium tumefaciens. Nature* **311**, 763–4.

Hopkins A. S., Murray N. E. & Brammar W. J. (1976) Characterization of λtrp-transducing bacteriophages made *in vitro. J. Mol. Biol.* **107**, 549–69.

Hopwood D. A., Bibb M. J., Chater K. F. & Kieser T. (1987) Plasmid and phage vectors for gene cloning and analysis in *Streptomyces. Methods Enzymol.* **153**, 116–66.

Hopwood D. A., Lydiate D. J., Malpartida F. & Wright H. M. (1985a) Conjugative sex plasmids of *Streptomyces. Basic Life Sci.* **30**, 615–34.

Hopwood D. A., Malpartida F., Kieser H. M., Iheda H., Duncan J., Fujii I., Rudd B. A. M., Floss H. G. & Omura S. (1985b) Production of "hybrid" antibiotics by genetic engineering. *Nature* **314**, 642–4.

Horsch R. B., Fraley R. T., Rogers S. G., Sanders P. R., Lloyd A. & Hoffmann N.

(1984) Inheritance of functional genes in plants. *Science* **223**, 496—8.

Horsch R. B., Fry J. E., Hoffmann N. L., Eicholtz D., Rogers S. G. & Fraley R. T. (1985) A simple and general method for transferring genes into plants. *Science* **227**, 1229—31.

Hosieth S. K. & Stocker B. A. D. (1981) Aromatic-dependent *Salmonella typhimurium* are non-virulent and effective as live vaccines. *Nature* **291**, 238—9.

Houghton M., Eaton M. A. W., Stewart A. G., Smith J. C., Doel S. M., Catlin G. H., Lewis H. M., Patel T. P., Emtage J. S., Carey N. H. & Porter A. G. (1980) The complete amino acid sequence of human fibroblast interferon as deduced using synthetic oligodeoxyribonucleotide primers of reverse transcriptase. *Nucleic Acids Res.* **8**, 2885—94.

Howell S. H., Walker L. L. & Dudley R. K. (1980) Cloned cauliflower mosaic virus DNA infects turnips (*Brassica rapa*). *Science* **208**, 1265—7.

Hsiao C. L. & Carbon J. (1981) Characterization of a yeast replication origin (ars2) and construction of stable minichromosomes containing cloned yeast centromere DNA (CEN 3). *Gene* **15**, 157—66.

Hu N. & Messing J. (1982) The making of strand-specific M13 probes. *Gene* **17**, 271—7.

Hu S.-L., Kosowski S. P. & Dalrymple J. M. (1986) Expression of AIDS virus envelope gene by a recombinant vaccinia virus. *Nature* **320**, 537—40.

Hughes S. & Kosick E. (1984) Mutagenesis of the region between *env* and *src* of the SR-A strain of Rous sarcoma virus for the purpose of constructing helper-independent vectors. *Virology* **136**, 89—99.

Hui A., Hayflick J., Dinkelspiel K. & de Boer H. A. (1984) Mutagenesis of the three bases preceding the start codon of the β-galactosidase mRNA and its effect on translation in *Escherichia coli*. *EMBO J.* **3**, 623—9.

Hull R. & Al-Hakim A. (1988) Nucleic acid hybridization in plant virus diagnosis and characterization. *Trends Biotechnol.* **6**, 213—18.

Hull R. & Covey S. N. (1983a) Replication of cauliflower mosaic virus DNA. *Sci. Prog.* **68**, 403—22.

Hull R. & Covey S. N. (1983b) Does cauliflower mosaic virus replicate by reverse transcription? *Trends Biol. Sci.* **8**, 119—21.

Humphries P., Old R., Coggins, L. W., McShane T., Watson C. & Paul J. (1978) Recombinant plasmids containing *Xenopus laevis* structural genes derived from complementary DNA. *Nucleic Acids Res.* **5**, 905—24.

Huttner K. M., Barbosa J. A., Scangos G. A., Pratchera D. D. & Ruddle F. H. (1981) DNA-mediated gene transfer without carrier DNA. *J. Cell. Biol.* **91**, 153—6.

Iida S., Meyer J., Bachi B., Stalhammer-Carlemalm M., Schrickel S., Bickle T. A. & Arber W. (1982) DNA restriction-modification genes of phage P1 and plasmid p15B. Structure and *in vitro* transcription. *J. Mol. Biol.* **165**, 1—18.

Ikemura T. (1981a) Correlation between the abundance of *Escherichia coli* transfer RNAs and the occurrence of the respective codons in its protein genes. *J. Mol. Biol.* **146**, 1—21.

Ikemura T. (1981b) Correlation between the abundance of *Escherichia coli* transfer RNAs and the occurrence of the respective codons in its protein genes. A proposal for a synonymous codon choice that is optimal for the *E. coli* translational system. *J. Mol. Biol.* **151**, 389—409.

Imanaka T., Takagaki K. & Aiba S. (1986) Construction of high, intermediate and low-copy promoter-probe plasmids for *Bacillus subtilis*. *Gene* **43**, 231—6.

Ingham P. (1988) The molecular genetics of embryonic pattern formation in

Drosophila. Nature **335**, 25−34.

Inouye H. & Beckwith J. (1977) Synthesis and processing of an *Escherichia coli* alkaline phosphatase precursor *in vitro. Proc. Natl. Acad. Sci. USA* **74**, 1440−4.

Iserentant D. & Fiers W. (1980) Secondary structure of mRNA and efficiency of translation initiation. *Gene* **9**, 1−12.

Ish-Horowicz D. & Burke J. F. (1981) Rapid and efficient cosmid cloning. *Nucleic Acids Res.* **9**, 2989−98.

Itakura K., Hirose T., Crea R., Riggs A. D., Heyneker H. L., Bolivar F. & Boyer H. W. (1977) Expression in *Escherichia coli* of a chemically synthesized gene for the hormone somatostatin. *Science* **198**, 1056−63.

Ito K., Bassford P. J. & Beckwith J. (1981) Protein localization in *E. coli*: is there a common step in the secretion of periplasmic and outer-membrane proteins? *Cell* **24**, 707−17.

Jabbar M. A., Sivasubramanian N. & Nayak D. P. (1985) Influenza viral (A/WSN/33) hemagglutinin is expressed and glycosylated in the yeast *Saccharomyces cerevisiae. Proc. Natl. Acad. Sci. USA* **82**, 2019−23.

Jablonski E., Moomaw E. W., Tullis R. H. & Ruth J. L. (1986) Preparation of oligodeoxynucleotide-alkaline phosphatase conjugates and their use as hybridization probes. *Nucleic Acids Res.* **14**, 6115−28.

Jackson D. A., Symons R. H. & Berg P. (1972) Biochemical method for inserting new genetic information into DNA of Simian virus 40: circular SV40 DNA molecules containing lambda phage genes and the galactose operon of *Escherichia coli. Proc. Natl. Acad. Sci. USA* **69**, 2904−9.

Jacob A. E., Cresswell J. M., Hedges R. W., Coetzee J. N. & Beringer J. E. (1976) Properties of plasmids constructed by *in vitro* insertion of DNA from *Rhizobium leguminosarum* or *Proteus mirabilis* into RP4. *Mol. Gen. Genet.* **147**, 315−23.

Jacobs E., Dewerchin M. & Boeke J. D. (1988) Retrovirus-like vectors for *Saccharomyces cerevisiae*: integration of foreign genes controlled by efficient promoters into yeast chromosomal DNA. *Gene* **67**, 259−69.

Jacobs W. R., Tuckman M. & Bloom B. R. (1987) Introduction of foreign DNA into mycobacteria using a shuttle plasmid. *Nature* **327**, 532−5.

Jaenisch R. (1988) Transgenic animals. *Science* **240**, 1468−74.

Jaenisch R. & Mintz B. (1974) Simian virus 40 DNA sequences in DNA of healthy adult mice derived from preimplantation blastocysts injected with viral DNA. *Proc. Natl. Acad. Sci. USA* **71**, 1250−4.

Janniere L., Niaudet B., Pierre E. & Ehrlich S. D. (1985) Stable gene amplification in the chromosome of *Bacillus subtilis. Gene* **40**, 47−55.

Jeffreys A. J., Brookfield J. F. Y. & Semenoff R. (1985b) Positive identification of an immigration test case using human DNA fingerprints. *Nature* **317**, 577−9.

Jeffreys A. J. & Morton D. B. (1987) DNA fingerprints of cats and dogs. *Animal Genet.* **18**, 1−15.

Jeffreys A. J., Wilson V. & Thein S. L. (1985a) Individual-specific 'fingerprints' of human DNA. *Nature* **316**, 76−9.

Jen G. C. & Chilton M. D. (1986) The right border region of pTiT37 T-DNA is intrinsically more active than the left border region in promoting T-DNA transformation. *Proc. Natl. Acad. Sci. USA* **83**, 3895−9.

Jensen J. S., Marcker K. A. Otten L. & Schell J. (1986) Nodule-specific expression of a chimaeric soybean leghaemoglobin gene in transgenic *Lotus corniculatus. Nature* **321**, 669−74.

Jimenez A. & Davies J. (1980) Expression of a transposable antibiotic resistance element in *Saccharomyces. Nature* **287**, 869–71.

Jones D. & Errington J. (1987) Construction of improved bacteriophage φ105 vectors for cloning by transfection in *Bacillus subtilis. J. Gen. Microbiol.* **133**, 483–92.

Jones I. M., Primrose S. B. & Ehrlich S. D. (1982) Recombination between short direct repeats in a *recA* host. *Mol. Gen. Genet.* **188**, 486–9.

Jones I. M., Primrose S. B., Robinson A. & Ellwood D. C. (1980) Maintenance of some Col E1-type plasmids in chemostat culture. *Mol. Gen. Genet.* **180**, 579–84.

Jones K. & Murray K. (1975) A procedure for detection of heterologous DNA sequences in lambdoid phage by *in situ* hybridization. *J. Mol. Biol.* **51**, 393–409.

Jones L., Ristow S., Yilma T. & Moss B. (1986) Accidental human vaccination with vaccinia virus expressing nucleoprotein gene. *Nature* **319**, 543.

Josephson S. & Bishop R. (1988) Secretion of peptides from *E. coli*: a production system for the pharmaceutical industry. *Trends Biotechnol.* **6**, 218–24.

Julius D. J., Blair L. C., Brake A. J., Sprague G. F. & Thurner J. (1983) Yeast alpha-factor is processed from a larger precursor polypeptide: the essential role of a membrane-bound dipeptidyl amino-peptidase. *Cell* **32**, 839–52.

Julius D., Scheckman R. & Thorner J. (1984) Glycosylation and processing of prepro-alpha-factor through the yeast secretory pathway. *Cell* **36**, 309–18.

Kan Y. W. & Dozy A. M. (1978) Polymorphisms of DNA sequence adjacent to human β-globin structural gene: relation to sickle mutation. *Proc. Natl. Acad. Sci. USA* **75**, 5631–5.

Kaper J. B., Lockman H., Baldini M. M. & Levine M. M. (1984) A recombinant live oral cholera vaccine. *Biotechnology* **1**, 345–9.

Karess R. E. & Rubin G. M. (1984) Analysis of P transposable element functions in *Drosophila. Cell* **38**, 135–46.

Karn J., Brenner S., Barnett L. & Cesareni G. (1980) Novel bacteriophage λ cloning vector. *Proc. Natl. Acad. Sci. USA* **77**, 5172–6.

Kaufman R. J. & Sharp P. A. (1982) Construction of a modular dihydrofolate reductase cDNA gene: analysis of signals utilized for efficient expression. *Mol. Cell. Biol.* **9**, 1304–19.

Kaufman R. J., Walsey L. C., Spiliotes A. J., Gossels S. D., Latt S. A., Larsen G. R. & Kay R. M. (1985) Coamplification and coexpression of human tissue-type plasminogen activator and murine dihydrofolate reductase sequences in Chinese hamster ovary cells. *Mol. Cell. Biol.* **5**, 1750–9.

Keggins K. M., Lovett P. S. & Duvall E. J. (1978) Molecular cloning of genetically active fragments of *Bacillus* DNA in *Bacillus subtilis* and properties of the vector plasmid pUB110. *Proc. Natl. Acad. Sci. USA* **75**, 1423–7.

Keilty S. & Rosenberg M. (1987) Constitutive function of a positively regulated promotor reveals new sequences essential for activity. *J. Biol. Chem.* **262**, 6389–95.

Kelley W. S., Chalmers K. & Murray N. E. (1977) Isolation and characterization of a λ*polA* transducing phage. *Proc. Natl. Acad. Sci. USA* **74**, 5632–6.

Kelly J. M. & Hynes M. J. (1987) Multiple copies of the *amd*S gene of *Aspergillus nidulans* cause titration of *trans*-activity regulatory proteins. *Current Genet.* **12**, 21–31.

Kelly T. J. & Smith H. O. (1970) A restriction enzyme from *Hemophilus influenzae*. II. Base sequence of the recognition site. *J. Mol. Biol.* **51**, 393–409.

Kemp D. J., Coppel R. L., Cowman A. F., Saint R. B., Brown G. V. & Anders R. F. (1983) Expression of *Plasmodium falciparum* blood-stage antigens in *Escherichia coli*: detection with antibodies from immune humans. *Proc.*

Natl. Acad. Sci. USA **80**, 3787–91.

Kempe T., Sundqvist W. I., Chow F. & Hu S. L. (1985) Chemical and enzymatic biotin-labeling of oligodeoxyribonucleotides. *Nucleic Acids Res.* **13**, 45–57.

Kessler C., Neumaier P. S. & Wolf W. (1985) Recognition sequences of restriction endonucleases and methylases—a review. *Gene* **33**, 1–102.

Khoury G. & Gruss P. (1983) Enhancer elements. *Cell* **33**, 313–14.

Kinashi H., Shimaji M. & Sakai A. (1987) Giant linear plasmids in *Streptomyces* which code for antibiotic biosynthesis genes. *Nature* **328**, 454–6.

Kingsman A. J., Clarke L., Mortimer R. K. & Carbon J. (1979) Replication in *Saccharomyces cerevisae* of plasmid pBR313 carrying DNA from the yeast *trp* 1 region. *Gene* **7**, 141–52.

Kingsman S. M. & Kingsman A. J. (1988) *Genetic Engineering: an introduction to gene analysis and exploitation in eukaryotes.* Blackwell Scientific Publications, Oxford.

Klee H., Montoya A., Horodyski F., Lichtenstein C. & Garfinkel D. (1984) Nucleotide sequence of the *tms* genes of the pTiA6NC octopine Ti plasmid: two gene products in plant tumorigenesis. *Proc. Natl. Acad. Sci. USA* **81**, 1728–32.

Kleid D. G., Yansura D., Small B., Dowbenko D., Moore D. M., Grubman M. J., McKercher P. D., Morgan D. O., Robertson B. H. & Bachrach H. L. (1981) Cloned viral protein vaccine for foot-and-mouth disease: responses in cattle and swine. *Science* **214**, 1125–9.

Klein T. M., Wolf E. D., Wu R. & Sanford J. C. (1987) High-velocity microprojectiles for delivering nucleic acids into living cells. *Nature* **327**, 70–3.

Klotsky R.-A. & Schwartz I. (1987) Measurement of *cat* expression from growth-rate-regulated promoters employing β-lactamase activity as an indicator of plasmid copy number. *Gene* **55**, 141–6.

Koekman B. P., Ooms G., Klapwijk P. M. & Schilperoort R. A. (1979) Genetic map of an octopine Ti plasmid. *Plasmid* **2**, 347–57.

Kogan S. C., Doherty M. & Gitschier J. (1987) An improved method for prenatal diagnosis of genetic diseases by analysis of amplified DNA sequences. *N. Engl. J. Med.* **317**, 985–90.

Kollias G., Wrighton N., Hurst J., Grosveld F. (1986) Regulated expression of human $^A\gamma$-, β-, and hybrid γβ-globin genes in transgenic mice: manipulation of the developmental expression patterns. *Cell* **46**, 89–94.

Kondo J. K. & McKay L. L. (1985) Gene transfer systems and molecular cloning in group N streptococci: a review. *J. Dairy Sci.* **68**, 2143–59.

Kondoh H., Yasuda K. & Okada T. S. (1983) Tissue-specific expression of a cloned chick δ-crystallin gene in mouse cells. *Nature* **301**, 440–2.

Kornberg A. (1980) *DNA Replication.* W. H. Freeman & Co., San Francisco.

Koshland D. & Bostein D. (1982) Evidence of post-translation translocation of β-lactamase across the bacterial inner membrane. *Cell* **30**, 893–902.

Koukolikova-Nicola Z., Shillito R. D., Hohn B., Wang K., Van Montagu M. & Zambryski P. (1985) Involvement of circular intermediates in the transfer of T-DNA from *Agrobacterium tumefaciens* to plant cells. *Nature* **313**, 191–6.

Kramer R. A., Cameron J. R. & Davis R. W. (1976) Isolation of bacteriophage λ containing yeast ribosomal RNA genes: screening by *in situ* RNA hybridization to plaques. *Cell* **8**, 227–32.

Kramer B., Kramer W. & Fritz H.-J. (1984a) Different base/base mismatches are corrected with different efficiencies by the methyl-directed DNA mismatch-repair system of *E. coli*. *Cell* **38**, 879–88.

Kramer W., Drutsa V., Jansen H.-W., Kramer B., Pflugfelder M. & Fritz H.-J.

(1984b) The gapped duplex DNA approach to oligonucleotide-directed mutation construction. *Nucleic Acids Res.* **12**, 9441–56.

Kreft J. & Hughes C. (1981) Cloning vectors derived from plasmids and phage of *Bacillus. Curr. Top. Microbiol. Immunol.* **96**, 1–17.

Krieg P. A. & Melton D. A. (1984a) Formation of the 3' end of histone mRNA by post-transcriptional processing. *Nature* **308**, 203–6.

Krieg P. A. & Melton D. A. (1984b) Functional messenger RNAs are produced by SP6 *in vitro* transcription of cloned cDNAs. *Nucleic Acids Res.* **12**, 7071.

Krieg P. A. & Melton D. A. (1985) Developmental regulation of a gastrula-specific gene injected into fertilized *Xenopus* eggs. *EMBO J.* **4**, 3464–71.

Krisch H. M. & Selzer G. B. (1981) Construction and properties of a recombinant plasmid containing gene *32* of bacteriophage T4D. *J. Mol. Biol.* **148**, 199–218.

Kroyer J. & Chang S. (1981) The promoter-proximal region of the *Bacillus licheniformis* penicillinase gene. Nucleotide sequence and predicted leader peptide sequence. *Gene* **15**, 343–7.

Kruger D. H., Schroeder C., Reuter M., Bogdarina I. G., Buryanov Y. I. & Bickle T. A. (1985) DNA methylation of bacterial viruses T3 and T7 by different DNA methylases in *Escherichia coli* K12 cells. *Eur. J. Biochem.* **150**, 323–30.

Kuehn M. R., Bradley A., Robertson E. J. and Evans M. J. (1987) A potential model for Lesch–Nyhan syndrome through introduction of HPRT mutations in mice. *Nature* **326**, 295–8.

Kukuruzinska M. A. & Robbins P. W. (1987) Protein glycosylation in yeast: transcript heterogeneity of the ALG7 gene. *Proc. Natl. Acad. Sci. USA* **84**, 2145–9.

Kunkel T. A. (1985) Rapid and efficient site-specific mutagenesis without phenotypic selection. *Proc. Natl. Acad. Sci. USA* **82**, 488–92.

Kuo C.-L. & Campbell J. L. (1983) Cloning of *Saccharomyces cerevisiae* DNA replication genes: isolation of the CDC8 gene and two genes that compensate for the cdc8–1 mutation. *Mol. Cell. Biol.* **3**, 1730–7.

Kurjan J. & Herskowitz I. (1982) Structure of a yeast pheromone (MF alpha). A putative alpha factor precursor contains four tandem copies of mature alpha factor. *Cell* **30**, 933–43.

Kurland C. G. (1987) Strategies for efficiency and accuracy in gene expression. 1. The major codon preference: a growth optimization strategy. *Trends Biochem. Sci.* **12**, 126–8.

Kuroda K., Hauser C., Rott R., Klenk H.-D. & Doerfler W. (1986) Expression of the influenza virus haemagglutinin in insect cells by a baculovirus vector. *EMBO J.* **5**, 1359–65.

Kurtz D. T. & Nicodemus C. F. (1981) Cloning of $\alpha_{2\mu}$ globulin cDNA using a high efficiency technique for the cloning of trace messenger RNAs. *Gene* **13**, 145–152.

Lacy E., Roberts S., Evans E. P., Burtenshaw M. D. & Constantini F. D. (1983) A foreign β-globin gene in transgenic mice: integration at abnormal chromosomal positions and expression in inappropriate tissues. *Cell* **34**, 343–58.

Lamond A. I. & Travers A. A. (1983) Requirement for an upstream element for optimal transcription of a bacterial tRNA gene. *Nature* **304**, 248–50.

Land H., Grey M., Hanser H., Lindenmaier W. & Schutz G. (1981) 5'-Terminal sequences of eucaryotic mRNA can be cloned with a high efficiency. *Nucleic Acids Res.* **9**, 2251–66.

Lane C. D., Colman A., Mohun T., Morser J., Champion J., Kourides I., Craig

R., Higgins S., James T. C., Appelbaum S. W., Ohlsson R. I., Pauch E., Houghton M., Matthews J. & Miflin B. J. (1980) The *Xenopus* oocyte as a surrogate secretory system. The specificity of protein export. *Eur. J. Biochem.* **111**, 225–35.

Langer P. R., Waldrop A. A. & Ward D. C. (1981) Enzymatic synthesis of biotin-labeled polynucleotides: novel nucleic acid affinity probes. *Proc. Natl. Acad. Sci. USA* **78**, 6633–7.

Langford C. J. & Gallwitz (1983) Evidence for an intron-contained sequence required for the splicing of yeast RNA polymerase II transcripts. **33**, 519–27.

Langford C. J., Klinz F.-J., Donath C. & Gallwitz D. (1984) Point mutations identify the conserved intron-contained TACTAAC box as an essential splicing signal sequence in yeast. *Cell* **36**, 645–53.

Langford C., Nellen W., Niessing J. & Gallwitz D. (1983) Yeast is unable to excise foreign intervening sequences from hybrid gene transcripts. *Proc. Natl. Acad. Sci. USA* **80**, 1496–500.

Lapeyre B. & Amalric F. (1985) A powerful method for the preparation of cDNA libraries: isolation of cDNA encoding a 100-kDal nucleolar protein. *Gene* **37**, 215–20.

Larsen J. E. L., Gerdes K., Light J. & Molin S. (1984) Low-copy-number plasmid cloning vectors amplifiable by derepression of an inserted foreign promoter. *Gene* **28**, 45–54.

Laskey R. A., Harland R. M. & Mechali M. (1983) Induction of chromosome replication during maturation of amphibian oocytes. In *Molecular Biology of Egg Maturation*. Ciba Symposium, pp. 25–36. Pitman, London.

Laski F. A., Rio D. C. & Rubin G. M. (1986) Tissue specificity of *Drosophila* P element transposition is regulated at the level of mRNA splicing. *Cell* **44**, 7–19.

Law M.-F., Byrne J. & Hawley P. M. (1983) A stable bovine papillomavirus hybrid plasmid that expresses a dominant selective trait. *Mol. Cell Biol.* **3**, 2110–15.

Leach D. R. F. & Stahl F. (1983) Viability of lambda phages carrying a perfect palindrome in the absence of recombination nucleases. *Nature* **305**, 448–51.

Leder P., Tiemeier D. & Enquist L. (1977) EK2 derivatives of bacteriophage lambda useful in the cloning of DNA from higher organisms: the λgt WES system. *Science* **196**, 175–7.

Lederberg S. & Meselson M. (1964) Degradation of non-replicating bacteriophage DNA in non-accepting cells. *J. Mol. Biol.* **8**, 623–8.

Lee F., Mulligan R., Berg P. & Ringold G. (1981) Glucocorticoids regulate expression of dihydrofolate reductase cDNA in mouse mammary tumour virus chimaeric plasmids. *Nature* **294**, 228–32.

Leemans J., Deblaere R., Willmitzer L., De Greve H., Hernalsteens J. P., Van Montagu M. & Schell J. (1982a) Genetic identification of functions of T_L–DNA transcripts in octopine crown galls. *EMBO J.* **1**, 147–52.

Leemans J., Langenakens J., De Greve H., Deblaere R., Van Montagu M. & Schell J. (1982b) Broad-host-range cloning vectors derived from the W-plasmid Sa. *Gene* **19**, 361–4.

Le Hegarat J. C. & Anagnostopoulos C. (1977) Detection and characterization of naturally occurring plasmids in *Bacillus subtilis*. *Mol. Gen. Genet.* **157**, 164–74.

Lehtovaara P., Ulmanen I. & Palva I. (1984) *In vivo* transcription initiation and termination sites of an α-amylase gene from *Bacillus amyloliquefaciens* cloned in *Bacillus subtilis*. *Gene* **30**, 11–16.

Lemmers M., De Beuckeleer M., Holsters M., Zambryski P., Hernalsteens J. P., Van Montagu M. & Schell J. (1980) Internal organization, boundaries and integration of Ti plasmid DNA in nopaline crown gall tumours. *J. Mol. Biol.* **144**, 353–76.

Lis J. T., Simon J. A. & Sutton C. A. (1983) New heat shock puffs and β-galactosidase activity resulting from transformation of *Drosophila* with an *hsp 70-lac Z* hybrid gene. *Cell* **35**, 403–10.

Lobban P. E. & Kaiser A. D. (1973) Enzymatic end-to-end joining of DNA molecules. *J. Mol. Biol.* **78**, 453–71.

Loenen W. A. M. & Brammar W. J. (1980) A bacteriophage lambda vector for cloning large DNA fragments made with several restriction enzymes **10**, 249–59.

Loison G., Findel A., Bernard S., Nguyen-Juilleret M., Marquet M., Rienl-Bellon N., Carvallo D., Guerra-Santos L., Brown S. W., Courtney M., Roitsch C. & Lemoine Y. (1988) Expression amd secretion in *S. cerevisiae* of biologically active leech hirudin. *Biotechnology* **6**, 72–7.

Lorz H., Baaker B. & Schell J. (1985) Gene transfer to cereal cells mediated by protoplast transformation. *Mol. Gen. Genet.* **199**, 178–182.

Lowe R. S., Keller P. M., Keeck B. J., Davison A. J., Whang Y., Morgan A. J., Kieff E. & Ellis R. W. (1987) Varicella-zoster as a live vector for the expression of foreign genes. *Proc. Natl. Acad. Sci. USA* **84**, 3896–900.

Lowy I., Pellicer A., Jackson J. F., Sim G. K., Silverstein S. & Axel R. (1980) Isolation of transforming DNA: Cloning of the hamster hprt gene. *Cell* **22**, 817–23.

Lu A. L., Clark S. & Modrich P. (1983) Methyl-directed repair of DNA base-pair mismatches *in vitro*. *Proc. Natl. Acad. Sci. USA* **80**, 4639–43.

Lusky M. & Botchan M. (1981) Inhibitory effect of specific pBR322 DNA sequences upon SV40 replication in simian cells. *Nature* **293**, 79–81.

Lydiate D. J., Malpartida F. & Hopwood D. A. (1985) The *Streptomyces* plasmid SCP2*: its functional analysis and development into useful cloning vectors. *Gene* **35**, 223–35.

McClarin J. A., Frederick C. A., Wang B.-C., Green P., Boyer H. W., Grable J. & Rosenberg J. M. (1986) Structure of the DNA-*Eco* RI endonuclease recognition complex at 3 A resolution. *Science* **234**, 1526–41.

McClelland M., Kessler L. G. & Bittner M. (1984) Site-specific cleavage of DNA at 8- and 10-base-pair sequences. *Proc. Natl. Acad. Sci. USA* **81**, 983–7.

McLaughlin J. R., Murray C. L. & Rabinowitz J. C. (1981) Unique features in the ribosome binding site sequence of the Gram-positive *Staphylococcus aureus* β-lactamase gene. *J. Biol. Chem.* **256**, 11283–91.

McPherson D. T. (1988) Codon preference reflects mistranslational constructs: a proposal. *Nucleic Acids Res.* **16**, 4111–20.

Ma H., Kunes S., Schatz P. J. & Botstein D. (1987) Plasmid construction by homologous recombination in yeast. *Gene* **58**, 201–16.

Maeda S., Kawai T., Obinata M., Fujiwara H., Horiuchi T., Saeki Y., Sato Y. & Furusawa M. (1985) Production of human α-interferon in silkworm using a baculovirus vector. *Nature* **315**, 592–4.

Mandel M. & Higa A. (1970) Calcium-dependent bacteriophage DNA infection. *J. Mol. Biol.* **53**, 159–62.

Maniatis T., Goodbourn S. & Fischer J. A. (1987) Regulation of inducible and tissue-specific gene expression. *Science* **236**, 1237–45.

Maniatis T., Hardison R. C., Lacy E., Lauer J., O'Connell C., Quon D., Sim G. K. & Efstratiadis A. (1978) The isolation of structural genes from libraries of eucaryotic DNA. *Cell* **15**, 687–701.

Maniatis T., Sim Gek Kee, Efstratiadis A. & Kafatos F. C. (1976) Amplification and characterization of a β-globin gene synthesized *in vitro*. *Cell* **8**, 163–82.

Mann R. S., Mulligan R. C. & Baltimore D. (1983) Construction of a retrovirus packaging mutant and its use to produce helper-free defective retrovirus. *Cell* **32**, 871–9.

Mannhaupt G., Pilz U. & Feldmann H. (1988) A series of shuttle vectors using chloramphenicol acetyltransferase as a reporter enzyme in yeast. *Gene* **67**, 287–94.

Mansour S. L., Thomas K. R. & Capecchi M. R. (1988) Disruption of the proto-oncogene int-2 in mouse embryo-derived stem cells: a general strategy for targeting mutations to non-selectable genes. *Nature* **336**, 348–52.

Mantell S. H., Matthews J. A. & McKee R. A. (1985) *Principles of Plant Biotechnology*. Blackwell Scientific Publications, Oxford.

Marini N., Hiiyanna K. T. & Benbow R. M. (1989) Differential replication of circular DNA molecules co-injected into early *Xenopus laevis* embryos. *Nucleic Acids Res.* **17**, 5793–808.

Marinus M. G., Carraway M., Frey A. Z., Brown L. & Arraj J. A. (1983) Insertion mutations in the *dam* gene of *Escherichia coli* K-12. *Mol. Gen. Genet.* **192**, 288–9.

Masu Y., Nakayama K., Tamaki H., Harada Y., Kuno M. & Nakanishi S. (1987) cDNA cloning of bovine substance-K receptor through oocyte expression system. *Nature* **329**, 836–8.

Masui Y., Mizuno T. & Inouye M. (1984) Novel high-level expression cloning vehicles: 10^4-fold amplification of *Escherichia coli* minor protein. *Biotechnology* **2**, 81–5.

Matzke A. J. M. & Chilton M.-D. (1981) Site-specific insertion of genes into T-DNA of the *Agrobacterium* tumour-inducing plasmid: an approach to genetic engineering of higher plant cells. *J. Mol. Appl. Genet.* **1**, 39–49.

Maxam A. M. & Gilbert W. (1977) A new method for sequencing DNA. *Proc. Natl. Acad. Sci. USA* **74**, 560–64.

Maxam A. M. & Gilbert W. (1980) Sequencing end-labelled DNA with base-specific chemical cleavages. *Methods Enzymol.* **65**, 499–560.

Meagher R. B., Tait R. C., Betlach M. & Boyer H. W. (1977) Protein expression in *E. coli* minicells by recombinant plasmids. *Cell* **10**, 521–36.

Mellon P., Parker V., Gluzman Y. & Maniatis T. (1981) Identification of DNA sequences required for transcription of the human α-globin gene in a new SV40 host-vector system. *Cell* **27**, 279–88.

Mellor J., Dobson M. J., Roberts M. A., Tuite M. F., Emtage J. S., White S., Lowe P. A., Patel T., Kingsman A. J. & Kingsman S. M. (1983) Efficient synthesis of enzymatically active calf chymosin in *Saccharomyces cerevisiae*. *Gene* **24**, 1–14.

Mellor J., Malim M., Gull K., Tuite M. F., McCready S., Sibbayawan T., Kingsman S. M. & Kingsman A. J. (1985) Reverse transcriptase activity and Ty RNA are associated with virus-like particles in yeast. *Nature* **318**, 583–6.

Melton D. A., Krieg P. A., Rebaglich M. R., Maniatis T., Zinn K. & Green M. R. (1984) Efficient *in vitro* synthesis of biologically active RNA and RNA hybridization probes from plasmids containing a bacteriophage SP6 promoter. *Nucleic Acids Res.* **12**, 7035–56.

Mercerau-Puijalon O., Royal A., Cami B., Garapin A., Krust A., Gannon F. & Kourilsky P. (1978) Synthesis of an ovalbumin-like protein by *Escherichia coli* K12 harbouring a recombinant plasmid. *Nature* **275**, 505–10.

Mermod N., Ramos J. L., Lehrbach P. R. & Timmis K. N. (1986) Vector for regulated expression of cloned genes in a wide range of Gram-negative bacteria. *J. Bacteriol.* **167**, 447–54.

Mertz J. E. & Berg P. (1974) Defective simian virus 40 genomes. Isolation and growth of individual clones. *Virology* **62**, 112–24.

Mertz J. E. & Gurdon J. B. (1977) Purified DNAs are transcribed after micro-injection into *Xenopus* oocytes. *Proc. Natl. Acad. Sci. USA* **74**, 1502–6.

Meselson M. & Yuan R. (1968) DNA restriction enzyme from *E. coli*. *Nature* **217**, 1110–14.

Messing J., Crea R. & Seeburg P. H. (1981) A system for shotgun DNA sequencing. *Nucleic Acids Res.* **9**, 309–21.

Messing J., Gronenborn B., Muller-Hill B. & Hofschneider P. H. (1977) Filamentous coliphage M13 as a cloning vehicle: insertion of a *Hin*d II fragment of the *lac* regulatory region in M13 replicative form *in vitro*. *Proc. Natl. Acad. Sci. USA* **74**, 3642–6.

Messing J. & Vieira J. (1982) A new pair of M13 vectors for selecting either DNA strand of double-digest restriction fragments. *Gene* **19**, 269–76.

Meyer P., Heidmann I., Forkmann G. & Saedler H. (1987) A new petunia flower colour generated by transformation of a mutant with a maize gene. *Nature* **330**, 677–8.

Michel B., Niaudet B. & Ehrlich S. D. (1982) Intramolecular recombination during plasmid transformation of *Bacillus subtilis* competent cells. *EMBO J.* **1**, 1565–71.

Michel B., Palla E., Niaudet B. & Ehrlich S. D. (1980) DNA cloning in *Bacillus subtilis*. III. Efficiency of random-segment cloning and insertional inactivation vectors. *Gene* **12**, 147–54.

Michelson A. M. & Orkin S. H. (1982) Characterizaton of the homopolymer tailing reaction catalyzed by terminal deoxynucleotidyl transferase. Implications for the cloning of cDNA. *J. Biol. Chem.* **256**, 1473–82.

Miller A. D., Law M. F. & Verma I. M. (1985) Generation of helper-free amphitropic retroviruses that transduce a dominant-acting, methotrexate-resistant dihydrofolate reductase gene. *Mol. Cell. Biol.* **5**, 431–7.

Miller C. A., Tucker W. T., Meacock P. A., Gustafsson P. & Cohen S. N. (1983) Nucleotide sequence of the partition locus of *Escherichia coli* plasmid pSC101. *Gene* **24**, 309–15.

Miller C. G., Strauch K. L., Kukral A. M., Miller J. L., Wingfield P. T., Mazzei G. J., Werlen R. C., Graber P. & Movva N. R. (1987) N-terminal methionine-specific peptidase in *Salmonella typhimurium*. *Proc. Natl. Acad. Sci. USA* **84**, 2718–22.

Miller O. K. & Temin H. M. (1983) High efficiency ligation and recombination of DNA fragments by vertebrate cells. *Science* **200**, 606–9.

Miller D. W., Safer P. & Miller L. K. (1987) An insect baculovirus host–vector system for high level expression of foreign genes. In *Genetic Engineering*, vol. 8, eds Setler J. K. & Hollaender A. Plenum Press, New York.

Miller H. I., Henzel W. J., Ridgeway J. B., Kuang W.-J., Chisholm V. & Liu C. C. (1989) Cloning and expression of a yeast ubiquitin-protein cleaving activity in *Escherichia coli*. *Biotechnology* **7**, 698–704.

Mitraki A. & King J. (1989) Protein folding intermediates and inclusion body formation. *Biotechnology* **7**, 690–7.

Miyajima A., Otsu K., Schreurs J., Bond M. W., Abrams J. S. & Arai K. (1986) Expression of murine and human granulocyte-macrophage colony-stimulating factors in *S. cerevisiae*: mutagenesis of the potential glycosylation sites. *EMBO J.* **5**, 1193–7.

Miyamoto C., Smith G. E., Farrell-Towt J., Chizzonite R., Summers M. D. & Ju G. (1985) Production of human *c-myc* protein in insect cells infected with baculovirus expression vector. *Mol. Cell. Biol.* **5**, 2860–5.

Moerschell R. P., Tsunasawa S. & Sherman F. (1988) Transformation of yeast with synthetic oligonucleotides. *Proc. Natl. Acad. Sci. USA* **85**, 524–8.

Moir A. & Brammar W. J. (1976) Use of specialized transducing phages in amplification of enzyme production. *Mol. Gen. Genet.* **149**, 87–99.

Moir D. T. & Dumais D. R. (1987) Glycosylation and secretion of human alpha-1-antitrypsin of yeast. *Gene* **56**, 209–17.

Montaya A. L., Chilton M.-D., Gordon M. P., Sciaky D. & Nester E. W. (1977) Octopine and nopaline metabolism in *Agrobacterium tumefaciens* and crown gall tumor cells: role of plasmid genes. *J. Bacteriol.* **129**, 101–7.

Moran C. P., Lang N., Le Grice S. F. J., Lee G., Stephens M., Sonenshein A. L., Pero J. & Losick R. (1982) Nucleotide sequences that signal the initiation of transcription and translation in *Bacillus subtilis*. *Mol. Gen. Genet.* **186**, 339–46.

Moriarty A. M., Hoyer B. H., Shih J. W.-K., Gerin J. L. & Hamer D. H. (1981) Expression of the hepatitis B virus surface antigen gene in cell culture by using a Simian virus 40 vector. *Proc. Natl. Acad. Sci. USA* **78**, 2606–10.

Morris C. E., Klement J. F. & McCallister W. T. (1986) Cloning and expression of the bacteriophage T3 RNA-polymerase gene. *Gene* **41**, 193–200.

Morris R. O. (1986) Genes specifying auxin and cytokinin biosynthesis in phytopathogens. *Ann. Rev. Plant Physiol.* **37**, 509–38.

Morrow J. F., Cohen S. N., Chang A. C. Y., Boyer H. W., Goodman H. M. & Helling R. B. (1974) Replication and transcription of eukaryotic DNA in *Escherichia coli*. *Proc. Natl. Acad. Sci. USA* **71**, 1743–7.

Moser R., Thomas R. M. & Gutte B. (1983) An artificial crystalline DDT-binding polypeptide. *FEBS Lett.* **157**, 247–51.

Moss B., Smith G. L., Gerin J. L. & Purcell R. H. (1984) Live recombinant vaccinia virus protects chimpanzees against hepatitis B. *Nature* **311**, 67–9.

Mottes M., Grandi G., Sgaramella V., Canosi U., Morelli G. & Trautner T. A. (1979) Different specific activities of the monomeric and oligomeric forms of plasmid DNA in transformation of *B. subtilis* and *E. coli*. *Mol. Gen. Genet.* **174**, 281–6.

Mountain A. (1989) Gene expression systems for *Bacillus subtilis*. (In press).

Muesing M., Tamm J., Shepard H. M. & Polisky B. (1981) A single base pair alteration is responsible for the DNA overproduction phenotype of a plsmid copy-number mutant. *Cell* **24**, 235–42.

Mulligan R. C. & Berg P. (1980) Expression of a bacterial gene in mammalian cells. *Science* **209**, 1422–7.

Mulligan R. C. & Berg P. (1981a) Factors governing the expression of a bacterial gene in mammalian cells. *Mol. Cell. Biol.* **1**, 449–59.

Mulligan R. C. & Berg P. (1981b) Selection for animal cells that express the *Escherichia coli* gene coding for xanthine–guanine phosphoribosyl-transferase. *Proc. Natl. Acad. Sci. USA* **78**, 2072–6.

Mulligan R. C., Howard B. H. & Berg P. (1979) Synthesis of rabbit β-globin in cultured monkey kidney cells following infection with SV40 β-globin recombinant genome. *Nature* **277**, 108–114.

Murray A. W., Schultes N. P. & Szostak J. W. (1986) Chromosome length controls mitotic chromosome segregation in yeast. *Cell* **45**, 529–536.

Murray A. W. & Szostak J. W. (1983a) Pedigree analysis of plasmid segregation in yeast. *Cell* **34**, 961–70.

Murray A. W. & Szostak J. W. (1983b) Construction of artificial chromosomes in yeast. *Nature* **300**, 189–93.

Murray A. W. & Szostak J. W. (1983c) Construction of artificial chromosomes in yeast. *Nature* **305**, 189–193.

Murray J. A. H. (1987) Bending the rules: the 2μ plasmid of yeast. *Mol. Microbiol.* **1**, 1–14.

Murray K. & Murray N. E. (1975) Phage lambda receptor chromosomes for DNA fragments made with restriction endonuclease III of *Haemophilus influenzae* and restriction endonuclease I of *Escherichia coli*. *J. Mol. Biol.* **98**, 551–64.

Murray M. J., Shilo B.-Z., Chiaho S., Cowing D., Hsu H. W. & Weinberg R. A. (1981) Three different human tumor cell lines contain different oncogenes. *Cell* **25**, 355–61.

Murray N. E. (1983) Phage lambda and molecular cloning. In *The Bacteriophage Lambda*, vol. 2, eds Hendrix R. W., Roberts J. W., Stahl F. W. & Weisberg R. A., Lambda II (Monograph No. 13). Cold Spring Harbor Laboratory, Cold Spring Harbor, New York.

Murray N. E., Brammar W. J. & Murray K. (1977) Lambdoid phages that simplify recovery of *in vitro* recombinants. *Mol. Gen. Genet.* **150**, 53–61.

Murray N. E., Bruce S. A. & Murray K. (1979) Molecular cloning of the DNA ligase gene from bacteriophage T4. II. Amplification and preparation of the gene product. *J. Mol. Biol.* **132**, 493–505.

Murray N. E. & Kelley W. S. (1979) Characterization of λ*pol A* transducing phages. Effective expression of the *E. coli pol A* gene. *Mol. Gen. Genet.* **175**, 77–87.

Myers R. M., Lerman L. S. & Maniatis T. (1985a) A general method for saturation mutagenesis of cloned DNA fragments. *Science* **229**, 242–7.

Myers R. M., Fischer S. G., Lerman L. S. & Maniatis T. (1985b) Nearly all single base substitutions in DNA fragments joined to a GC-clamp can be detected by denaturing gradient gel electrophoresis. *Nucleic Acids Res.* **13**, 3131–45.

Myers R. M. & Tjian R. (1980) Construction and analysis of simian virus 40 origins defective in tumor antigen binding and DNA replication. *Proc. Natl. Acad. Sci. USA* **77**, 6491–5.

Nasmyth K. A. & Reed S. I. (1980) Isolation of genes by complementation in yeast. Molecular cloning of a cell-cycle gene. *Proc. Natl. Acad. Sci. USA* **77**, 2119–23.

Nathans J. & Hogness D. S. (1983) Isolation, sequence analysis, and intron–exon arrangement of the gene encoding bovine rhodopsin. *Cell* **34**, 807–14.

Neugebauer K., Sprengel R. & Schaller H. (1981) Penicillinase from *Bacillus licheniformis*. Nucleotide sequence of the gene and implications for the biosynthesis of a secretory protein in a Gram-positive bacterium. *Nucleic Acids Res.* **9**, 2577–88.

Neumann E., Schaefer-Ridder M., Wang Y. & Hofschneider P. H. (1982) Gene transfer into mouse lymphoma cells by electroporation in high electric fields. *EMBO J.* **1**, 841–5.

Newman A. J., Linn T. G. & Hayward R. S. (1979) Evidence for cotranscription of the RNA polymerase genes *rpo*BC with a ribosome protein of *Escherichia coli*. *Mol. Gen. Genet.* **169**, 195–204.

Newport J. & Kirschner M. (1982a) A major developmental transition in early *Xenopus* embryos: I. Characterization and timing of cellular changes at the midblastula stage. *Cell* **30**, 675–86.

Newport J. & Kirschner M. (1982b) A major developmental transition in early

Xenopus embryos: II. Control of the onset of transcription. *Cell* **30**, 687−96.

Ng R. & Abelson J. (1980) Isolation and sequence of the gene for actin in *Saccharomyces cerevisiae*. *Proc. Natl. Acad. Sci. USA* **77**, 3912−16.

Niaudet B., Goze A. & Ehrlich S. D. (1982) Insertional mutagenesis in *Bacillus subtilis*. Mechanism and use in gene cloning. *Gene* **19**, 277−84.

Nicholson W., Park Y. & Chambliss G. (1987) Catabolite-repression-resistant mutants of the *Bacillus subtilis* α-amylase promoter affect transcription levels and are in an operator-like sequence. *J. Mol. Biol.* **198**, 609−18.

Norrander J., Kempe T. & Messing J. (1983) Construction of improved M13 vectors using oligodeoxynucleotide-directed mutagenesis. *Gene* **27**, 101−6.

Novick R. P., Clowes R. C., Cohen S. N., Curtiss R., Datta N. & Falkow S. (1976) Uniform nomenclature for bacterial plasmids: a proposal. *Bact. Rev.* **40**, 168−89.

Nugent M. E., Primrose S. B. & Tacon W. C. A. (1983) The stability of recombinant DNA. *Dev. Ind. Microbiol.* **24**, 271−85.

Obukowicz M. G., Perlak F. J., Bolten S. L., Kusano-Kretzmer K., Mayer E. J. & Watrud L. S. (1987) IS50L as a non-self transposable vector used to integrate the *Bacillus* thuringiensis delta-endotoxin gene into the chromosome of root-colonizing pseudomonads. *Gene* **51**, 91−6.

O'Hare K., Benoist C. & Breathnach R. (1981) Transformation of mouse fibroblasts to methotrexate resistance. *Proc. Natl. Acad. Sci. USA* **78**, 1527−31.

O'Hare K. & Rubin G. M. (1983) Structures of P transposable elements and their sites of insertion and excision in the *Drosophila melanogaster* genome. *Cell* **34**, 25−35.

Okamoto T., Fujita Y. & Irie R. (1985) Interspecific protoplast fusion between *Streptococcus cremoris* and *Streptococcus lactis*. *Agric. Biol. Chem.* **49**, 1371−6.

Okayama H. & Berg P. (1982) High-efficiency cloning of full-length cDNA. *Mol. Cell. Biol.* **2**, 161−70.

Old J. M., Ward R. H. T., Petrov M., Karagozlu F., Modell B. & Weatherall D. J. (1982a) First trimester diagnosis for haemoglobinopathies: a report of 3 cases. *Lancet* **2**, 1413−16.

Old R. W., Woodland H. R., Ballantine J. E. M., Aldridge T. C., Newton C. A., Bains W. A. & Turner P. C. (1982b) Organization and expression of cloned histone gene clusters from *Xenopus laevis* and *X. borealis*. *Nucleic Acids Res.* **10**, 7561−80.

Olivera B. M., Hall Z. W. & Lehman I. R. (1968) Enzymatic joining of polynucleotides. V. A DNA adenylate intermediate in the polynucleotide joining reaction. *Proc. Natl. Acad. Sci. USA* **61**, 237−44.

Olsen R. H., DeBusccher G. & McCombie W. R. (1982) Development of broad-host-range vectors and gene banks. Self-cloning of the *Pseudomonas aeruginosa* PAO chromosome. *J. Bacteriol.* **150**, 60−9.

Olson M. V. (1981) Applications of molecular cloning to *Saccharomyces*. In *Genetic Engineering*, eds Setlow J. K. & Hollaender A. Plenum Press, New York.

Ooms G., Hooykaas P. J. J., Moolenaar G. & Schilperoort R. A. (1981) Crown gall plant tumours of abnormal morphology induced by *Agrobacterim tumefaciens* carrying mutated octopine Ti plasmids: analysis of T-DNA functions. *Gene* **14**, 33−50.

Ooms G., Klapwijk P. M., Poulis J. A. & Schilperoort R. A. (1980) Characterization of Tn904 insertion in octopine Ti plasmid mutants of *Agrobacterium tumefaciens*. *J. Bacteriol.* **144**, 82−91.

Orkin S. H. (1982) Genetic diagnosis of the foetus. *Nature* **296**, 202−3.

Orkin S. H., Little P. F. R., Kazazian H. H. & Boehm C. (1982) Improved detection of the sickle mutation by DNA analysis. *N. Engl. J. Med.* **307**, 32–6.

Orr-Weaver T. L., Szostak J. W. & Rothstein R. L. (1981) Yeast transformation: a model system for the study of recombination. *Proc. Natl. Acad. Sci. USA* **78**, 6354–8.

Ostrowski M. C., Richard-Foy H., Wolford R. G., Berard D. S. & Hager G. L. (1983) Glucocorticoid regulation of transcription at an amplified episomal promoter. *Mol. Cell. Biol.* **3**, 2045–57.

Ou C. Y., Kwok S., Mitchell S. W., Mack D. H., Sninsky J. J., Krebs J. W., Feorino P., Warfield D. & Schochetman G. (1988) DNA amplification for direct detection of HIV-1 in DNA of peripheral blood mononuclear cells. *Science* **239**, 295–7.

Palmiter R. D. & Brinster R. L. (1986) Germ-line transformation of mice. *Ann. Rev. Genet.* **20**, 465–99.

Palmiter R. D., Brinster R. L., Hammer R. E., Trumbauer M. E., Rosenfeld M. G., Birnberg N. C. & Evans R. M. (1982a) Dramatic growth of mice that develop from eggs microinjected with metallothionein-growth hormone fusion genes. *Nature* **300**, 611–15.

Palmiter R. D., Chen H. Y. & Brinster R. L. (1982b) Differential regulation of metallothionein-thymidine kinase fusion genes in transgenic mice and their offspring. *Cell* **29**, 701–10.

Palmiter R. D., Chen H. Y., Messing A. & Brinster R. L. (1985) SV40 enhancer and large T-antigen are instrumental in development of choroid plexus tumors in transgenic mice. *Nature* **316**, 457–60.

Palmiter R. D., Norstedt G., Gelinas R. E., Hammer R. E. & Brinster (1983) Metallothionein-human GH fusion genes stimulate growth of mice. *Science* **222**, 809–14.

Palva I., Lehtovaara P., Kääriäinen L., Sibakov M., Cantell K., Schein C. H., Kashiwagi K. & Weissmann C. (1983) Secretion of interferon by *Bacillus subtilis*. *Gene* **22**, 229–35.

Palva I., Pettersson R. F., Nalkkinnen N., Lehtovaara P., Sarvas M., Söderlund H., Takkinen K. & Kääriäinen L. (1981) Nucleotide sequence of the promoter and NH_2-terminal signal peptide region of the α-amylase gene from *Bacillus amyloliquefaciens*. *Gene* **15**, 43–51.

Palva M. & Ranki M. (1985) Microbial diagnosis by nucleic acid sandwich hybridization. *Clin. Lab. Med.* **5**, 475–90.

Palva I., Sarvas M., Lehtovaara P., Sibakov M. & Kääriäinen L. (1982) Secretion of *Escherichia coli* β-lactamase from *Bacillus subtilis* by the aid of α-amylase signal sequence. *Proc. Natl. Acad. Sci. USA* **79**, 5582–6.

Panasenko S. M., Cameron J. R., Davis R. W. & Lehman I. R. (1977) Five hundredfold overproduction of DNA ligase after induction of a hybrid lambda lysogen contructed *in vitro*. *Science* **196**, 188–9.

Paszkowski J., Baur M., Bogucki A. & Potrykus I. (1988) Gene targeting in plants. *EMBO J.* **7**, 4021–6.

Paszkowski J., Shillito R. D., Saul M., Mandak V., Hohn T., Hohn B. & Potrykus I. (1984) Direct gene transfer to plants. *EMBO J.* **3**, 2717–22.

Paterson B. M., Roberts B. E. & Kuff E. L. (1977) Structural gene identification and mapping by DNA.mRNA hybrid-arrested cell-free translation. *Proc. Natl. Acad. Sci. USA* **74**, 4370–4.

Patzer E. J., Nakamura G. R., Hershberg R. D., Gregory T. J., Crowley C., Levinson A. D. & Eichberg J. W. (1986) Cell culture derived recombinant HBsAg is highly immunogenic and protects chimpanzees from infection with hepatitis B virus. *Biotechnology* **4**, 630—6.

Peden K. W. C. (1983) Revised sequence of the tetracycline-resistance gene of pBR322. *Gene* **22**, 277—80.

Peijnenburg A. A. C. M., Bron S. & Venema G. (1987) Structural plasmid instability in recombination- and repair-deficient strains of *Bacillus subtilis*. *Plasmid* **17**, 167—70.

Penttila M. E., Andre L., Lehtovaara P., Bailey M., Teeri T. T. & Knowles J. K. C. (1988) Efficient secretion of two fungal cellobiohydrolases by *Saccharomyces cerevisiae*. *Gene* **63**, 103—12.

Peralta E. G., Hellmiss R. & Ream W. (1986) *Overdrive*, a T-DNA transmission enhancer on the *A. tumefaciens* tumour-inducing plasmid. *EMBO J.* **5**, 1137—42.

Perkus M. E., Piccini A., Lipinskas B. R. & Paoletti E. (1985) Recombinant vaccinia virus: immunization against multiple pathogens. *Science* **229**, 981—4.

Perry L. J. & Wetzel R. (1984) Disulfide bond engineered into T4 lysozyme: stabilization of the protein toward thermal inactivation. *Science* **226**, 555—7.

Perucho M., Goldfarb M., Shimizu K., Lama C., Fogh J. & Wigler M. (1981) Human-tumor-derived cell lines contain common and different transforming genes. *Cell* **27**, 467—76.

Perucho M., Hanahan D., Lipsich L. & Wigler M. (1980a) Isolation of the chicken thymidine kinase gene by plasmid rescue. *Nature* **285**, 207—10.

Perucho M., Hanahan D. & Wigler M. (1980b) Genetic and physical linkage of exogenous sequences in transformed cells. *Cell* **22**, 309—17.

Peschke U., Beuck V., Bujard H., Gentz R. & Le Grice S. (1985) Efficient utilisation of *Escherichia coli* transcriptional signals in *Bacillus subtilis*. *J. Mol. Biol.* **186**, 547—55.

Pfeiffer P. & Hohn T. (1983) Involvement of reverse transcription in the replication of cauliflower mosaic virus. A detailed model and test of some aspects. *Cell* **33**, 781—9.

Pittius C. W., Hennighausen L., Lee E., Westphal H., Nicols E., Vitale J. & Gordon K. (1988) A milk protein gene promoter directs expression of human tissue plasminogen activator cDNA to the mammary gland in transgenic mice. *Proc. Natl. Acad. Sci. USA* **85**, 5874—8.

Plaskon R. R. & Wartell R. M. (1987) Sequence distributions associated with DNA curvature are found upstream of strong *E. coli* promoters. *Nucleic Acids Res.* **15**, 785—96.

Plasterk R. H. A., Simon M. I. & Barbour A. G. (1985) Transposition of structural genes to an expression sequence on a linear plasmid causes antigenic variation in the bacterium *Borrelia hermsii*. *Nature* **318**, 257—63.

Pohlner J., Halter R., Beyreuther K. & Meyer T. F. (1987) Gene structure and extracellular secretion of *Neisseria gonorrhoeae* IgA protease. *Nature* **325**, 458—62.

Polsky-Cynkin R., Parsons G. H., Allerdt L., Landes G., Davis G. & Rashtchian A. (1985) Use of DNA immobilized on plastic and agarose supports to detect DNA by sandwich hybridization. *Clin. Chem.* **31**, 1438—43.

Potrykus I., Paszkowski J., Saul M. W., Petruska J. & Shillito R. D. (1985a) Molecular and general genetics of a hybrid foreign gene introduced into tobacco by direct gene transfer. *Mol. Gen. Genet.* **199**, 169—77.

Potrykus I., Saul M. W., Petruska J., Paszkowski J. & Shillito R. (1985b) Direct gene transfer to cells of a graminaceous monocot. *Molec. Gen. Genet.* **199**, 183—8.

Potter H., Weir L. & Leder P. (1984) Enhancer-dependent expression of human K immunoglobulin genes introduced into mouse pre-B lymphocytes by electroporation. *Proc. Natl. Acad. Sci. USA* **81**, 7161–5.

Poustka A. & Lehrach H. (1986) Jumping libraries and linking libraries: the next generation of molecular tools in mammalian genetics. *Trends Genet.* **2**, 174–9.

Poustka A., Pohl T. M., Barlow D. P., Frischauf A. M. & Lehrach H. (1987) Construction and use of human chromosome jumping libraries from *Not* I-digested DNA. *Nature* **325**, 353–5.

Powell-Abel P., Nelson R. S., De B. Hoffmann N., Rogers S. G., Fraley R. T. & Beachy R. N. (1986) Delay of disease development in transgenic plants that express the tobacco mosaic virus coat protein gene. *Science* **232**, 738–43.

Pratt J. M., Boulnois G. J., Darby V., Orr E., Wahle E. & Holland I. B. (1981) Identification of gene products programmed by restriction endonuclease DNA fragments using an *E. coli in vitro* system. *Nucleic Acids Res.* **9**, 4459–74.

Primrose S. B., Derbyshire P., Jones I. M., Nugent M. E. & Tacon W. C. A. (1983) Hereditary instability of recombinant DNA molecules. In *Bioactive Microbial Products 2: Development and Production*, eds Nisbet L. J. & Winstanley D. J., pp. 63–77. Academic Press, London.

Primrose S. B. & Ehrlich S. D. (1981) Isolation of plasmid deletion mutants and a study of their instability. *Plasmid* **6**, 193–201.

Pursel V. G., Pinkert C. A., Miller K. F., Bolt D. J., Campbell R. G., Palmiter R. D., Brinster R. L. & Hammer R. E. (1989) Genetic engineering of livestock. *Science* **244**, 1281–8.

Purvis I. J., Bethany A. J. E., Santiago T. C., Coggins J. R., Duncan K., Eason R. & Brown A. J. P. (1987a) The efficiency of folding of some proteins is increased by controlled rates of translation *in vivo*: a hypothesis. *J. Mol. Biol.* **193**, 413–17.

Purvis I. J., Bethany A. J. E., Loughlin L. & Brown A. J. P. (1987b) The effects of alterations within the 3′ untranslated region of the pyruvate kinase messenger RNA upon its stability and translation in *Saccharomyces cerevisiae*. *Nucleic Acids Research* **15**, 7951–62.

Putney S. D., Herlihy W. C. & Schimmel P. (1983) A new troponin T and cDNA clones for 13 different muscle proteins, found by shotgun sequencing. *Nature* **302**, 718–21.

Puyet A., Sandoval M., López P., Aguilar A., Martin J. F. & Espinosa M. (1987) A simple medium for rapid regeneration of *Bacillus subtilis* protoplasts transformed with plasmid DNA. *FEBS Microbiol. Lett.* **40**, 1–5.

Radloff R., Bauer W. & Vinograd J. (1967) A dye-bouyant-density method for the detection and isolation of closed circular duplex DNA: the closed circular DNA in HeLa cells. *Proc. Natl. Acad. Sci. USA* **57**, 1514–21.

Radman M. & Wagner R. (1984) Effects of DNA methylation on mismatch repair, mutagenesis, and recombination in *Escherichia coli. Curr. Top. Microbiol. Immunol.* **108**, 23–28.

Raleigh E. A., Murray N. E., Revel H., Blumenthal R. M., Westaway D., Reith A. D., Rigby P. W. J., Elhai J. & Hanahan D. (1988) Mcr A and Mcr B restriction phenotypes of some *E. coli* strains and implications for gene cloning. *Nucleic Acids Res.* **16**, 1563–75.

Raleigh E. A. & Wilson G. (1986) *Escherichia coli* K-12 restricts DNA containing 5-methylcytosine. *Proc. Natl. Acad. Sci. USA* **83**, 9070–4.

Ramshaw I. A., Andrew M. E., Phillips S. M., Boyle D. B. & Coupar B. E. H. (1987) Recovery of immunodeficient mice from a vaccinia virus/IL-2 recombinant infection. *Nature* **329**, 545–6.

Ranki M., Palva A., Virtanen M., Laaksonen N. & Soderlund H. (1983) Sandwich hybridization as a convenient method for the detection of nucleic acids in crude samples. *Gene* **21**, 77–85.

Rassoulzadegan M., Binetruy B. & Cuzin F. (1982) High frequency of gene transfer after fusion between bacteria and eukaryotic cells. *Nature* **295**, 257–9.

Ratzkin B. & Carbon J. (1977) Functional expression of cloned yeast DNA in *Escherichia coli*. *Proc. Natl. Acad. Sci. USA* **74**, 487–91.

Razin A. & Riggs A. D. (1980) DNA methylation and gene function. *Science* **210**, 604–10.

Renart J. & Sandoval I. V. (1984) Western blots. *Methods Enzymol.* **104**, 455–60.

Renz M. & Kurz C. (1984) A colorimetric method for DNA hybridization. *Nucleic Acids Res.* **12**, 3435–44.

Richards J. E., Gilliam T. C., Cole J. L., Drumm M. L., Wasmuth J, J., Gusella J. F. & Collins F. S. (1988) Chromosome jumping from D4S10 (G8) toward the Huntington disease gene. *Proc. Natl. Acad. Sci. USA* **85**, 6437–41.

Richards K. E., Guilley H. & Jonard G. (1981) Further characterization of the discontinuities in cauliflower mosaic virus DNA. *FEBS Lett.* **134**, 67–70.

Rigby P. W. J., Dieckmann M., Rhodes C. & Berg P. (1977) Labelling deoxyribonuclic acid to high specific activity *in vitro* by nick translation with DNA polymerase I. *J. mol. Biol.* **113**, 237–51.

Rio D. C., Clark S. G. & Tjian R. (1985) A mammalian host–vector system that regulates expression and amplification of transfected genes by temperature induction. *Science* **227**, 23–28.

Robbins D. M., Ripley S., Henderson A. & Axel R. (1981) Transforming DNA integrates into the host chromosome. *Cell* **23**, 29–39.

Roberts R. J. (1984) Restriction and modification enzymes and their recognition sequences. *Nucleic Acids Res.* **12**, 167–91.

Roberts T. M., Bikel I., Yocum R. R., Livingston D. M. & Ptashne M. (1979a) Synthesis of simian virus 40 t antigen in *Escherichia coli*. *Proc. Natl. Acad. Sci. USA* **76**, 5596–600.

Roberts T. M., Kacich R. & Ptashne M. (1979b) A general method for maximizing the expression of a cloned gene. *Proc. Natl. Acad. Sci. USA* **76**, 760–4.

Roberts T. M., Swanberg S. L., Poteete A., Riedal G. & Backman K. (1980) A plasmid cloning vehicle allowing a positive selection for inserted fragments. *Gene* **12**, 123–7.

Robinson M., Lilley R., Little S., Emtage J. S., Yarranton G., Stephens P., Millican A., Eaton M. & Humphreys G. (1984) Codon usage can affect efficiency of translation of genes in *Escherichia coli*. *Nucleic Acids Res.* **12**, 6663–71.

Rodriguez R. L., West R. W., Heyneker H. L., Bolivar P. & Boyer H. W. (1979) Characterizing wild-type and mutant promoters of the tetracycline resistance gene in pBR313. *Nucleic Acids Res.* **6**, 3267–87.

Roeder R. G. (1974) Multiple forms of DNA-dependent RNA polymerase in *X. laevis*. *J. Biol. Chem.* **249**, 249–56.

Rogers J., Goedert M. & Wilson P. M. (1988) An extra sequence in the lambda EMBL3 polylinker. *Nucleic Acids Res.* **16**, 1633.

Rood J. I., Sneddon M. K. & Morrison J. F. (1980) Instability in *tyr*R strains of plasmids carrying the tyrosine operon: isolation and characterization of plasmid derivatives with insertions or deletions. *J. Bacteriol.* **144**, 552–9.

Rosamond J., Endlich B. & Linn S. (1979) Electron microscopic studies of the mechanism of action of the restriction endonuclease of *Escherichia coli* B. *J. Mol. Biol.* **129**, 619—35.

Rose M., Casadaban M. J. & Botstein D. (1981) Yeast genes fused to β-galactosidase in *Escherichia coli* can be expressed normally in yeast. *Proc. Natl. Acad. Sci. USA* **78**, 2460—4.

Rose M. D., Novick P., Thomas J. H., Botstein D. & Fink G. R. (1987) A *Saccharomyces cerevisiae* genomic plasmid bank based on a centromere-containing shuttle vector. *Gene* **60**, 237—43.

Rosteck P. R. Jr & Hershberger C. L. (1983) Selective retention of recombinant plasmids coding for human insulin. *Gene* **25**, 29—38.

Rothstein R. J., Lau L. F., Bahl C. P., Narang S. A. & Wu R. (1979) Synthetic adaptors for cloning DNA. *Methods Enzymol.* **68**, pp. 98—109.

Roychoudhury R., Jay E. & Wu R. (1976) Terminal labelling and addition of homopolymer tracts to duplex DNA fragments by terminal deoxynucleotidyl transferase. *Nucleic Acids Res.* **3**, 863—77.

Rubin E. M., Wilson G. A. & Young F. E. (1980) Expression of thymidylate synthetase activity in *Bacillus subtilis* upon integration of a cloned gene from *Escherichia coli*. *Gene* **10**, 227—35.

Rubin G. M., Kidwell M. G. & Bingham P. M. (1982) The molecular basis of P—M hybrid dysgenesis: the nature of induced mutations. *Cell* **29**, 987—94.

Rubin G. M. & Spradling A. C. (1982) Genetic transformation of *Drosophila* with transposable element vectors. *Science* **218**, 348—53.

Rubin G. M. & Spradling A. (1983) Vectors for P element-mediated gene transfer in *Drosophila*. *Nucleic Acids Res.* **11**, 6341—51.

Rubin J. S., Joyner A. L., Bernstein A. & Whitmore G. F. (1983) Molecular identification of a human DNA repair gene following DNA-mediated gene transfer. *Nature* **306**, 206—8.

Rudolph H. & Hinnen A. (1987) The yeast PHO5 promoter: phosphate control elements and sequences mediating mRNA start-site selection. *Proc. Natl. Acad. Sci. USA* **84**, 1340—4.

Rusconi S. & Schaffner W. (1981) Transformation of frog embryos with a rabbit β-globin gene. *Proc. Natl. Acad. Sci. USA* **78**, 5051—5.

Russell D. R. & Bennett G. N. (1982) Construction and analysis of *in vivo* activity of *E. coli* promoter hybrids and promoter mutants that alter the −35 to −10 spacing. *Gene* **20**, 231—43.

Ryoji M. & Worcel A. (1984) Chromatin assembly in *Xenopus* oocytes: *in vivo* studies. *Cell* **37**, 21—32.

Sabin E. A., Lee-Ng C. T., Shuster J. R. & Barr P. J. (1989) High-level expression and *in vivo* processing of chimeric ubiquitin fusion proteins in *Saccharomyces cerevisiae*. *Biotechnology* **7**, 705—9.

Sakai R. K., Gelford D. H., Stoeffel S., Scharf S. J., Higuchi R., Horn G. T., Mullis K. B. & Erlich H. A. (1988) Primer-directed enzymatic amplification of DNA with a thermostable DNA polymerase. *Science* **239**, 487—91.

Sakai R. K., Scharf S., Faloona F., Mullis K. B., Horn G. T., Erlich H. A. & Arnheim N. (1985) Enzymatic amplification of β-globin genomic sequences and restriction site analysis for diagnosis of sickle cell anaemia. *Science* **230**, 1350—4.

Saloman F., Deblaere R., Leemans J., Hernalsteens J. P., Van Montagu M. & Schell J. (1984) Genetic identification of functions of TR-DNA transcripts in octopine crown galls. *EMBO J.* **3**, 141—6.

Sambrook J., Rodgers L., White J. & Getling M. J. (1985) Lines of BPV-transformed murine cells that constitutively express influenza virus haemagglutinin. *EMBO J.* **4**, 91–103.

Sancar A., Hack A. M. & Rupp W. D. (1979) Simple method for identification of plasmid-coded proteins. *J. Bacteriol.* **137**, 692–3.

Sanger F., Coulson A. R., Hong G. F., Hill D. F. & Peterson G. B. (1982) Nucleotide sequence of bacteriophage λ DNA. *J. Mol. Biol.* **162**, 729–73.

Sanger F., Nicklen S. & Coulson A. R. (1977) DNA sequencing with chain-terminating inhibitors. *Proc. Natl. Acad. Sci. USA* **74**, 5463–7.

Sarver N., Gruss P., Law M.-F., Khoury G. & Howley P. M. (1981a) Bovine papilloma virus deoxyribonucleic acid: a novel eucaryotic cloning vector. *Mol. Cell. Biol.* **1**, 486–96.

Sarver N., Gruss P., Law M.-F., Khoury G. & Howley P. M. (1981b) Rat insulin gene covalently linked to bovine papilloma virus DNA is expressed in transformed mouse cells. In *Developmental Biology Using Purified Genes*, eds Brown D. & Fox C. R., ICN-UCLA Symposia on molecular and cellular biology, vol. 23. Academic Press, New York.

Scahill S. J., Devos R., Van der Heyden J. & Fiers W. (1983) Expression and characterisation of the product of a human immune interferon cDNA gene in Chinese hamster ovary cells. *Proc. Natl. Acad. Sci. USA* **80**, 4654–8.

Scalenghe F., Turco E., Edstrom J. E., Pirotta V. & Melli M. (1981) Micro-dissection and cloning of DNA from a specific region of *Drosophila melanogaster* polytene chromosomes. *Chromosoma* **82**, 205–16.

Scangos G. A., Huttner K. M., Juricek D. K. & Ruddle F. H. (1981) DNA-mediated gene transfer in mammalian cells: molecular analysis of unstable transformants and their progression to stability. *Mol. Cell. Biol.* **1**, 111–20.

Schafer W., Gorz A. & Kahl G. (1987) T-DNA integration and expression in a monocot crop plant after induction of *Agrobacterium*. *Nature* **327**, 529–31.

Schaffner W. (1980) Direct transfer of cloned genes from bacteria to mammalian cells. *Proc. Natl. Acad. Sci. USA* **77**, 2163–7.

Scheidereit C., Greisse S., Westphal H. M. & Beato M. (1983) The glucocorticoid receptor binds to defined nucleotide sequences near the promoter of mouse mammary tumour virus. *Nature* **304**, 749–52.

Schein C. H. & Noteborn M. H. M. (1988) Formation of soluble recombinant proteins in *Escherichia coli* is favored by lower growth temperature. *Bio-technology* **6**, 291–4.

Schell J. & Van Montagu M. (1977) The Ti-plasmid of *Agrobacterium tumefaciens*, a natural vector for the introduction of fix genes in plants. In *Genetic Engineering for Nitrogen Fixation*, ed. Hollaender A., pp. 159–79. Plenum Press, New York.

Scheller R. H., Dickerson R. E., Boyer H. W., Riggs A. D. & Itakura K. (1977) Chemical synthesis of restriction enzyme recognition sites useful for cloning. *Science* **196**, 177–80.

Scherzinger E., Bagdasarian M. M., Scholz P., Lurz R., Ruchert B. & Bagdasarian M. (1984) Replication of the broad host range plasmid RSF1010: requirement for three plasmid-encoded proteins. *Proc. Natl. Acad. Sci. USA* **81**, 654–8.

Schimke R. T. (1984) Gene amplification in cultured animal cells. *Cell* **37**, 705–13.

Schimke R. T., Kaufman R. J., Alt F. W. & Kellems R. F. (1978) Gene amplification and drug resistance in cultured murine cells. *Science* **202**, 1051–5.

Schleif R. F. & Wensink P. C. (1981) *Practical Methods in Molecular Biology*. Springer-Verlag, Berlin.

Schmidhauser T. J., Ditta G. & Helinski D. R. (1988) Broad-host-range plasmid cloning vectors for Gram-negative bacteria. In *Vectors. A survey of molecular cloning vectors and their uses*, eds Rodriguez R. L. & Denhardt D. T., pp.

287–332. Butterworths, London.

Schmidhauser T. J., Filutowicz M. & Helinski D. R. (1983) Replication of derivatives of the broad host range plasmid RK2 in two distantly related bacteria. *Plasmid* **9**, 325–30.

Scholnick S. B., Morgan B. A. & Hirsh J. (1983) The cloned Dopa decarboxylase gene is developmentally regulated when reintegrated into the *Drosophila* genome. *Cell* **34**, 37–45.

Scholz. P., Haring V., Wittman-Liebold B., Ashman K., Bagdasarian M. & Scherzinger E. (1989) Complete nucleotide sequence and gene organization of the broad-host-range plasmid RSF1010. *Gene* **75**, 271–88.

Schoner R. G., Ellis L. F. & Schoner B. E. (1985) Isolation and purification of protein granules from *Escherichia coli* cells overproducing bovine growth hormone. *Biotechnology* **3**, 151–4.

Schoner R. G., Williams D. M. & Lovett P. S. (1983) Enhanced expression of mouse dihydrofolate reductase in *Bacillus subtilis*. *Gene* **22**, 47–57.

Schroder G., Waffenschidt S., Weiler E. W. & Schroder J. (1984) The T-region of Ti plasmids codes for an enzyme synthesizing indole-3-acetic acid. *Eur. J. Biochem.* **138**, 387–91.

Schumann W. (1979) Construction of an *Hpa* I and *Hin*d II plasmid vector allowing direct selection of transformants harboring recombinant plasmids. *Mol. Gen. Genet.* **174**, 221–4.

Schwartz D. C. & Cantor C. R. (1984) Separation of yeast chromosomal-sized DNAs by pulsed field gradient gel electrophoresis. *Cell* **37**, 67–75.

Sciaky D., Montoya A. L. & Chilton M. D. (1978) Fingerprints of *Agrobacterium* Ti plasmids. *Plasmid* **1**, 238–54.

Seeburg P. H., Shine J., Martial J. A., Baxter J. D. & Goodman H. M. (1977) Nucleotide sequence and amplification in bacteria of the structural gene for rat growth hormone. *Nature* **270**, 486–94.

Seed B. (1983) Purification of genome sequences from bacteriophage libraries by recombination and selection *in vivo*. *Nucleic Acids Res.* **8**, 2427–45.

Seelke R., Kline B., Aleff R., Porter R. D. & Shields M. S. (1987) Mutations in the *rec*D gene of *Escherichia coli* that raise the copy number of certain plasmids. *J. Bacteriol.* **169**, 4841–4.

Sgaramella V. (1972) Enzymatic oligomerization of bacteriophage P22 DNA and of linear simian virus 40 DNA. *Proc. Natl. Acad. Sci. USA* **69**, 3389–93.

Shah D. M., Horsch R. B., Klee H. J., Kishare G. M., Winter J. A., Turner N. E., Hironaka C. M., Sanders P. R., Gasser C. S., Aykent S., Siegel N. R., Rogers S. G. & Fraley R. T. (1986) Engineering herbicide tolerance in transgenic plants. *Science* **233**, 478–81.

Sharp P. M. & Bulmer M. (1988) Selective differences among translation termination codons. *Gene* **63**, 141–5.

Shaw C. H., Leemans J., Shaw C. H., Van Montague M. & Schell J. (1983) A general method for the transfer of cloned genes to plant cells. *Gene* **23**, 315–30.

Shaw C. H., Watson M. D., Carter G. H. & Shaw C. H. (1984) The right hand copy of the nopaline Ti-plasmid 25 bp repeat is required for tumour formation. *Nucleic Acids Res.* **12**, 6031–41.

Shepard H. M., Yelverton E. & Goeddel D. V. (1982) Increased synthesis in *E. coli* of fibroblast and leukocyte interferons through alterations in ribosome binding sites. *DNA* **1**, 125–31.

Shih M. F., Arsenakis M., Tiollais P. & Roizman B. (1984) Expression of hepatitis B virus S gene by herpes simplex virus type 1 vectors carrying α- and β-regulated gene chimeras. *Proc. Natl. Acad. Sci. USA* **81**, 5867–70.

Shimamoto K., Terada R., Izawa T. & Fujimoto H. (1989) Fertile transgenic rice plants regenerated from transformed protoplasts. *Nature* **338**, 274–6.

Shimotohno K. & Temin H. M. (1981) Formation of infectious progeny virus after insertion of herpes simplex thymidine kinase gene into DNA of an avian retrovirus. *Cell* **26**, 67–77.

Shimotohno K. & Temin H. M. (1982) Loss of intervening sequences in genomic mouse α-globin DNA inserted in an infectious retrovirus vector. *Nature* **299**, 265–8.

Shine J. & Dalgarno L. (1975) Determinant of cistron specificity in bacterial ribosomes. *Nature* **254**, 34–8.

Shine J., Fettes I., Lan N. C. Y., Roberts J. L. & Baxter J. D. (1980) Expression of cloned β-endorphin gene sequences by *Escherichia coli*. *Nature* **285**, 456–61.

Shorenstein R. G. & Losick R. (1973) Comparative size and properties of the sigma subunits of ribonucleic acid polymerase from *Bacillus subtilis* and *Escherichia coli*. *J. Biol. Chem.* **248**, 6170–3.

Short J. M., Fernandez J. M., Sorge J. A. & Huse W. D. (1988) λZAP: a bacteriophage lambda expression vector with *in vivo* excision properties. *Nucleic Acids Res.* **16**, 7583–600.

Simmler M. O., Johnsson C., Petit C., Rouyer F., Vergnaud G. & Weissenbach J. (1987) Two highly polymorphic minisatellites from the pseudoautosomal region of the human sex chromosomes. *EMBO J.* **6**, 963–9.

Simon L. D., Randolph B., Irwin N. & Binkowski G. (1983) Stabilization of proteins by a bacteriophage T4 gene cloned in *Escherichia coli*. *Proc. Natl. Acad. Sci. USA* **80**, 2059–62.

Simons J. P., Wilmut I., Clark A. J., Archibald A. L., Bishop J. O. & Lathe R. (1988) Gene transfer into sheep. *Biotechnology* **6**, 179–83.

Simons R. W., Houman F. & Kleckner N. (1987) Improved single and multicopy *lac*-based cloning vectors for protein and operon fusions. *Gene* **53**, 85–96.

Simpson R., O'Hara P., Lichtenstein C., Montoya A. L., Kwok K., Gordon M. P. & Nester E. W. (1982) The DNA from A6S/Z tumor contains scrambled Ti plasmid sequence near its junction with plant DNA. *Cell* **29**, 1005–14.

Singh H., Bieker J. J. & Dumas L. B. (1982) Genetic transformation of *Saccharomyces cerevisiae* with single-stranded circular DNA vectors. *Gene* **20**, 441–9.

Singh M., Heaphy S. & Gait M. J. (1986) Oligonucleotide-directed misincorporation mutagenesis on single-stranded DNA templates. *Prot. Eng.* **1**, 75–6.

Sinn E., Muller W., Pattengale P., Tepler I., Wallace R. & Leder P. (1987) Coexpression of MMTV/v-Ha-*ras* and MMTV/c-*myc* genes in transgenic mice: synergistic action of oncogenes *in vivo*. *Cell* **49**, 465–75.

Skalka A. & Shapio L. (1976) *In situ* immunoassays for gene translation products in phage plaques and bacterial colonies. *Gene* **1**, 65–79.

Skogman G., Nilsson J. & Gustafsson P. (1983) The use of a partition locus to increase stability of tryptophan–operon-bearing plasmids in *Escherichia coli*. *Gene* **23**, 105–15.

Smith D. F., Searle P. F. & Williams J. G. (1979) Characterization of bacterial clones containing DNA sequences derived from *Xenopus laevis*. *Nucleic Acids Res.* **6**, 487–506.

Smith E. F. & Townsend C. O. (1907) A plant-tumor of bacterial origin. *Science* **25**, 671–3.

Smith G. E., Summers M. D. & Fraser M. J. (1983) Production of human β-interferon in insect cells infected with a baculovirus expression vector. *Mol. Cell. Biol.* **3**, 2156–65.

Smith G. L., Mackett M. & Moss B. (1983a) Infectious vaccinia virus recombinants that express hepatitis B virus surface antigen. *Nature* **302**, 490–5.

Smith G. L., Murphy B. R. & Moss B. (1983b) Construction and characterization of an infectious vaccinia virus recombinant that expresses the influenza hemagglutinin gene and induces resistance to influenza virus infection in hamsters. *Proc. Natl. Acad. Sci. USA* **80**, 7155–9.

Smith H., Bron S., Van Ee J. & Venema G. (1987) Construction and use of signal sequence selection vectors in *Escherichia coli* and *Bacillus subtilis. J. Bacteriol.* **169**, 3321–8.

Smith H. O. & Nathans D. (1973) A suggested nomenclature for bacterial host modification and restriction systems and their enzymes. *J. Mol. Biol.* **81**, 419–23.

Smith H. O. & Wilcox K. W. (1970) A restriction enzyme from *Hemophilus influenzae*. I. Purification and general properties. *J. Molec. Biol.* **51**, 379–91.

Smith J., Cook E., Fotheringham I., Pheby S., Derbyshire R., Eaton M. A. W., Doel M., Lilley D. M. J., Patel T., Lewis H. & Bell L. D. (1982) Chemical synthesis and cloning of gene for human β-urogastrone. *Nucleic Acids Res.* **10**, 4467–82.

Smith J. C., Derbyshire R. B., Cook E., Dunthorne L., Viney J., Brewer S. J., Sassenfeld H. M. & Bell L. D. (1984) Chemical synthesis and cloning of a poly(arginine)-coding gene fragment designed to aid polypeptide purification. *Gene* **32**, 321–7.

Smithies O., Gregg R. G., Boggs S. S., Koralewski M. A. & Kucherlapati R. (1985) Insertion of DNA sequences into the human β-globin locus by homologous recombination. *Nature* **317**, 230–4.

Sorge J., Wright D., Erdman V. D. & Cutting A. E. (1984) Amphotropic retrovirus vector system for human cell gene transfer. *Mol. Cell. Biol.* **4**, 1730–7.

Southern E. M. (1975) Detection of specific sequences among DNA fragments separated by gel electrophoresis. *J. Mol. Biol.* **98**, 503–17.

Southern E. M. (1979a) Measurement of DNA length by gel electrophoresis. *Anal. Biochem.* **100**, 319–23.

Southern E. M. (1979b) Gel electrophoresis of restriction fragments. *Methods Enzymol.* **68**, 152–76.

Southern P. J. & Berg P. (1982) Transformation of mammalian cells to antibiotic resistance with a bacterial gene under the control of the SV40 early region promoter. *J. Mol. Appl. Genet.* **1**, 327–41.

Spoerel N., Herlich P. & Bickle T. A. (1979) A novel bacteriophage defence mechanism: the anti-restriction protein. *Nature* **278**, 30–4.

Spradling A. C. & Rubin G. M. (1982) Transposition of cloned P elements into *Drosophila* germ line chromosomes. *Science* **218**, 341–7.

Spradling A. C. & Rubin G. M. (1983) The effect of chromosomal position on the expression of the *Drosophila* xanthine dehydrogenase gene. *Cell* **34**, 47–57.

Sprague K. V., Faulds D. H. & Smith G. R. (1978) A single base-pair change creates a *chi* recombinational hotspot in bacteriophage λ. *Proc. Natl. Acad. Sci. USA* **75**, 6182–6.

Stachel S. E., Messens E., Van Montagu M. & Zambryski P. (1985) Identification of the signal molecules produced by wounded plant cells that activate T-DNA transfer in *Agrobacterium tumefaciens. Nature* **318**, 624–9.

Stachel S. E., Timmerman B. & Zambryski P. (1986) Generation of single stranded T-DNA molecules during the initial stages of T-DNA transfer from *Agrobacterium tumefaciens* to plant cells. *Nature* **322**, 706–11.

Stachel S. E. & Zambryski P. (1986) *VirA* and *virG* control the plant-induced activation of the T-DNA transfer process in *A. tumefaciens*. *Cell* **46**, 325–33.

Stahl D. A., Flesher B., Mansfield H. R. & Montgomery L. (1988) Use of phylogenetically based hybridization probes for studies of ruminal microbial ecology. *Appl. Env. Microbiol.* **54**, 1079–84.

Stallcup M. R., Sharrock W. J. & Rabinowitz J. C. (1974) Ribosome and messenger specificity in protein synthesis by bacteria. *Biochem. Biophys. Res. Commun.* **58**, 92–8.

Stark M. J. R. (1987) Multicopy expression vectors carrying the *lac* repressor gene for regulated high-level expression of genes in *Escherichia coli*. *Gene* **51**, 255–67.

Stassi D. L. & Lacks S. A. (1982) Effect of strong promoters on the cloning in *Escherichia coli* of DNA fragments from *Streptococcus pneumoniae*. *Gene* **18**, 319–28.

Staudt L. M., Clerc R. G., Singh H., LeBowitz J. H., Sharp P. A. & Baltimore D. (1988) Cloning of a lymphoid-specific cDNA encoding a protein binding the regulatory octamer DNA motif. *Science* **241**, 577–9.

Steitz J. A. (1979) Genetic signals and nucleotide sequences in messenger RNA. In *Biological Regulation and Development. 1. Gene Expression*, ed. Goldberger R. F., pp. 349–99. Plenum Press, New York.

Steller H. & Pirotta V. (1985) A transposable P vector that confers selectable G418 resistance to *Drosophila* larvae. *EMBO J.* **4**, 167–71.

Stevis P. E. & Ho N. W. Y. (1987) Positive selection vectors based on xylose utilization suppression. *Gene* **55**, 67–74.

Stewart T. A., Pattengale P. K. & Leder P. (1984) Spontaneous mammary adenocarcinomas in transgenic mice that carry and express MTV/myc fusion genes. *Cell* **38**, 627–37.

Stinchcomb D. T., Mann C. & Davis R. W. (1982) Centromeric DNA from *Saccharomyces cerevisiae*. *J. Mol. Biol.* **158**, 157–79.

Stinchcomb D. T., Struhl K. & Davis R. W. (1979) Isolation and characterization of a yeast chromosomal replicator. *Nature* **282**, 39–43.

Stockhaus J., Eckes P., Rocha-Sosa M., Schell J. & Willmitzer L. (1987) Analysis of *cis*-active sequences involved in the leaf-specific expression of a potato gene in transgenic plants. *Proc. Natl. Acad. Sci. USA* **84**, 7943–7.

Stoker N. G., Fairweather N. F. & Spratt B. G. (1982) Versatile low-copy-number plasmid vectors for cloning in *Escherichia coli*. *Gene* **18**, 335–41.

Storb U., O'Brien R. L., McMullen M. D., Gollahon K. A. & Brinster R. L. (1984) High expression of cloned immunoglobulin kappa gene in transgenic mice is restricted to B-lymphocytes. *Nature* **310**, 238–48.

Stormo G. D., Schneider T. D. & Gold L. M. (1982) Characterization of translational initiation sites in *E. coli*. *Nucleic Acids Res.* **10**, 2971–96.

Storms R. K., McNeil J. B., Khandekar P. S., An G., Parker J. & Friesen J. D. (1979) Chimaeric plasmids for cloning of deoxyribonucleic acid sequences in *Saccharomyces cerevisiae*. *J. Bacteriol.* **140**, 73–82.

Strueli M., Hall A., Boll W., Stewart W. E. II, Nagata S. & Weissmann C. (1982) Target cell specificity of two species of human interferon-α produced in *Escherichia coli* and of hybrid molecules derived from them. *Proc. Natl. Acad. Sci. USA* **78**, 2848–52.

Struhl K. (1983) The new yeast genetics. *Nature* **305**, 391–7.

Struhl K., Cameron J. R. & Davis R. W. (1976) Functional genetic expression of eukaryotic DNA in *Escherichia coli*. *Proc. Natl. Acad. Sci. USA* **73**, 1471–5.

Struhl K. & Davis R. D. (1980) A physical, genetic and transcriptional map of the cloned his3 gene region of *Saccharomyces cerevisiae*. *J. Mol. Biol.* **136**, 309–32.

Struhl K., Stinchcomb D. T., Scherer S. & Davis R. W. (1979) High-frequency transformation of yeast: autonomous replication of hybrid DNA molecules. *Proc. Natl. Acad. Sci. USA* **76**, 1035–9.

Stueber D. & Bujard H. (1982) Transcription from efficient promoters can interfere with plasmid replication and diminish expression of plasmid specified genes. *EMBO J.* **1**, 1399–404.

Stunnenberg H. G. & Birnstiel M. L. (1982) Bioassay for components regulating eukaryotic gene expression: a chromosomal factor involved in the generation of histone mRNA 3'-termini. *Proc. Natl. Acad. Sci. USA* **79**, 6201–4.

Subramani S., Mulligan R. & Berg P. (1981) Expression of mouse dihydrofolate reductase complementary deoxyribonucleic acid in simian virus 40 vectors. *Mol. Cell. Biol.* **1**, 854–64.

Summers D. K. & Sherratt D. J. (1984) Multimerization of high copy number plasmids causes instability: Col E1 encodes a determinant essential for plasmid monomerization and stability. *Cell* **36**, 1097–103.

Sussman D. J. & Milman G. (1984) Short-term, high-efficiency expression of transfected DNA. *Mol. Cell. Biol.* **4**, 1641–3.

Sutcliffe J. G. (1979) Complete nucleotide sequence of the *Escherichia coli* plasmid pBR322. *Cold Spring Harbor Symp. Quant. Biol.* **43** (1), 77–90.

Sveda M. M. & Lai C.-J. (1981) Functional expression in primate cells of cloned DNA coding for the hemagluttinin surface glycoprotein of influenza virus. *Proc. Natl. Acad. Sci USA* **78**, 5488–92.

Sweeney G. E. & Old R. W. (1988) *Trans*-activation of transcription, from promoters containing immunoglobulin gene octamer sequences, by myeloma cell mRNA in *Xenopus* oocytes. *Nucleic Acids Res.* **16**, 4903–13.

Swift G. H., Hammer R. E., MacDonald R. J. & Brinster R. L. (1984) Tissue specific expression of the rat pancreatic elastase 1 gene in transgenic mice. *Cell* **38**, 639–46.

Swyryd E. A., Seaver S. & Stark G. R. (1974) N-(phosphonacetyl)-L-aspartate, a potent transition state analog inhibitor of aspartate transcarbamylase, blocks proliferation of mammalian cells in culture. *J. Biol. Chem.* **249**, 6945–50.

Syvanen A.-C., Laaksonen M. & Soderlund H. (1986) Fast quantitation of nucleic acid hybrids by affinity hybrid collection. *Nucleic Acids Res.* **14**, 5037–48.

Szostak J. W. & Blackburn E. H. (1982) Cloning yeast telomeres on linear plasmid vectors. *Cell* **29**, 245–55.

Szybalska E. H. & Szybalski W. (1962) Genetics of human cell lines, IV. DNA-mediated heritable transformation of a biochemical trait. *Proc. Natl. Acad. Sci. USA* **48**, 2026–34.

Tabin C. J., Hoffman J. W., Groff S. P. & Weinberg R. A. (1982) Adaption of a retrovirus as a eucaryotic vector transmitting the herpes simplex virus thymidine kinase gene. *Mol. Cell. Biol.* **2**, 426–36.

Tacon W. C. A., Bonass W. A., Jenkins B. & Emtage J. S. (1983) Plasmid expression vectors containing tryptophan promoter transcriptional regulons lacking the attenuator. *Gene* **23**, 255–65.

Tacon W., Cary N. & Emtage S. (1980) The construction and characterization of plasmid vectors suitable for the expression of all DNA phases under the control of the *E. coli* tryptophan promoter. *Mol. Gen. Genet.* **177**, 427–38.

Tait R. C., Close T. J., Lundquist R. C., Hagiya M., Rodriguez R. L. & Kado

C. I. (1983) Construction and characterization of a versatile broad host range DNA cloning system for Gram-negative bacteria. *Biotechnology* **1**, 269–75.

Tait R. C., Lundquist R. C. & Kado C. I. (1982) Genetic map of the crown gall suppressive *Inc*W plasmid pSa. *Mol. Gen. Genet.* **186**, 10–15.

Takagi H., Morinaga Y., Tsuchiya M., Ikemura H. & Inouye M. (1988) Control of folding of proteins secreted by a high expression secretion vector, pIN-111-ompA: 16-fold increase in production of active subtilisin E in *Escherichia coli*. *Biotechnology* **6**, 948–50.

Takamatsu N., Ishikawa M., Meshi T. & Okada Y. (1987) Expression of bacterial chloramphenicol acetyltransferase gene in tobacco plants mediated by TMV-RNA. *EMBO J.* **6**, 307–11.

Talmadge K. & Gilbert W. (1980) Construction of plasmid vectors with unique *Pst*I cloning sites in a signal sequence coding region. *Gene* **21**, 235–41.

Talmadge K. & Gilbert W. (1982) Cellular location affects protein stability in *Escherichia coli*. *Proc. Natl. Acad. Sci. USA* **79**, 1830–3.

Talmadge K., Stahl S. & Gilbert W. (1980) Eukaryotic signal sequence transports insulin antigen in *Escherichia coli*. *Proc. Natl. Acad. Sci. USA* **77**, 3369–73.

Taniguchi T., Guarente L., Roberts T. M., Kimelman D., Douhan J. & Ptashne M. (1980) Expression of the human fibroblast interferon gene in *Escherichia coli*. *Proc. Natl. Acad. Sci. USA* **77**, 5230–3.

Tartaglia J. & Paoletti E. (1988) Recombinant vaccinia virus vaccines *Trends Biotechnol.* **6**, 43–6.

Taylor J. W, Schmidt W., Cosstick R., Okruszek A. & Eckstein F. (1985) The use of phosphorothioate-modified DNA in restriction enzyme reactions to prepare niched DNA. *Nucleic Acids Res.* **13**, 8749–64.

Tchen P., Fuchs R. P. P., Sage E. & Leng M. (1984) Chemically modified nucleic acids as immunodetectable probes in hybridization experiments. *Proc. Natl. Acad. Sci. USA* **81**, 3466–70.

Teem J. L. & Rosbash M. (1983) Expression of a β-galactosidase gene containing the ribosomal protein 51 intron is sensitive to the *rna* 2 mutation of yeast. *Proc. Natl. Acad. Sci. USA* **80**, 4403–7.

Tenover F. C. (1988) Diagnostic deoxyribonucleic acid probes for infectious diseases. *Clin. Microbiol. Rev.* **1**, 82–101.

Tepfer D. (1984) Transformation of several species of higher plants by *Agrobacterium rhizogenes*: sexual transmission of the transformed genotype and phenotype. *Cell* **37**, 959–67.

Thomas C. M., Stalker D. M., Guiney D. G. & Helinski D. R. (1979) Essential regions for the replication and conjugal transfer of the broad host range plasmid RK2. In *Plasmids of Medical, Environmental and Commercial Importance*, ed. Timmis K. N. & Puhler A., pp. 375–85. Elsevier/North Holland Biomedical Press, Amsterdam.

Thomas C. M., Stalker D. M. & Helinski D. R. (1981) Replication and incompatibility properties of segments of the origin region of replication of the broad host range plasmid RK2. *Mol. Gen. Genet.* **81**, 1–7.

Thomas K. R. & Cappechi M. R. (1986) Introduction of homologous DNA sequences into mammalian cells induces mutations in the cognate gene. *Nature* **324**, 34–8.

Thomas K. R., Folger K. R. & Cappechi M. R. (1986) High frequency targeting of genes to specific sites in the mammalian genome. *Cell* **44**, 419–28.

Thomas M., Cameron J. R. & Davis R. W. (1974) Viable molecular hybrids of bacteriophage lambda and eukaryotic DNA. *Proc. Natl. Acad. Sci. USA* **71**, 4579–83.

Thomas P. S. (1980) Hybridization of denatured RNA and small DNA fragments transferred to nitrocellulose. *Proc. Natl. Acad. Sci. USA* **77**, 5201–5.

Thomashow M. F., Nutter R., Montoya A. L., Gordon M. P. & Nester E. W. (1980) Integration and organization of Ti plasmid sequences in crown gall tumours. *Cell* **19**, 729–39.

Tikchonenko T. I., Karamov E. V., Zavizion B. A. & Naroditsky B. S. (1978) *Eco*RI* activity: enzyme modification or activation of accompanying endonuclease? *Gene* **4**, 195–212.

Tilghman S. M., Tiemeier D. C., Polsky F., Edgell M. H., Seidman J. G., Leder A., Enquist L. W., Norman B. & Leder P. (1977) Cloning specific segments of the mammalian genome: bacteriophage λ containing mouse globin and surrounding gene sequences. *Proc. Natl. Acad. Sci. USA* **74**, 4406–10.

Timberlake W. E. & Marshall M. A. (1988) Genetic regulation of development in *Aspergillus nidulans*. *Trends Genet.* **4**, 162–9.

Tomizawa J.-I. & Itoh T. (1981) Plasmid ColE1 incompatibility determined by interaction of RNA I with primer transcript. *Proc. Natl. Acad. Sci. USA* **78**, 6096–100.

Tomizawa J.-I. & Itoh T. (1982) The importance of RNA secondary structure in ColE1 primer formation. *Cell* **31**, 575–83.

Tommassen J., van Tol H. & Lugtenberg B. (1983) The ultimate localization of an outer membrane protein of *Escherichia coli* K-12 is not determined by the signal sequence. *EMBO J.* **2**, 1275–9.

Towbin H., Staehelin T. & Gordon J. (1979) Electrophoretic transfer of proteins from polyacrylamide gels to nitrocellulose sheets: procedure and some applications. *Proc. Natl. Acad. Sci. USA* **76**, 4350–4.

Traboni C., Cortese R., Cilibert G. & Cesarini G. (1983) A general method to select M13 clones carrying base pair substitution mutants constructed *in vitro*. *Nucleic Acids Res.* **11**, 4229–39.

Tschopp J. F., Sverlow G., Kosson R., Craig W. & Grinna L. (1987) High-level secretion of glycosylated invertase in the methylotrophic yeast, *Pichia pastoris*. *Biotechnology* **5**, 1305–8.

Tschumper G. & Carbon J. (1983) Copy number control by a yeast centromere. *Gene* **23**, 221–32.

Tschumper G. & Carbon J. (1987) *Saccharomyces cerevisiae* mutants that tolerate centromere plasmids at high copy number. *Proc. Natl. Acad. Sci. USA* **84**, 7203–7.

Tuite M. F., Dobson M. J., Roberts N. A., King R. M., Burke D. C., Kingsman S. M. & Kingsman A. J. (1982) Regulated high efficiency expression of human interferon-alpha in *Saccharomyces cerevisiae*. *EMBO J.* **1**, 603–8.

Turgeon R., Wood H. N. & Braun A. C. (1976) Studies on the recovery of crown gall tumor cells. *Proc. Natl. Acad. Sci. USA* **73**, 3562–4.

Turner N. E., O'Connell K. M., Nelson R. S., Sanders P. R., Beachey R. N., Fraley R. T. & Shah D. M. (1987) Expression of alfalfa mosaic virus coat protein gene confers cross-protection in transgenic tobacco and tomato plants. *EMBO J.* **6**, 1181–8.

Twigg A. J. & Sherratt D. (1980) Trans-complementable copy-number mutants of plasmid ColE1. *Nature* **283**, 216–18.

Uchimaya H., Fushimi T., Hashimoto H., Harada H., Syono K. & Sugawara Y. (1986) Expression of a foreign gene in callus derived from DNA-treated protoplasts of rice (*Oryza sativa*). *Mol. Gen. Genet.* **204**, 204–7.

Uhlin B. E., Molin S., Gustafsson P. & Nordstrom K. (1979) Plasmids with temperature-dependent copy number for amplification of cloned genes and their products. *Gene* **6**, 91–106.

Uhlin B. E., Schweickart V. & Clark A. J. (1983) New runaway−replication−plasmid cloning vectors and suppression of runaway replication by novobiocin. *Gene* **22**, 255−65.

Upshall A. (1986) Filamentous fungi in biotechnology. *Biotechniques* **4**, 158−66.

Upshall A., Kumar A. A., Bailey M. C., Parker M. D., Favreau M. A., Lewison K. P., Joseph M. L., Maraganore J. M. & McKnight G. L. (1988) Secretion of active human tissue plasminogen activator from the filamentous fungus *Aspergillus nidulans. Biotechnology* **5**, 1301−4.

Urdea M. S., Running J. A., Horn T., Clyne J., Ku L., & Warner B. D. (1987) A novel method for the rapid detection of specific nucleotide sequences in crude biological samples without blotting or radioactivity: application to the analysis of hepatitis B virus in human serum. *Gene* **61**, 253−64.

Vaeck M., Reynaerts A., Hofte H., Jansens S., De Beuckeleer M., Dean C., Zabeau M., van Montagu M. & Leemans J. (1987) Transgenic plants protected from insect attack. *Nature* **328**, 33−7.

Valenzuela P., Medina A., Rutter W. J., Ammerer G. & Hall B. D. (1982) Synthesis and assembly of hepatitis B virus surface antigen particles in yeast. *Nature* **298**, 347−50.

Van Larbeke N., Engler G., Holsters M., van den Elsacker S., Zaenen I., Schilperoort R. A. & Schell J. (1974) Large plasmid in *Agrobacterium tumefaciens* essential for crown gall-inducing ability. *Nature* **252**, 169−70.

van Randen J. & Venema G. (1984) Direct plasmid transfer from replica-plated *E. coli* colonies to competent *B. subtilis* cells. Identification of an *E. coli* clone carrying the *his*H and *tyr*A genes of *B. subtilis. Mol. Gen. Genet.* **195**, 57−61.

Van Sluys M. A., Tempe J. & Fedoroff N. (1987) Studies on the introduction and mobility of the maize *Activator* element in *Arabidopsis thaliana* and *Daucus carota. EMBO J.* **6**, 3881−9.

Vapnek D., Hautala J. A., Jacobson J. W., Giles N. H. & Kushner S. R. (1977) Expression in *Escherichia coli* K12 of the structural gene for catabolic dehydroquinase of *Neurospora crassa. Proc. Natl. Acad. Sci. USA* **74**, 3508−12.

Vassart G., Georges M., Monsieur R., Brocas H., Lequarre A. S. & Christophe D. (1987) A sequence in M13 phage detects hypervariable minisatellites in human and animal DNA. *Science* **235**, 683−4.

Velten J., Fukada K. & Abelson J. (1976) *In vitro* construction of bacteriophage λ and plasmid DNA molecules containing DNA fragments from bacteriophage T4. *Gene* **1**, 93−106.

Venema G. (1979) Bacterial transformation. *Adv. Microbiol. Physiol.* **19**, 245−331.

Verspierin P., Cornelissen A. W. C. A., Thuong N. T., Helene C. & Toulme J. J. (1987) An acridine-linked oligodeoxynucleotide targeted to the common 5′ end of trypanosome mRNAs kills cultured parasites. *Gene* **61**, 307−15.

Vieira J. & Messing J. (1982) The pUC plasmids, an M13mp7-derived system for insertion mutagenesis and sequencing with synthetic universal primers. *Gene* **19**, 259−68.

Vieira J. & Messing J. (1987) Production of single-stranded plasmid DNA. *Method Enzymol.* **153**, 3−11.

Villa-Komaroff L., Efstratiadas A., Broome S., Lomedico P., Tizard R., Naber S. P., Chick W. L. & Gilbert W. (1978) A bacterial clone synthesizing proinsulin. *Proc. Natl. Acad. Sci. USA* **75**, 3727−31.

Vinson C. R., LaMarco K. L., Johnson P. F., Landschulz W. H. & McKnight S. L. (1987) *In situ* detection of sequence-specific DNA binding activity specified by a recombinant bacteriophage. *Genes Dev.* **2**, 801–6.

Vogel J., Hinrichs S. H., Reynolds R. K., Luciw P. A. & Jay G. (1988) The HIV *tat* gene induces dermal lesions resembling Kaposis sarcoma in transgenic mice. *Nature* **335**, 606–11.

Wahle E. & Kornberg A. (1988) The partition locus of plasmid pSC101 is a specific binding site for DNA gyrase. *EMBO J.* **7**, 1889–95.

Wahl G. M., de Saint Vincent B. R. & DeRose M. L., (1984) Effect of chromosomal position on amplification of transfected genes in animal cells. *Nature* **307**, 516–20.

Walker M. D., Edlund T., Boulet A. M. & Rutter W. J. (1983) Cell-specific expression controlled by the 5'-flanking region of insulin and chymotrypsin genes. *Nature* **306**, 557–61.

Wallace, R. B., Johnson P. F., Tanaka S., Schold M., Itakura K. & Abelson J. (1980) Directed deletion of a yeast transfer RNA intervening sequence. *Science* **209**, 1396–400.

Wallace R. B., Schold M., Johnson M. J., Dembek P. & Itakura K. (1981) Oligonucleotide directed mutagenesis of the human β-globin gene: a general method for producing specific point mutations in cloned DNA. *Nucleic Acids Res.* **9**, 3647–56.

Walmsey M. E. & Patient R. K. (1987) Highly efficient β-globin transcription in the absence of both a viral enhancer and erythroid factors. *Development* **101**, 815–27.

Wasylyk B. (1988) Enhancers and transcription factors in the control of gene expression. *Biochim. Biophys. Acta* **951**, 17–35.

Watson J. D. (1972) Origin of concatameric T7 DNA. *Nature New Biol.* **239**, 197–201.

Watson B., Currier T. C., Gordon M. P., Chilton M.-D. & Nester E. W. (1975) Plasmid requirement for virulence of *Agrobacterium tumefaciens*. *J. Bacteriol.* **123**, 255–64.

Weatherall D. J. (1985) *The New Genetics and Clinical Practice* 2nd edn. Oxford University Press, Oxford.

Weatherall D. J. & Clegg J. B. (1982) Thalassemia revisited. *Cell* **29**, 7–9.

Weatherall D. J. & Old J. M. (1983) Antenatal diagnosis of the haemoglobin disorders by analysis of foetal DNA. *Mol. Biol. Med.* **1**, 151–5.

Weck P. K., Apperson S., Stebbing N., Gray P. W., Leung D., Shepard H. M. & Goeddel D. V. (1982) Antiviral activities of hybrids of two major human leukocyte interferons. *Nucleic Acids Res.* **9**, 6153–66.

Wei C.-M., Gibson P., Spear P. G. & Scolnick E. M. (1981) Construction and isolation of a transmissible retrovirus containing the *src* gene of Harvey Sarcoma Virus and the thymidine kinase gene of herpes simplex virus type 1. *J. Virol.* **39**, 935–44.

Weiler E. W. & Schroder J. (1987) Hormone genes and crown gall disease. *TIBS* **12**, 271–5.

Weiss R., Teich N., Varmus H. & Coffin J. (1985) *RNA tumour viruses* 2nd edn. Cold Spring Harbor Laboratory, Cold Spring Harbor, New York.

Wells J. A. & Estell D. A. (1988) Subtilisin—an enzyme designed to be engineered. *Trends Biochem.* **13**, 291–7.

Wells J. A., Vasser M. & Powers D. B. (1985) Cassette mutagenesis: an efficient method for generation of multiple mutations at defined sites. *Gene* **34**, 315–23.

Wensink P. C., Finnegan D. J., Donelson J. E. & Hogness D. S. (1974) A system for mapping DNA sequences in the chromosomes of *Drosophila melano-*

gaster. Cell **3**, 315—25.

Wetzel R., Kleid D. G., Crea R., Heyneber H. L., Yansura D. G., Hirose T., Kraszewski A., Riggs A. D., Itakura K. & Goeddel D. V. (1981) Expression in *Escherichia coli* of a chemically synthesized gene for a "mini-C" analog of human proinsulin. *Gene* **16**, 63—71.

Wetzel R., Perry L. J., Baase W. A. & Becktel W. J. (1988) Disulphide bonds and thermal stability in T4 lysozyme. *Proc. Natl. Acad. Sci. USA* **85**, 401—5.

White F. F. & Nester E. W. (1980) Relationship of plasmids responsible for hairy root and crown gall tumorigenicity. *J. Bacteriol.* **144**, 710—20.

Widera G., Gautier F., Lindenmaier W. & Collins J. (1978) The expression of tetracycline resistance after insertion of foreign DNA fragments between the *Eco* RI and *Hin*d III sites of the plasmid cloning vector pBR322. *Mol. Gen. Genet.* **163**, 301—5.

Wigler M., Perucho M., Kurtz D., Dana S., Pellicer A., Axel R. & Silverstein S. (1980) Transformation of mammalian cells with an amplifiable dominant-acting gene. *Proc. Natl. Acad. Sci. USA* **77**, 3567—70.

Wigler M., Silverstein S., Lee L. S., Pellicer A., Cheng Y. C. & Axel R. (1977) Transfer of purified herpes virus thymidine kinase gene to cultured mouse cells. *Cell* **11**, 223—32.

Wigler M., Sweet R., Sim G. K., Wold B., Pellicer A., Lacy E., Maniatis T., Silverstein S. & Axel R. (1979) Transformation of mammalian cells with genes from procaryotes and eucaryotes. *Cell* **16**, 777—85.

Wilkinson A. J., Fersht A. R., Blow D. M., Carter P. & Winter G. (1984) A large increase in enzyme-substrate affinity by protein engineering. *Nature* **307**, 187—8.

Williams D. A., Lemischka I. R., Nathan D. G. & Mulligan R. C. (1984) Introduction of new genetic material into pluripotent haemotopoietic stem cells of the mouse. *Nature* **310**, 476—80.

Williams D. C., Van Frank R. M., Muth W. L. & Burnett J. P. (1982) Cytoplasmic inclusion bodies in *Escherichia coli* producing biosynthetic human insulin proteins. *Science* **215**, 687—8.

Williams D. M., Duvall E. J. & Lovett P. S. (1981a) Cloning restriction fragments that promote expression of a gene in *Bacillus subtilis*. *J. Bacteriol.* **146**, 1162—5.

Williams D. M., Schoner R. G., Duvall E. J., Preis L. H. & Lovett P. S. (1981b) Expression of *Escherichia coli trp* genes and the mouse dihydrofolate reductase gene cloned in *Bacillus subtilis*. *Gene* **16**, 199—206.

Williams J. G. & Lloyd M. M. (1979) Changes in the abundance of polyadenylated RNA during slime mould development measured using cloned molecular hybridization probes. *J. Mol. Biol.* **129**, 19—35.

Williamson R., Eskdale J., Coleman D. V., Niazi M., Loeffler F. E. & Modell B. (1981) Direct gene analysis of chorionic villi: a possible technique for first trimester diagnosis of haemoglobinopathies. *Lancet* **2**, 1127.

Willmitzer L., Dhaese P., Schreier P. H., Schmalenbach W., Van Montagu M. & Schell J. (1983) Size, location & polarity of transferred DNA encoded transcripts in nopaline crown gall tumours: common transcripts in octopine and nopaline tumours. *Cell* **32**, 1045—6.

Willmitzer L., Simons G. & Schell J. (1982) The Ti DNA in octopine crown gall tumours codes for seven well-defined polyadenylated transcripts. *EMBO J.* **1**, 139—46.

Wilson C., Cross G. S. & Woodland H. R. (1986) Tissue-specific expression of actin genes injected into *Xenopus* embryos. *Cell* **47**, 589—99.

Wilson G. G. & Murray N. E. (1979) Molecular cloning of the DNA ligase gene from bacteriophage T4. I. Characterization of the recombinants. *J. Mol. Biol.* **132**, 471–91.

Wilson J. M., Jefferson D. M., Chowdhury J. R., Novikoff P. M., Johnston D. E. & Mulligan R. C. (1988a) Retrovirus-mediated transduction of adult hepatocytes. *Proc. Natl. Acad. Sci. USA* **85**, 3014–18.

Wilson J. M., Johnston D. E., Jefferson D. M. & Mulligan R. C. (1988b) Correction of the genetic defect in hepatocytes from the Watanabe heritable hyperlipidemic rabbit. *Proc. Natl. Acad. Sci. USA* **85**, 4421–5.

Windass J. D., Worsey M. J., Pioli E. M., Pioli D., Barth P. T., Atherton K. T., Dart E. C., Byrom D., Powell K. & Senior P. J. (1980) Improved conversion of methanol to single-cell protein by *Methylophilus methylotrophus*. *Nature* **287**, 396–401.

Winter G., Fersht A. R., Wilkinson A. J., Zoller M. & Smith M. (1982) Redesigning enzyme structure by site-directed mutagenesis: tyrosyl tRNA synthetase and ATP binding. *Nature* **299**, 756–8.

Winter J. A., Wright R. L. & Gurley W. B. (1984) Map locations of five transcripts homologous to TR-DNA in tobacco and sunflower crown gall tumours. *Nucleic Acids Res.* **12**, 2391–2406.

Wood C. R., Boss M. A., Kenlen J. H., Calvert J. E., Roberts N. A. & Emtage J. S. (1988) The synthesis and *in vivo* assembly of functional antibodies in yeast. *Nature* **314**, 446–9.

Woolston C. J., Covey S. N., Penswick J. R. & Davies J. W. (1983) Aphid transmission and a polypeptide are specified by a defined region of the cauliflower mosaic virus genome. *Gene* **23**, 15–23.

Wu R., Bahl C. P. & Narang S. A. (1978) Chemical synthesis of oligonucleotides. *Prog. Nucleic Acid Res. Mol. Biol.* **21**, 101–38.

Wullems G. J., Molendijk L., Ooms G. & Schilperoort R. (1981a) Differential expression of crown gall tumor markers in transformants obtained after *in vitro Agrobacterium tumefaciens* induced transformation of cell wall regenerating protoplasts derived from *Nicotiana tabacum*. *Proc. Natl. Acad. Sci. USA* **78**, 4344–8.

Wullems G. J., Molendijk L., Ooms G. & Schilperoort R. A. (1981b) Retention of tumor markers in F1 progeny plants formed from *in vitro* induced octopine and nopaline tumor tissues. *Cell* **24**, 719–28.

Wyman A. R., Wolfe L. B. & Botstein D. (1985) Propagation of some human DNA sequences in bacteriophage lambda vectors requires mutant *Escherichia coli* hosts. *Proc. Natl. Acad. Sci. USA* **82**, 2880–4.

Yadav N. S., Vanderleyden J., Bennet D., Barnes W. M. & Chilton M.-D. (1982) Short direct repeats flank the T-DNA on a nopaline Ti plasmid. *Proc. Natl. Acad. Sci. USA* **79**, 6322–26.

Yagi Y., McLellan T., Frez W. & Clewell D. (1978) Characterization of a small plasmid determining resistance to erythromycin, lincomycin and vernamycin B_α in a strain of *Streptococcus sanguis* isolated from dental plaque. *Antimicrob. Agents Chemother.* **13**, 884–7.

Yanisch-Perron C., Vieira J. & Messing J. (1985) Improved M13 phage cloning vectors and host strains: nucleotide sequences of the M13 mp18 and pUC19 vectors. *Gene* **33**, 103–19.

Yanofsky C. & Kolter R. (1982) Attenuation in amino acid biosynthetic operons. *Ann. Rev. Genet.* **16**, 113–34.

Yanofsky M. F., Porter S. G., Young C., Albright L. M., Gordon M. P. & Nester E. W. (1986) The *vir D* operon of *Agrobacterium tumefaciens* encodes a site-specific endonuclease. *Cell* **47**, 471–7.

Yansura D. G. & Henner D. J. (1984) Use of the *Escherichia coli lac* repressor and operator to control gene expression in *Bacillus subtilis*. *Proc. Natl. Acad. Sci. USA* **81**, 439−43.

Yarranton G. T., Wright E., Robinson M. K. & Humphreys G. O. (1984) Dual-origin plasmid vectors whose origin of replication is controlled by the coliphage lambda promoter P_L. *Gene* **28**, 293−300.

Yasuda S. & Takagi T. (1983) Overproduction of *Escherichia coli* replication proteins by the use of runaway-replication plasmids. *J. Bacteriol.* **154**, 1153−61.

Yoshizumi H. & Ashikari T. (1987) Expression, glycosylation and secretion of fungal hydrolases in yeast. *Trends Biotech.* **5**, 277−81.

Young B. D., Birnie G. D. & Paul J. (1976) Complexity and specificity of polysomal poly $(A)^+$ RNA in mouse tissues. *Biochemistry* **15**, 2823−8.

Young F. E. (1987) DNA Probes. Fruits of the new biotechnology. *J. Am. Med. Assoc.* **258**, 2404−6.

Young R. A. & Davis R. W. (1983) Efficient isolation of genes by using antibody probes. *Proc. Natl. Acad. Sci. USA* **80**, 1194−8.

Zaballos A., Salas M. & Mellado R. P. (1987) A set of expression plasmids for the synthesis of fused and unfused polypeptides in *Escherichia coli*. *Gene* **58**, 67−76.

Zaenen I., Van Larbeke N., Teuchy H., Van Montagu M. & Schell J. (1974) Super-coiled circular DNA in crown-gall inducing *Agrobacterium* strains. *J. Mol. Biol.* **86**, 109−127.

Zambryski P., Depicker A., Kruger H. & Goodman H. (1982) Tumor induction by *Agrobacterium tumefaciens*: analysis of the boundaries of T-DNA. *J. Mol. Appl. Genet.* **1**, 361−70.

Zambryski P., Holsters M., Kruger K., Depicker A., Schell J., Van Montagu M. & Goodman H. (1980) Tumor DNA structure in plant cells transformed by *A. tumefaciens*. *Science* **209**, 1385−91.

Zambryski P., Joos H., Genetello C., Leemans J., Van Montagu M. & Schell J. (1983) Ti plasmid vector for the introduction of DNA into plant cells without alteration of their normal regeneration capacity. *EMBO J.* **2**, 2143−50.

Zimmerman S. B. & Pheiffer B. (1983) Macromolecular crowding allows blunt end ligation by DNA ligases from rat liver or *Escherichia coli*. *Proc. Natl. Acad. Sci. USA* **80**, 5852−6.

Zimmerman U. & Vienken J. (1983) Electric field induced cell to cell fusion. *J. Membr. Biol.* **67**, 165−82.

Zinder N. D. & Boeke J. D. (1982) The filamentous phage (Ff) as vectors for recombinant DNA−a review. *Gene* **19**, 1−10.

Zinn K., Di Maio D. & Maniatis T. (1983) Identification of two distinct regulatory regions adjacent to the human B-interferon gene. *Cell* **34**, 865−79.

Zoller M. J. & Smith M. (1983) Oligonucleotide-directed mutagenesis of DNA fragments cloned into M13 vectors. *Methods Enzymol.* **100**, 468−500.

Index